Räume – Grenzen – Hybriditäten

Im Kontext des Aufbrechens von Eindeutigkeiten durch kulturelle und soziale Hybridisierungen, durch die partielle Verschmelzung und Neuordnung von lokal, regional, national und global verlieren Grenzen wie auch räumliche Bezüge an Klarheit und Verbindlichkeit. Supranationalstaatlich einzuordnende Migrationsströme wirken bis in lokale Räume. Die Metropolisierung der Welt folgt globalen Mustern und manifestiert sich zugleich lokal sehr unterschiedlich. Und auch Kulturalität, Diversität und Gender entziehen sich einer einfach verortbaren Betrachtungsebene. Übergreifend gehen mit unterschiedlich gearteten Umbrüchen vielfältige und differenzierte In- und Exklusionsprozesse einher. Neue Grenzen werden gezogen, in Frage gestellt und verändert – physisch, kulturell, sozial, politisch, mental etc. –, wobei gleichzeitig ehemals eindeutig scheinende Unterscheidungsmuster unscharf werden. Es entstehen Felder eines *in-betweens* und sowohl-als-auchs, in dem Hybridisierungen stattfinden. Diese Uneindeutigkeiten lösen mitunter Angst aus und führen wiederum zu einem Gefühl der Sehnsucht nach Eindeutigkeit. Mit der Reihe „Räume – Grenzen – Hybriditäten" im Verlag Springer VS wird ein inter- und transdisziplinäres Forum geboten, das sich an Soziologie, Kultur- und Literaturwissenschaften, Geographie, Border Studies und weitere inhaltlich tangierte Disziplinen richtet. Ziel ist es, komplexe Veränderungsprozesse in aktueller ebenso wie in historischer Perspektive zu fokussieren sowie Grenzziehungen und gleichzeitig Hybridisierungen zu konzeptionalisieren. Die übergreifende Klammer bildet ein konstruktivistischer Zugang, mit dem die ‚Gemachtheit' und Wandelbarkeit von Räumen, Grenzen, Kulturen betont und analysiert wird.

Spaces—Borders—Hybridities
Within the context of the breakdown of unequivocality through cultural and social hybridizations, through the partial merging and reorganization of the local, regional, national, and global, borders as well as spatial references are losing their clarity and binding character. Supranational migration flows also have an impact on local areas. The metropolization of the world follows global patterns and manifests itself very differently on a local level. And culturality, diversity, and gender also elude an analysis level which can be located easily. Overall, diverse and differentiated inclusion and exclusion processes go hand in hand with different types of upheavals. New borders are drawn, questioned, and changed—physically, culturally, socially, politically, mentally, etc.—while at the same time, patterns of distinction that once appeared to be clear, become blurred. Fields of in-betweens and both/ands arise, in which hybridizations take place. These ambiguities sometimes trigger fear and, in turn, lead to a feeling of longing for unequivocality.
The series "Spaces—Borders—Hybridities" from Springer VS offers an interdisciplinary and transdisciplinary forum aimed at sociology, cultural and literary studies, geography, border studies and other related disciplines. The aim is to focus on complex processes of change from a contemporary as well as historical perspective and to conceptualize borderings and hybridizations. The overarching theme is a constructivist approach, in which the "fabrication" and changeability of spaces, borders, and cultures is emphasized and analyzed.

Albert Rossmeier

San Diego's Hybrid Urban Borderlands

An Urban Landscape- and Border-Theoretical Approach to the Inner-Ring Redevelopment of America's Finest City

Springer VS

Albert Rossmeier
Bad Füssing, Germany

ISSN 2662-1932 ISSN 2662-1940 (electronic)
Räume – Grenzen – Hybriditäten
ISBN 978-3-658-42666-8 ISBN 978-3-658-42667-5 (eBook)
https://doi.org/10.1007/978-3-658-42667-5

This Springer VS imprint is published by the registered company Springer Fachmedien Wiesbaden GmbH, part of Springer Nature.
The registered company address is: Abraham-Lincoln-Str. 46, 65189 Wiesbaden, Germany

Foreword

In 2015, Albert Rossmeier became part of our research team at the University of Applied Sciences Weihenstephan-Triesdorf in Freising and has accompanied our research programs ever since. After completing his master's degree at the University of Tübingen, he started his dissertation on neighborhood change in San Diego, California—and he did this as an external candidate from the United States, which required a lot of perseverance, courage, dedication and determination. These qualities, in turn, align remarkably well with the author of the work presented here. Over the years, particularly his broad interest and complex thinking in the fields of (urban) landscape, energy transition, discourse theory and border studies made him a valuable research partner.

In the present monograph, published in the Springer VS book series "Spaces—Borders—Hybridities," Albert Rossmeier focuses on neighborhood change processes in the course of the recent urban redevelopment in the central communities of San Diego. The author approaches the development in selected inner-ring neighborhoods through a uniquely combined urban landscape and urban border lens that focuses on individual perceptions, experiences and perspectives. Studying the various ways in which the demographics, aesthetics and ways of living are changing in the central neighborhoods, the author addresses a research gap in the related fields of urban landscape and urban border studies pertaining to the redevelopment trend at the inner-city edges and the first rings around the downtowns of major cities in the US and Europe. Beyond the pertinent thematic dedication, Albert Rossmeier successfully develops the interdisciplinary framework of *hybrid urban borderlands*, which offers a remarkable triangulation of urban landscape and urban border-theoretical perspectives on the temporal hybridity and ambiguity of transforming neighborhoods and their edges. Accordingly, the present study is an informative insight into recent trends regarding neighborhood redevelopment that is of particular relevance to scholars of urban phenomena, planners, developers and residents alike.

We wish the readers much joy in reading this rewarding work that brings together landscape research and border studies on an international level. Allow yourself to be captivated by theoretical-conceptual and empirical insights!

Tübingen, Germany — Olaf Kühne
Saarbrücken, Germany — Florian Weber

Acknowledgments

This work has already opened many doors for me and has given the last three years significant structure. However, the completion of this dissertation project would not have been possible without the help of a number of people.

First, I would like to thank all of my interviewees who helped me gain exciting and helpful insights about the San Diego-Tijuana cross-border metropolis. Additionally, I would like to thank Thomas Herman (San Diego State University) and Kate Swanson (Dalhousie University, previously SDSU) who became my local contact persons and took time to support my project in various ways. In particular, I would like to express my special thanks to my two doctoral advisors Olaf Kühne and Florian Weber. It was Olaf who got me interested in geography as well as the Southern California region in my early years as a landscape architecture student. He has been instrumental in shaping my academic career and has actively supported and motivated me over the years. I am also very grateful to Florian Weber for all the pleasant and successful collaborations as well as the permanent professional support. He has always found time, an open ear, and helpful words. Additionally, I would like to thank my fellow student, colleague, and, above all, friend Nora Crossey. The numerous conversations and cheerful moments contributed significantly to the success of this and several other study and research projects.

I would especially like to thank my dear parents, Albert and Gertraud Rossmeier, and my sister Anna Rossmeier. Without them, I would not have been able to realize the decision to move to California, nor would I have been able to complete this work. I thank you from the bottom of my heart for the uninterrupted support and belief in my goals. Furthermore, my heartfelt thanks go to the people who have become family over the years, Natalia Brdys and Johannes Eisner. Today, I look back on many significant moments in more than ten years together.

Many thanks for that and for the special cohesion that has always been of great importance on my personal as well as professional path.

Last but not least, I would like to thank my dear husband, Gabriel Rossmeier, who has caringly supported me at all times and in every phase of this work. Your positive and cordial nature motivates me and has kept me looking forward, even in the challenging phases of this work.

Los Angeles
November 2022

Albert Rossmeier

Abstract

This study aims primarily for a wider understanding of the recent redevelopment processes that emerged several decades ago in San Diego's central city and now gradually spread over the downtown edges into the inner-ring neighborhoods. Perspectively situated in the fields of urban landscape and urban border studies, the research project outlines how the eastward 'redevelopment wave' in San Diego contests socialized neighborhood (boundary) perceptions by transforming the former first-tier suburbs from outdated and disinvested communities into dense 'urban villages' and trendy places to be. This undertaking enriches the various disciplines that are concerned with the recent course of 'post-suburbia,' for yet only a modest number of studies have shed light on the manifold process of inner-ring neighborhood redevelopment. Taking this gap as a starting point, the study applied a phenomenology-oriented social constructivist methodology to trace how the subjective constructions and everyday life meanings but also the demographics and the physical foundation of the different communities changed over time. Using a mixed methods approach, the research has shown that the decadelong redevelopment of San Diego's central neighborhoods is following a spatiotemporal sequence and taking place in varying extents throughout the inner ring. Large commercial corridors connecting the redeveloped neighborhoods in the west with the newly developing low-income communities in the east act as 'development axes' along which new businesses, college-educated residents, development projects, and symbolics progress eastward, promoting structural, social, and economic juxtapositions, among others. The research shows that these (temporary) hybridizations enhance the divergence and plurality of individual place perceptions and thus the ambiguity of b/ordering and othering processes in the redeveloping inner ring. The resulting phenomena of 'cross-border hybridity' are perforating, dissolving, or shifting socialized, linear neighborhood boundaries into areas that are simultaneously part of the one and the other neighborhood. In the present study, these rather undefined or stretched border areas have been referred to as *hybrid urban borderlands*. This notion is a novel conceptual approach that has been tested in the present research and can be deemed a promising lens for future studies on neighborhood change, urban redevelopment, and socio-spatial re-interpretation beyond the context of San Diego.

Zusammenfassung

Diese Studie strebt primär ein breiteres Verstehen der jüngeren Sanierungsprozesse von San Diego an, welche vor mehreren Jahrzehnten im Zentrum begonnen haben und aktuell nach und nach über die Innenstadtgrenzen in die Nachbarschaften des inneren Rings vordringen. Perspektivisch situiert in den Feldern der Urban Landscape und Urban Border Studies zeichnet das Forschungsprojekt nach, wie die ostwärts vordringende 'Welle der Sanierung' in San Diego sozialisierte Wahrnehmungen von Nachbarschaften und ihren Grenzen herausfordert – durch die Transformation der ehemaligen First-Tier Suburbs von veralteten und vernachlässigten Communities in dichte 'Urban Villages' und angesagte Orte. Dieses Vorhaben bereichert die verschiedenen Disziplinen, die sich mit dem jüngeren Entwicklungskurs von 'Post-Suburbia' beschäftigen, denn bisher haben nur wenige Studien die vielfältigen Prozesse der Reurbanisierung von Nachbarschaften im inneren Ring ins Licht gerückt. Ausgehend von dieser Forschungslücke verwendet die vorliegende Studie eine phänomenologisch orientierte, sozialkonstruktivistische Methodologie, um nachzuzeichnen, wie sich die subjektiven Konstruktionen und sozialisierten Alltagsbedeutungen, aber auch die Demographie und die physischen Grundlagen der verschiedenen Nachbarschaften über die Zeit verändert haben. Mit Hilfe eines Mixed Methods Ansatzes konnte die Forschung zeigen, dass die jahrzehntelange Sanierung der zentralen Nachbarschaften San Diegos einer raumzeitlichen Sequenz folgt und sich in unterschiedlichem Ausmaße im inneren Ring niederschlägt. Große Gewerbekorridore verbinden die bereits sanierten Nachbarschaften im Westen mit Gebieten im Osten, die erst am Anfang der Sanierung stehen und agieren als ‚Entwicklungsachsen,' an welchen neue Geschäfte, Bauprojekte sowie dazugehörige Symboliken ostwärts vordringen und – unter anderen – strukturelle, soziale und ökonomische Gegensätze befördern. Die Forschung zeigt, dass diese (vorübergehenden) Hybridisierungen die Divergenz und Pluralität von individuellen Ortswahrnehmungen begünstigt und damit auch zur Uneindeutigkeit von B/ordering und Othering Prozessen im sich entwickelnden inneren Ring beiträgt. Die sich dabei ergebenden Phänomene grenzüberschreitender Hybridität perforieren, verschieben oder weichen die sozialisierten, linearen Nachbarschaftsgrenzen zu Zonen auf, die gleichzeitig Teil der einen und der anderen Nachbarschaft sind. Diese eher uneindeutigen und gedehnten Grenzgebiete wurden in der vorliegenden Forschung als *hybrid urban borderlands* bezeichnet. Dieser Begriff stellt einen neuen, konzeptuellen Ansatz dar, welcher in der vorliegenden Studie erstmals getestet wurde und abschließend als vielversprechende Linse für zukünftige Untersuchungen zu Nachbarschaftswandel, städtischen Sanierungsprozessen und sozialräumlicher Reinterpretation jenseits des Kontexts von San Diego vorgeschlagen wird.

Contents

List of Figures

List of Tables

Introduction: Crossing the Inner-City Boundaries—Urban Redevelopment in San Diego and its Progression from the Center into the Inner Ring

1

With its close and entangled everyday life ties across the U.S.-Mexico border, the international borderland of San Diego-Tijuana is home to more than 5 million inhabitants. As the largest western cross-border metropolitan region, it stretches over 6,200 square miles (roughly 16,058 square kilometer) from the military base Camp Pendleton and the City of Oceanside in the northwest, the mountain town of Julian in the northeast over the central and southern jurisdictions of San Diego County and the three Mexican municipalities of Tijuana, Playas de Rosarito in the southwest, and Tecate in the southeast (Herzog and Sohn 2017). Beyond the particularly strong 'bottom-up' networks—the social connections and organizations—the physically bordered region is locally tied together through 'top-down' structures—multilevel cross-border governance approaches and guidelines as well as an array of planning institutions—that deal with the shared natural environment, markets, production and manufacturing relations, as well as with tourism in the popular and rapidly growing region (cf. Mendoza and Dupeyron 2020; Peña 2007; Roßmeier 2020a; on multilevel cross-border governance see generally Crossey and Weber 2020; Hooghe and Marks 2003).

The U.S.-American part of the bi-national urban agglomeration (see Figure 1.1) constitutes of a spread-out, horizontal cluster of various low-density commercial centers, job magnets, and residential communities, reaching from the shoreline of the Pacific Ocean up the canyons and the mesa tops to the wooded Cuyamaca mountains and the arid badlands of the Anza-Borrego desert (Pryde 2014). Omitting the history of the region's various Native American bands, which dates back several thousand years (outlined in chapter 4), the majority of the communities and structures in the settlement are less than 100 years old. Representing classic sunbelt city development, San Diego's sprawling growth pattern throughout the 20th century mirrored the success of the private automobile and was spatially defined by the development of a sustained web of large freeways. Thus,

A. Rossmeier, *San Diego's Hybrid Urban Borderlands*,
Räume – Grenzen – Hybriditäten, https://doi.org/10.1007/978-3-658-42667-5_1

it is no surprise that the various communities throughout the county surpassed the City of San Diego in terms of population number as early as the 1940 s—a trend that has intensified since then and turned the young region inside out (illustrated in Figure 4.2). Therefore, despite standing in the historically cast shadow of the larger and more populous mega-metropolis of Los Angeles to its north, California's most southern urban region has been studied intensively for its early as well as vast suburban development dynamics (among many Davis et al. 2003; Ford 2005, 2014; Killory 1993; Kühne and Schönwald 2015a; Lynch and Appleyard 1974; Mirkowich 1941; Pryde and Stutz 2014; Roßmeier 2019).

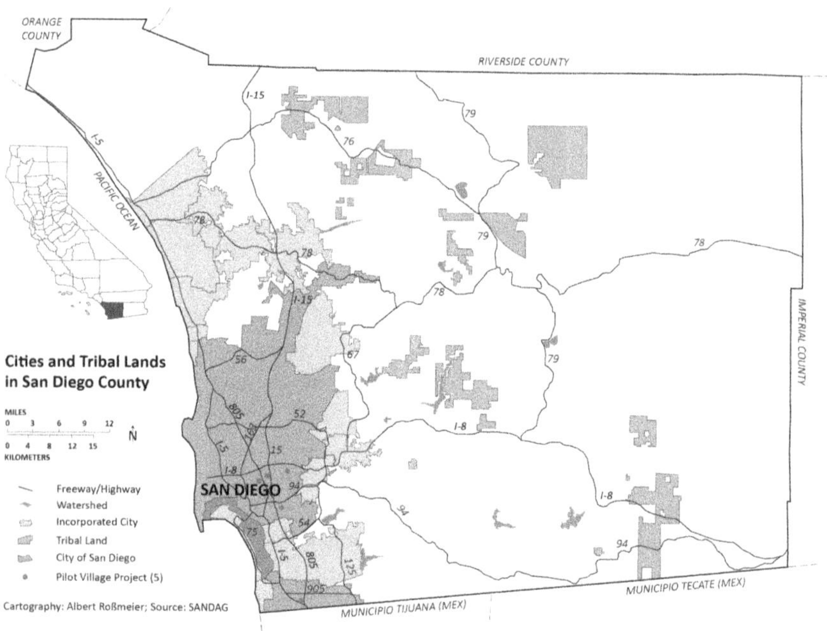

Figure 1.1 Localization of San Diego County, it's Cities, and Tribal Lands. (The map shows San Diego County and localizes it within the State of California. Adjacent counties on the U.S. side and municipalities on the Mexican side are shown in italics. Source: Own cartography and design)

Generally, more often than not, suburbanism—broadly understood as the decentralization of housing, consumption, and work—has been addressed from a quite negative standpoint in urban studies by highlighting various challenges and undesirable impacts concerning transportation, the structure of the city and of social everyday life, land and energy consumption, and relatedly greenhouse gas emissions and regional sustainability goals. Early research work on suburbanization in the U.S. dates back to the 1950 s (Bello 1958; Schnore 1957; Whyte 1958) and has faced increasing diversification in terms of thematic focus, perspective, and case study areas in the following decades (basic presentations among others by Duany et al. 2000; Jackson 1985; Stilgoe 1988; Teaford 2016 [1986]). On the one hand, suburbia and suburbanism have been discussed within a large body of research work due to the variety of fields and disciplines that are concerned with the local socio-spatial products and the dynamics of what has become a 'global process' (Herzog 2015). These fields include planning, architecture, urban as well as landscape studies, geography, and history, among others, which have examined suburbanization within a focus on, for instance, its historical onset and its rural aspects and influences (Gailing 2015; Sandul 2014), its planning and aesthetic qualities and regional consequences—addressed, for instance, by means of the notion of *Zwischenstadt* or 'in-between city' in the German planning and urban design discourse (Sieverts 2003; cf. Aring 1999; Vicenzotti and Qviström 2018)—or through highlighting its seemingly unsorted and hybrid combination of urban, rural, and landscape aspects (Kühne 2012, 2019; Soja 1993), as applied in the present study (illustrated in detail in section 2.2.4).

On the other hand, starting in the 1990 s but particularly since the 2000 s, the settlement fringes have been studied intensively for their gradual complexification and diversification in terms of socio-economic and racial or ethnic composition, functional orientation, as well as architectural appearance and structure (early recognition and discussion by Masotti and Hadden 1973, 1974). As a result, several authors have called for "more subtle frameworks in order better to [sic!] understand the structure of contemporary metropolitan areas" (Hanlon et al. 2006, p. 2129; cf. Airgood-Obrycki et al. 2020), particularly also in regard to the contrast between aging, older settlements in closer vicinity to the historic central business district—the inner-ring suburbs—and newer, more peripheral developments in the outer rings (Kruse and Sugrue 2006), as it will be discussed in the following. Reacting to the calls for typification and categorization, various forms and styles of suburban settlements have been identified. These range from seemingly endless job and commercial magnets along large freeways—referred to as edge or edgeless cities (Garreau 1992; Lang 2003; Lang et al. 2009)—to smaller fenced or walled and thus exclusive neighborhoods—the gated communities and enclaves (Breitung 2012; Liao et al. 2019; Villamizar-Santamaría 2019; Webster et al. 2002)—and from shrinking or decaying forms of suburbs (Hesse 2010; Hollander et al. 2018; Lucy and Phillips 2000; Sarzynski and Vicino 2019) within the process of the 'suburbanization of poverty' (Hochstenbach and Musterd 2018) to an overall state of autonomy from the inner-city area, termed as post-suburbia (Hanlon et al. 2009; Herzog 2019; Phelps 2018; Phelps et al. 2010; Phelps and Wu 2011; Teaford 2011, 2020 [1997]; Tzaninis and Boterman 2018; in the German-language discourse also discussed by Aring 1999; Burdack and Hesse 2006; Menzl 2014).

The notion of post-suburbia has been developed in order to overcome the outdated and "simplistic city-suburban dichotomy" (Bourne 1996a, p. 163) and to acknowledge the complexity and diversity that has emerged over the decades outside of the dense urban cores (Charmes and Keil 2015, p. 581). This conceptual turn follows the assumption that formerly suburban and peripheral spaces can at any time gain centrality and urbanity through the adaption of a typically urban architectural and structural foundation, densification and infill development, the diversification of the catalog of (land) uses, and in this course also the emergence of 'urban lifestyles' (Phelps 2017, p. 3; Soja 2014). Thus, in many metropolitan regions of the world, suburban communities have centralized and grown, while others have faced decline or even decay. In this process of post-suburbanization, communities overcome the state in which they were 'sub'-ordinated to the 'urban' inside (Bourne 1996a, p. 163), they undo traditional and hierarchical boundaries between inside and outside and evolve into a hybrid in-between version that is somewhat central but at the same time peripheral. Kühne (2019; cf. 2012) refers to this process as *urban/rural hybridization*, a development in which various influences blend and regroup through the renegotiation of differentiations, identities, and boundaries. As a result, the ambiguous post-suburban landscape develops into an unordered pastiche-like mosaic of different land uses, groups, and densities (discussed in detail in section 2.2.4).

This is particularly the case in San Diego, which has already begun to suburbanize in the late 19th century, mushrooming into numerous communities and forming several rings around its downtown, and which has—until now—not stopped to expand its edges (Kühne and Schönwald 2015a; Kühne and Weber 2019; F. Weber and Kühne 2017). However, along with the classic 20th century sub- und post-suburbanization, the City of San Diego has also faced the common and widely discussed consequences of the 'white flight' phenomenon: the decade-long processes of urban blight and downtown decay (see generally Beauregard 2003, 2006). Particularly since World War II and the following decade, as San Diego began to grow rapidly, the city was in need for fast and efficient housing solutions in order to accommodate the vast amount of new residents, which have arrived and settled in the militarized harbor town (Eddy 1993; Killory 1993; see Figure 4.2). The early construction and success of suburban neighborhoods and commercial centers beyond the downtown boundaries and the peripheral development of military housing in the 1940 s and'50 s have promoted the phase of decay and the central city's loss of significance for large parts of the white suburban population. In the course of the suburban growth, not only in San Diego but throughout the U.S., ethnic minorities and immigrants with low stocks of economic and symbolic capital have been 'trapped' inside the urban cores and the immediate neighborhoods surrounding it, due to segregative zoning and housing mortgage restrictions (referred to as redlining, see Hillier 2003). In the case of San Diego, not only the downtown but also the inner-ring suburbs and the industrialized central waterfront areas of East Village and Barrio Logan have already begun to decline during the Great Depression era, when the streetcars and the commencing mass motorization brought the white and affluent home buyers farther and farther away to the prestigious new neighborhoods at the city edges (Lynch and Appleyard 1974; cf. Appleyard and Stepner 2018).

In the following decades, the decline of central neighborhoods around the urban cores has become a trend in many larger U.S. metropolises, which is now referred to as the 'downfall of the inner-ring suburbs' (Airgood-Obrycki 2019; Cooke 2013; Cooke and Denton 2015; Hanlon 2008; Vicino 2008b). Throughout the U.S., starting in the 1960 s and particularly in the'70 s, first-tier neighborhoods "have generally experienced a lack of population growth, diminished economic status of residents, an aging population and housing stock, and uncompetitive labor force" (Vicino 2008a, p. 555; cf. Hanlon and Vicino 2007; Orfield 2002). This trend has perpetuated in many major metropolises throughout the 1980 s and'90 s and created inner-ring communities of predominantly immigrant and elderly population with increasing poverty levels and outdated commercial, public, and housing infrastructure (Puentes and Warren 2006). Accordingly, Hanlon (2010, p. 4) phrases drastically: "Once symbolic of the American dream, some [inner-ring communities] have now become America's nightmare." In addition to the attested complexity, diversity, and success of the sprawling post-suburban communities on the settlement fringes, the decline of the city centers and the inner-ring suburbs has intensified the contrast of 'inside' and 'outside' and thus also the urban/rural hybridity of the postmodern metropolises (Puentes and Orfield 2002, p. 4; Vicino 2008b, p. 512).

In the second half of the 20th century, as a counterreaction to the widespread phenomenon of downtown decay, policy makers, planners, and developers in many cities of the U.S. forged extensive and costly redevelopment plans (Herzog 2010). Within a modern planning approach, many of these revitalization projects were designed to raze historic vice areas and skid rows in the city centers and replace the dilapidated building stock with new commercial structures, such as shopping centers or major sports arenas. The aim was to construct new scenes of consumption and entertainment, to destroy the immaterial 'boundaries of fear' between the 'orderly' suburban communities and the 'chaotic' urban cores in order to draw the affluent white suburbanites back into the once thriving downtown areas. In the course of these efforts, the urban transient and homeless population, the immigrants, and the elderly low-income residents have been pushed further into the underutilized industrial and warehouse quarters at the downtown fringes as well as into the inner ring. This was particularly the case in San Diego, where the industrialized community of East Village has historically served as a 'containment area' (Kayzar 2006, pp. 153–158) for undesirable land uses and their users in the course of several redevelopment projects (illustrated in detail in section 5.4.1). In downtown San Diego, the displacement of the former residents and low-cost businesses has already begun during the Wilson-era in the 1970 s with the resolution for the major revitalization project of the Horton Plaza Mall and the Gaslamp Quarter, which targeted the former redlight district Stingaree and its adjoining area (Erie et al. 2010, 2011; Ervin 2008, 2007; Kayzar 2016; Rumpf 2016). However, despite the extensive redesign of more than 20 city blocks, the success of the costly project was mitigated by the continued efforts of bankers, development firms, and large landowners throughout the 1980 s and'90 s to develop the northern sections of the county into large commercial and job centers as well as idyllic neighborhoods (Davis 2003).

Nevertheless, at the turn of the century, several factors in San Diego have begun to change the familiar situation and mindset of endless growth and sprawl into the hinterland. Particularly in the years of 2003 and 2007, the increasing number and severance of natural catastrophes at the so-called wildland-urban interface of San Diego County has demonstrated the limits and risks of dispersed and decentralized housing (Erie et al. 2011, pp. 239–242; on the wildland-urban interface in the U.S. see generally Radeloff et al. 2005). With some of the highest death tolls in the history of California, the Cedar, Oak, and Paradise wildfires in 2003 and the Witch fire in 2007 have destroyed vast amounts of land and thousands of structures in San Diego (CAL FIRE 2022; cf. Pryde 2014, pp. 40–42). Besides the imminent fire danger, at the start of the millennium, developable land in the county has become scarce for the first time in the city's history. "For the most part, however, San Diego's planning strategy did little to address these knotty problems" (Erie et al. 2011, p. 141) and continued its expansive, land consuming development approach. Yet, soaring housing prices—particularly since the 1990 s (Griffin and Weeks 2014, p. 72)—and the mortgage crisis paired with the onset of the global recession in 2008 (Hesse 2008a, p. 229) began to redirect homebuyers and renters alike from the fringes toward the more affordable inside.

Furthermore, the beginning social reorientation and reconsideration of central neighborhoods is not only tied to the region's lack of affordable housing but also to the emergence of postmodern 'urban lifestyles' in San Diego and the increasing valorization of dense, walkable environments. Newcomers in the central areas "are often attracted by the unique cultural experiences that older urban neighborhoods seem to provide. In contrast to the dreary predictability of suburbs and upscale quarters, [San Diego's] low-income urban neighborhoods are perceived as bohemian, unique, gritty, multi-cultural, authentic, sociable, and community-oriented—cultural attributes that have become attractive [...] in a postmodern era of increasingly fluid identities" (Joassart-Marcelli 2021, p. 156). Generally, these perceptions are the result of the social shift to the postmodern value principle, which foregrounds and prioritizes emotionality and aesthetics and breaks with the modern scheme of functional-cognitive interpretation, which resonates rather with stereotypical suburban life and the urban morphology of separated functions and uses.

In international urban studies, geography, and sociology, this remarkable turning point has been discussed as a renunciation of the predictable and ordered everyday life in suburbia, which goes along with decreasing birth and marriage rates, higher educational levels and occupation rates of women, as well as other shifts in the job and housing market (Gallagher 2014; Siedentop et al. 2018; in the German-language discourse see Häußermann 2009). Besides this interpretation, the beginning social reorientation toward the centers and the rising number of redevelopment projects in the U.S. but also in Europe have been traced by means of the notion of social and physical reurbanization (Bourne 1996b; in Germany, among others, discussed by Brake and Herfert 2012; Danielzyk and Priebs 2012; Herfert and Osterhage 2012; Hesse 2008b; Kühne and Schönwald 2015b; Matthes 2014). According to Frank (2018, 2020), this is not only taking place through the postmodernization or urbanization of typically suburban lifestyles or the physical redevelopment of downtown areas but is particularly leading to a process of 'inner-city suburbanization' in the course of arriving households from suburbia, which carry their more or less suburban lifestyles and spatial demands into the city centers.

Despite this turnaround, the initial phases of the social and spatial reurbanization in the city centers have not necessarily led to an immediate 'resurgence' of the decayed inner-ring suburbs—neither in San Diego nor in many of the other sprawling U.S. metropolises (Hudnut 2003). In the 1970 s and'80 s, in the course of the redevelopment projects in the downtown areas, lower-income residents, social service providers, and low-cost businesses have been displaced, which fostered "increases in poverty, unemployment, and neighborhood disinvestment" (S. Lee and Leigh 2007, p. 148) in numerous inner-ring suburbs throughout the U.S. Thus, the first-tier neighborhoods have been left behind and continued their 'downfall' during times when the downtown areas had already begun to receive renewed political and private attention in the form of prestigious and costly redevelopment projects. This has particularly taken place in the neighborhoods on San Diego's downtown edge and in its inner ring, which deteriorated further in the 1980 s and'90 s during the realization of the large-scale downtown redevelopment projects of Horton Plaza Mall, the Gaslamp Quarter, as well as the Marina District and the Convention Center.

However, in recent years, this development trajectory has changed drastically in several metropolises. More and more cases are now showing a different expansion pattern that follows and supersedes the extensive displacement from the inner-city into the first ring. After several decades of downtown redevelopment, spatial limitations in the urban cores as well as changing planning policies toward 'smart growth' are now promoting a progression of the development and investment dynamics from the developed downtown areas into the outdated inner-ring neighborhoods. Over time, public planners and civic leaders began to recognize the dilapidated inner rings for their "unique set of assets" (Puentes and Warren 2006, p. 1; cf. S. Lee and Leigh 2005; Puentes and Orfield 2002), their centrality, historicity, density, and thus high walkability. In San Diego, this has led to the adoption of the 'City of Villages' strategy in the city's General Plan (CofSD 2018), an ambitious flagship objective that contrasts strongly with the decadelong planning approach of low-density, suburban sprawl. This new urban policy aims for infill development in the inner-ring neighborhoods and their transformation into attractive mixed-use, dense, transit-oriented, and walkable 'urban villages' (see section 5.2.1). Additionally, developers, realtors, and other private stakeholders began to emphasize the uniqueness and competitiveness of inner-ring suburbs compared to the peripheral neighborhoods and initiated redevelopment and new construction in the first tier (Sweeney and Hanlon 2017). Given Southern California's increasingly competitive and exclusive real estate market, the pressure on historically disinvested areas in San Diego, such as the eastern downtown fringe and the inner-ring neighborhoods, grew particularly high since the turn of the millennium. Thus, the ongoing public and private redevelopment efforts in San Diego's downtown "have begun to cross the central-city boundary" (Charles 2013, p. 1505) and are now reaching the historically disinvested communities in the first tier, leading to manifold social, spatial, and functional neighborhood changes.

Summing up this complex development concisely, the public, private, and social 'rediscovery' of the inner-ring suburbs constitutes a novel trend, which is currently reversing and undoing the previous state of decline and decay in the inner ring. Despite taking place not only in San Diego but also in other large metropolitan regions in the U.S. (among others see Charles and Peiser 2019; Hanlon and Airgood-Obrycki 2018; Hudnut 2003; Puentes and Warren 2006) and in Europe (studies in suburban Paris conducted by F. Weber 2019; F. Weber and Kühne 2017, 2020), so far only a modest number of studies have recognized and shed light on the manifold process of inner-ring neighborhood redevelopment. Within varying research interests, initial studies on the redevelopment dynamics in San Diego's downtown and the surrounding areas stem from Cantor and Rosentraub (2012), E. E. Delgado and Swanson (2019), Joassart-Marcelli (2021), Joassart-Marcelli and Bosco (2020), Kühne and Weber (2019), Kühne and Schönwald (2015a), and Le Texier (2007), including own precursory work (Roßmeier 2019, 2020b; Roßmeier and Weber 2021). However, the complexity of the changing social realities in the central neighborhoods of San Diego and the underlying spatial pattern of inner-ring redevelopment have not yet been addressed sufficiently. The present study takes this gap as a starting point for a comprehensive research that aims for a wider understanding of the manifold transformations on San Diego's downtown fringe and in its inner-ring neighborhoods. This undertaking will not only illuminate the various social meanings and interpretations of the neighborhood changes in San Diego but enrich the broad array of research fields and disciplines that are concerned with the current and future course of post-suburbia.

Within the aim of approaching the outlined transformations in San Diego in a comprehensive way, the present study requires a suitable methodology and perspective, which allows to go beyond mere descriptions of physical rearrangements, border locations, zoning updates, or changes of settlement density. Instead, it is crucial to consider the complex interplay of subjective meanings, social construction and interpretation processes, socio-spatial and demographic developments, as well as functional and aesthetic-symbolic updates. Above all, this brings the residents of San Diego's inner-ring neighborhoods—the individual voices, their experiences, and perceptions—in the center of the research interest. Thus, on the one hand, the study is drawing on phenomenological thinking and non- or more-than-representational theories within the aim of tracing individual experiences and feelings and uncovering subjective meanings (see generally Hitzler and Eberle 2004; Husserl 1970 [1954]; Merleau-Ponty 1962; Schütz 1970; within a landscape focus see Duineveld et al. 2017; Law 2009; Tilley 1994; Waterton 2019; Wylie 2003, 2019). By means of this methodology it becomes possible to “dissolve the boundaries between objects and space” (M. Jones 2009, p. 491; cf. Bosco 2015; Graham and Healey 1999). This enables the present analysis to take the subject’s individual meanings and experiences of the changing physical foundation of the neighborhoods and urban landscape into consideration without counteracting the methodological achievements of the anti-essential turn (outlined in detail in section 2.1.1).

On the other hand, the present analysis is embedded in a social constructivist methodology (Berger and Luckmann 1966; Burr 1995, 1998; Flick 2004; Weinberg 2009), which directs the research focus on the constructional character of individual perceptions and interpretations of the redeveloping neighborhoods (presented in detail in section 2.1.2). From this perspective, reality is a product of everyday social interaction and not a condition of given facts. Thus, the social constructivist methodology allows to “take a critical stance toward our taken-for-granted ways of understanding the world, including ourselves,” (Burr 2005, p. 2; cf. Gergen 1994, 1999; Schütz and Luckmann 1973) our own believes and ways of seeing the world, the sprawling city, and the neighborhood. This conception of social reality entails a particular awareness for aspects of processuality, plurality, and individuality and is connectable with post-structuralist and relational thinking (cf. Murdoch 2006). On the basis of a combined social constructivist and post-structuralist or relational methodology, it becomes possible to abandon (traditional planning-theoretical) “product-oriented physical” (Davoudi and Strange 2009, p. 32) approaches to space and embrace a “process-oriented and social conception” (Davoudi and Strange 2009, p. 32) instead. This perspectival orientation allows the present analysis to approach neighborhoods, places, and their images—and more generally also urban landscape—as constantly evolving constructs and outcomes of complex social interactions. Accordingly, the subjective perceptions and interpretations of San Diego’s evolving inner ring move into the analytical focus.

Beyond the methodological framework, the present analysis will gain perspectival guidance through a conceptual triangulation of urban landscape and border theory. Throughout its history, landscape theory has been enriched by various schools of thinking and has evolved in the course of several epochs, major paradigm shifts, and science-theoretical turns (cf. Duineveld et al. 2017, p. 375). Thus, the field of (urban) landscape research has become an endeavor of various disciplines, each of which contributes to the rich, yet ambiguous conceptual pool of landscape meanings and readings (see section 2.2). In light of the versatile conceptual history of landscape, among many others, social constructivist and phenomenological but also relational and post-structuralist approaches have emerged over time (see exemplary B. Anderson and Wylie 2009; Greider and Garkovich 1994; P. Howard et al. 2019; Kühne 2019; Waterton 2013), which provide a perspectival backdrop for the presently applied notion of landscape. Within the given analytical focus on processuality and changeability in San Diego's inner ring and against the background of the phenomenology-oriented social constructivist methodology, the present study will approach urban landscape as an ongoing individual *practice of spatial identity creation* (cf. Berr 2020, p. 190; drawing from Riehl 1996 and their idea of the 'landscape/scenic eye'). Thus, the notion of urban landscape will be used to highlight the practices of constructing, negotiating, and conferring identity to space in San Diego's inner ring. In addition to this understanding, also the idea of *(socio-)spatial hybridization* and pluralization, adopted from the postmodern landscape-theoretical focus on urban/rural hybridization (cf. Kühne 2019), will be integrated into the present landscape notion. Thus, it becomes possible to highlight spatial overlaps, aspects of (temporal) ambiguity, and other pastiche-like phenomena that arise in the processes of neighborhood change. For as mentioned above, the trend of inner-ring redevelopment seems to take place within a socio-spatial progression of growth perspectives, investment dynamics, and demographic conditions from the profitable but spatially limited downtown areas into the disinvested and outdated inner-ring neighborhoods. In other words, the analytical focus needs to be directed on processes of neighborhood expansion and the reinterpretation of spaces and neighborhoods. For a presumption of the study is that these processes are most likely not taking place *inside* the neighborhoods but particularly along the neighborhood *edges*, where they adjoin each other and potentially begin to overlap.

Besides the landscape-theoretical approach, the present research perspective is drawing on key realizations from the interdisciplinary field of border studies (see section 2.3). Similarly to landscape studies, border theory and the research field's object of investigation have evolved and diversified strongly during the 20th century and the various science-theoretical turns of the social sciences (F. Weber et al. 2020). Along with this evolution, also the border methodologies have changed fundamentally (Gerst and Krämer 2021b). Until the 1960 s and 70's, border research has mostly been conducted through positivist-territorial and essentialist perspectives *on* the border or *from the two sides* it separates, which meant to highlight its demarcating function (Gerst and Krämer 2021a, pp. 123–126). Subsequently, the shift to post-structuralist and relational thinking has led to a fruitful diversification of the research perspectives and border notions, which began to emphasize cross-border networks and governance structures (on the San Diego-Tijuana region see Herzog and Sohn 2014; on cross-border cooperation in Europe see Nienaber and Wille 2020; F. Weber 2022). However, particularly the emergence of social constructivist thinking, which has led to the "processual shift" (Brambilla et al. 2015, p. 1) of border studies, as well as the notion of *borders everywhere* (Balibar 2004; cf. Lyon 2005; Rumford 2012, p. 894) have changed the understanding of demarcations fundamentally. Instead of a simple look *on* or *over* the border, it became important to *see like a border*, which is a perspective that highlights the plurality, individuality, and processuality of borders (Gerst and Krämer 2021a, pp. 130–132; Rumford 2012, 2014). Thus, the processual turn, the notion of 'borders everywhere,' and the idea of 'seeing like a border' have helped to diversify the object of investigation from fixed diverging lines in space to *social processes of b/ordering and othering* (Brambilla et al. 2015, p. 2; van Houtum and van Naerssen 2002, p. 126; Iossifova 2020) or, in the sense of the present urban landscape notion, to *processes of inclusion and exclusion* through the social construction of spatial identity. In this course, the large body of border theory has also been transposed from the national border to other spatial levels, such as cities and neighborhoods, which has created the starting point for the present approach (on urban border research see among many Fauser 2017; Roßmeier 2020b; Roßmeier and Weber 2021; Sohn and Lara-Valencia 2013; Spierings 2012; F. Weber and Roßmeier 2020). Furthermore, in this context, it is also important to highlight the distinction of material and immaterial demarcations, which are each of strong everyday life relevance and will be taken into account in the terminology of the present study (outlined in section 2.3.1).

In addition to the understanding of borders as processes, the methodological perspective *into* the border (Gerst and Krämer 2021a, pp. 129–130) has been adopted in the present perspectival triangulation of urban landscape and border theories. This unique view allows for the conceptual transformation of border lines into three-dimensional *borderlands* that stretch demarcations into complex networks of social, political, economic, and environmental relations, among others (Anzaldúa 2012 [1987]; Pavlakovich-Kochi et al. 2004; Szytniewski et al. 2020). The resulting (urban) borderland concept traces the idea of abandoning the modern and polar dichotomy of the two different sides of a demarcation and argues for the existence of a unique *third space* instead (on hybridity and the third space see Bhabha 2000, 2012; Easthope 1998; Mitchell 1997). Accordingly, urban borderlands are "not just the fixed borders or boundaries around homogeneous territories—they are the negotiated, maintained and, occasionally, celebrated spaces in-between the different. These are spaces that allow for the emergence of alternatives" (Iossifova 2015, p. 104). Thus, the methodological perspective *into* the border and the (urban) borderland concept highlight aspects of connectivity and further the understanding of borders as spaces of coexistence

(Iossifova 2019). More importantly, however, the urban borderland concept is connectable to the ideas of plurality and socio-spatial hybridity, which have been adopted into the present notion of urban landscape. Furthermore, the perspective *into* the border denies the linear morphology of demarcations and confers the border a certain spatiality, which ultimately resembles (urban) landscape.

On the basis of these outlines, it becomes possible to triangulate the two developed notions of urban landscape and urban borders/borderlands by means of their conceptual focus on individuality, processuality, and hybridity. The result of this is the conceptualization of *hybrid urban borderlands* (presented in section 2.4), which will be applied and tested in the following analysis. On the basis of this perspectival-conceptual approach, it becomes possible to trace the processes of neighborhood expansion at the downtown edges and in between the adjacent neighborhoods or, in other words, the novel trend of inner-ring redevelopment. For, as presupposed, the spatial progression of redevelopment processes in San Diego is most likely commencing along the edges of adjacent, yet dissimilar neighborhoods, where it leads to 'border stretches,' overlaps, and socio-spatial hybridization between 'inside/outside,' 'here/there,' or simply between downtown and inner ring. These processes are believed to confer the concerned neighborhood boundaries a temporal, transitional spatiality—through the stretching of a differentiation line into a hybrid urban borderland—and thus also a certain socio-spatial and interpretational inaccuracy and ambiguity.

Taking these considerations as a starting point, the main objective of the following social constructivist empirical analysis is to further the understanding of the inner-ring redevelopment trend by examining how the neighborhood changes are socially perceived and interpreted. Thus, it becomes possible to trace in how far the redevelopment processes are going along with practices of socio-spatial identity (re-)creation and (re-)negotiation of San Diego's urban landscape, which means, for instance, in how far the outdated and disinvested neighborhoods are socially reinterpreted into trendy places to be. On top of this, it is of interest how the neighborhood changes and the (re-)interpretations are leading to community overlaps and new social and spatial inclusions or how these social processes might initiate new forms of exclusion and distinction in the city through new identity constructions and processes of b/ordering and othering. The phenomenological component of the present methodology allows to extend the analytical concern to individual neighborhood change experiences and the ways in which the everyday life meanings of the (physical foundations of the) neighborhoods are evolving. It is of interest, how the meanings of existing or new infrastructure emerge and develop, how the appearance of new construction projects and other updates in the neighborhoods is experienced, and thus also how socialized neighborhood images change over time. What are the expectations, the doubts and the fears, or the everyday thoughts and encounters of the residents in San Diego's downtown and inner-ring? How are their routines and (b/ordering) practices adapting in the course of the redevelopment and are spatial demands of new and old residents potentially getting into conflict? Finding answers to these questions is not only of importance for the residents of the central neighborhoods and their collective understanding of the ongoing redevelopment trend. Above this, answering the outlined questions within a focus on processuality and plurality of social reality is also beneficial for the determination and creation of future planning objectives in San Diego. Thus, these answers are also addressed to a variety of decision makers, planners, and public leaders. Given the scarcity of developable land in the county and the political urge to create a denser region, the private efforts to develop the neighborhoods, as well as the imposed transformation for the residents, understanding the trend of inner-ring redevelopment is crucial. The present study has taken up this complex aim and approaches the questions above by means of a comprehensive, multi-methodical research design, which draws on a mix of qualitative methods, quantitative data, as well as innovative representational-graphical illustrations.

Lastly, the introduction closes with a presentation of the structure of the present work. In the subsequent section, the operationalization of the phenomenological (2.1.1) and the social constructivist (2.1.2) methodology will be outlined in detail. In a next step, the present landscape approach will be presented. First, the historical formation of landscape geography will be illustrated in section 2.2.1, before the social constructivist approach to landscape as well as phenomenological variants of constructivist landscape approaches will be outlined in section 2.2.2 and 2.2.3. The subsection on landscape closes with an illustration of the sensitivity of the social constructivist landscape notion for aspects of urban/rural hybridity and spatial pastiches (2.2.4). Following the outline and the operationalization of the present urban landscape notion, the second component of the perspectival-conceptual approach will be introduced: the notion of borders. In order to be able to take the everyday life relevance of both material (physical) and immaterial (social processes, political lines, or aesthetic-symbolic) demarcations into account, it is crucial to develop a suitable terminology—this will be done in section 2.3.1. Subsequently, the historical genesis of the interdisciplinary research field of border studies will be outlined in detail (section 2.3.2). In addition to this, the section will discuss the evolution of common themes and concepts that have emerged over time. Lastly, the transposition of scale within the border studies field will be outlined, which forms the conceptual starting point for the present consideration of urban b/ordering processes and hybrid urban borderlands (section 2.3.3). After the presentation of the two key notions, it will be possible to triangulate the urban landscape and the border-theoretical approach into the conceptualization of hybrid urban borderlands (see section 2.4). For this endeavor, it is important to begin with a detailed outline of the borderland notion and a discussion of the relational connection of borders and space (section 2.4.1). In the following subsection (2.4.2), existing approaches and concepts that aim for the connection of (urban) landscape and border studies will be presented in order to assess common connecting points and overlaps, their conceptual qualities, as well as their suitability for the present study. Lastly, the conceptualization of hybrid urban borderlands will be elaborated. Following the methodological and conceptual-perspectival part of the work, in section 2.5, the central research questions will be deduced. The chapter will close with an initial critical reflection of the present methodology and theoretical foundation (section 2.6).

Subsequently, the multi-methodical approach of the present study will be presented in chapter 3. However, in order to approach the posed research questions, it will be crucial to outline the difficulties and the present social constructivist approach of 'bounding neighborhoods.' Accordingly, section 3.1 will discuss the present understanding of neighborhoods and neighborhood boundaries. In a next step, the neopragmatic approach to triangulating different methods and their methodologies will be presented. This approach allows the present study to draw simultaneously on qualitative methods, quantitative survey data, as well as on various graphic, rather representational illustrations (see section 3.2). Subsequently, the various methods will be presented: the two types of semi-structured and unstructured interviews in section 3.3.1, the participant short-term observations or phenomenological neighborhoods walks in section 3.3.2, the cartographic and illustration approach in section 3.3.3, and finally the approach of supplementing the outlined methods with quantitative survey data will be discussed in section 3.3.4. Lastly, the chapter will demonstrate the present evaluation approach—the qualitative content analysis—which allows to bring the different data together (section 3.4).

After the presentation of the methodical and the evaluation approach, the reader will be brought back to San Diego within an outline of the region's settlement history, which starts several thousand years ago with the migration of various Native American tribes into the present cross-border metropolis of San Diego-Tijuana (chapter 4). In subsequence to the illustrated chapters, it will be possible to present the empirical results and take a close look at the various neighborhoods of San Diego. The presentation will follow the direction of the discovered spatial pattern of a 'redevelopment wave' that moves through San

Diego's inner ring neighborhoods, initially starting in downtown and in Uptown San Diego (section 5.1), progressing over to North Park and east into Mid-City (section 5.2), as well as south into Greater Golden Hill (section 5.3), and now even further south from East Village (section 5.4.3) toward Southeastern San Diego and Barrio Logan (section 5.4.4), which are San Diego's "'borderlands' where gentrification pressures, negotiation about belonging, and threats of displacement are most intense" (Joassart-Marcelli 2021, p. 154). Each of the empirical subchapters will start with a historical and quantitative/demographic neighborhood overview and will close with an urban landscape and border-theoretical conclusion, which serves as a summary and theory-based conclusion of the presented results. At the end of the empirical presentation, section 5.5 will serve as a final discussion. There, the results will be outlined in detail against the background of the underlying methodology and theoretical foundation of the study.

The present work will close with a conclusion and a reflection on the presented results (chapter 6). This chapter is split into two section, the first part is dealing with methodological, methodical, as well as perspectival reflections and will address limitations and highlight the analytical scope of the present methodical and methodological triangulation. The second part of the conclusion will summarize the main results of the analysis and highlight the key realization for San Diego and for the trend of inner-ring redevelopment. Additionally, this section will point toward other possible case study areas that can possibly benefit from the application of the hybrid urban borderland conceptualization and constitute suitable contexts to test the conceptualizations' analytical scope further.

References

Airgood-Obrycki, W. (2019). Suburban status and neighbourhood change. *Urban Studies, 56* (14), pp. 2935–2952.

Airgood-Obrycki, W., Hanlon, B. & Rieger, S. (2020). Delineate the U.S. suburb: An examination of how different definitions of the suburbs matter. *Journal of Urban Affairs.*

Anderson, B. & Wylie, J. (2009). On Geography and Materiality. *Environment and Planning A: Economy and Space, 41* (2), pp. 318–335.

Anzaldúa, G. (2012 [1987]). *Borderlands/La frontera. The New Mestiza.* San Francisco: Aunt Lute Books.

Appleyard, B. & Stepner, M. (2018). Toward the Dreams and Realities of Temporary Paradise? Lynch and Appleyard's Look at the Special Landscape of San Diego/Tijuana. *Journal of the American Planning Association, 84* (3–4), pp. 230–236.

Aring, J. (1999). *Suburbia—Postsuburbia—Zwischenstadt. Die jüngere Wohnsiedlungsentwicklung im Umland der großen Städte Westdeutschlands und Folgerungen für die regionale Planung und Steuerung*. Hannover: Verl. der ARL.

Balibar, É. (2004). *We, the people of Europe. Reflections on transnational citizenship*. Princeton: Princeton University Press.

Beauregard, R. A. (2003). *Voices of Decline. The Postwar Fate of US Cities*. New York: Routledge.

Beauregard, R. A. (2006). *When America became suburban*. Minneapolis: U of Minnesota Press.

Bello, F. (1958). The City and the Car. In W. H. Whyte (Ed.), *The Exploding Metropolis* (Vol. 146, pp. 32–61). Berkley: University of California Press.

Berger, P. L. & Luckmann, T. (1966). *The social construction of reality*. New York: Anchor.

Berr, K. (2020). Visuality, Aesthetics, and Landscape: For the Clarification and Self-Awareness of Constructivist Landscape Research. In D. Edler, C. Jenal & O. Kühne (Eds.), *Modern Approaches to the Visualization of Landscapes* (pp. 189–215). Wiesbaden: Springer VS.

Bhabha, H. K. (2000). *Die Verortung der Kultur*. Tübingen: Stauffenburg.

Bhabha, H. K. (2012). Über kulturelle Hybridität: Tradition und Übersetzung. In A. Babka & G. Posselt (Eds.), *Über kulturelle Hybridität. Tradition und Übersetzung* (pp. 17–57). Wien: Turia + Kant.

Bosco, F. J. (2015). Actor-network theory, networks, and relational geographies. In S. C. Aitken & G. Valentine (Eds.), *Approaches to human geography. Philosophies, theories, people and practices* (2 ed., pp. 150–162). Los Angeles: Sage.

Bourne, L. S. (1996a). Reinventing the Suburbs: Old Myths and New Realities. *Progress in Planning, 46* (3), pp. 163–184.

Bourne, L. S. (1996b). Reurbanization, Uneven Urban Development, and the Debate on New Urban Forms. *Urban Geography, 17* (8), pp. 690–713.

Brake, K. & Herfert, G. (Eds.). (2012). *Reurbanisierung: Materialität und Diskurs in Deutschland*: Springer-Verlag.

Brambilla, C., Laine, J., Scott, J. W. & Bocchi, G. (2015). Introduction: Thinking, mapping, acting and living borders under contemporary globalisation. In C. Brambilla, J. Laine, J. W. Scott & G. Bocchi (Eds.), *Borderscaping: Imaginations and practices of border making* (pp. 1–9). Burlington: Ashgate.

Breitung, W. (2012). Enclave Urbanism in China: Attitudes Towards Gated Communities in Guangzhou. *Urban Geography, 33* (2), pp. 278–294.

Burdack, J. & Hesse, M. (2006). Reife, Stagnation oder Wende? Perspektiven zu Suburbanisierung, Post-Suburbia und Zwischenstadt: Ein Überblick zum Stand der Forschung. *Berichte zur deutschen Landeskunde, 80* (4), pp. 381–399.

Burr, V. (1995). *An Introdcution to Social Constructionism*. New York: Routledge.

Burr, V. (1998). Overview: Realism, Relativism, Social Constructionism and Discourse. In I. Parker (Ed.), *Social Constructionism, Discourse and Realism* (pp. 13–26). London: Sage.

Burr, V. (2005). *Social constructionism*. London: Routledge.

CAL FIRE. (2022). Stats and Events. The California Department of Forestry and Fire Protection. 20. September 2022.

Cantor, M. B. & Rosentraub, M. S. (2012). A ballpark and neighborhood change: Economic integration, a recession, and the altered demography of San Diego's Ballpark District after eight years. *City, Culture and Society, 3*, pp. 219–226.

Charles, S. L. (2013). Understanding the determinants of single-family residential redevelopment in the inner-ring suburbs of Chicago. *Urban Studies, 50* (8), pp. 1505–1522.

Charles, S. L. & Peiser, R. B. (2019). Inner-Ring Suburban Retrofit and Neighborhood Change in the Post-Suburban Era. In T. Banerjee & A. Loukaitou-Sideris (Eds.), *The New Companion to Urban Design* (pp. 1–13). London: Routledge.

Charmes, E. & Keil, R. (2015). The politics of post-suburban densification in Canada and France. *International Journal of Urban and Regional Research, 39* (3), pp. 581–602.

CofSD, P. D. (2018). General Plan. City of San Diego Planning Department. https://www.sandiego.gov/planning/programs/genplan#genplan. Accessed: 15. January 2022.

Cooke, T. J. (2013). Residential Mobility of the Poor and the Growth of Poverty in Inner-Ring Suburbs. *Urban Geography, 31* (2), pp. 179–193.

Cooke, T. J. & Denton, C. (2015). The suburbanization of poverty? An alternative perspective. *Urban Geography, 36* (2), pp. 300–313.

Crossey, N. & Weber, F. (2020). Zur Konstitution multipler *Borderlands* im Zuge der Frankreichstrategie des Saarlandes. In F. Weber, C. Wille, B. Caesar & J. Hollstegge (Eds.), *Geographien der Grenzen. Räume—Grenzen—Hybriditäten* (pp. 145–166). Wiesbaden: Springer VS.

Danielzyk, R. & Priebs, A. (2012). Suburbanisierung-angesichts von Reurbanisierungstendenzen ein Phänomen "von gestern"? In W. Schenk, M. Kühn, M. Leibenath & S. Tzschaschel (Eds.), *Suburbane Räume als Kulturlandschaften* (pp. 25–57). Hannover: Verl. d. ARL.

Davis, M. (2003). The Next Little Dollar: The Private Governments of San Diego. In M. Davis, K. Mayhew & J. Miller (Eds.), *Under the Perfect Sun the San Diego Tourists Never See* (pp. 17–144). New York: The New Press.

Davis, M., Mayhew, K. & Miller, J. (Eds.). (2003). *Under the Perfect Sun. The San Diego Tourists Never See*. New York: The New York Press.

Davoudi, S. & Strange, I. (2009). Space and place in the twentieth century planning: an analytical framework and a historical review. In S. Davoudi & I. Strange (Eds.), *Conceptions of Space and Place in Strategic Spatial Planning* (pp. 7–42). London: Routledge.

Delgado, E. E. & Swanson, K. (2019). *Gentefication* in the barrio: Displacement and urban change in Southern California. *Journal of Urban Affairs*, pp. 1–16.

Duany, A., Plater-Zyberk, E. & Speck, J. (2000). *Suburban Nation: The Rise of Sprawl and the Decline of the American Dream*. New York: North Press Point.

Duineveld, M., Van Assche, K. & Beunen, R. (2017). Re-Conceptualising political landscapes after the material turn: a typology of material events. *Landscape Research, 42* (4), pp. 375–384.

Easthope, A. (1998). Bhabha, hybridity and identiy. *Textual Practice, 12* (2), pp. 341–348.

Eddy, L. (1993). War comes to San Diego. *The Journal of San Diego History, 39* (1 & 2), p. n.pag.

Erie, S. P., Kogan, V. & MacKenzie, S. A. (2010). Redevelopment, San Diego Style: The Limits of Public—Private Partnerships. *Urban Affairs Review, 45* (5), pp. 644–678.

Erie, S. P., Kogan, V. & MacKenzie, S. A. (2011). *Paradise plundered: Fiscal Crisis and Governance Failures in San Diego*. Stanford: Stanford University Press.

Ervin, J. C. (2007). Reinventing downtown San Diego: A Spatial and Cultural Analysis of the Gaslamp Quarter. *The Journal of San Diego History, 53* (4), pp. 188–217.

Ervin, J. C. (2008). San Diego's Urban Trophy: Horton Plaza Redevelopment Project. *Southern California Quaterly, 90* (4), pp. 419–453.

Fauser, M. (2017). The Emergence of Urban Border Spaces in Europe. *Journal of Borderlands Studies, 34* (4), pp. 605–622.

Flick, U. (2004). Constructivism. In U. Flick, E. von Kardoff & I. Steinke (Eds.), *A Companion to Qualitative Research* (pp. 88–94). London: SAGE.

Ford, L. (2005). *Metropolitan San Diego: How Geography and Lifestyle Shape a New Urban Environment*. Philadelphia: University of Pennsylvania Press.

Ford, L. (2014). The Visions of the Builders. The Historical Evolution of the San Diego Cityscape. In P. R. Pryde (Ed.), *San Diego. An Introduction to the Region* (pp. 175–190). San Diego: Sunbelt Publications.

Frank, S. (2018). Inner-City Suburbanization–No Contradiction in Terms. Middle-Class Family Enclaves are Spreading in the Cities. *Raumforschung und Raumordnung | Spatial Research and Planning, 76* (2), pp. 123–132.

Frank, S. (2020). Suburbane Räume und Lebensweisen: Von der Peripherie ins Zentrum—und nicht zurück. In I. Breckner, A. Göschel & U. Matthiesen (Eds.), *Stadtsoziologie und Stadtentwicklung. Handbuch für Wissenschaft und Praxis* (pp. 257–268). Baden-Baden: Nomos.

Gailing, L. (2015). Die Transformation suburbaner Räume in westlichen Gesellschaften und die Perspektive der sozialwissenschaftlichen Landschaftsforschung. In O. Kühne, K. Gawroński & J. Hernik (Eds.), *Transformation und Landschaft: Die Folgen sozialer Wandlungsprozesse auf Landschaft* (pp. 83–93). Wiesbaden: Springer VS.

Gallagher, L. (2014). *The End of the Suburbs: Where the American Dream is Moving*. New York: Penguin.

Garreau, J. (1992). *Edge city: Life on the New Frontier*. New York: Anchor.

Gergen, K. (1994). *Realities and relationships: Soundings in social construction*. Cambridge: Harvard University Press.

Gerst, D. & Krämer, H. (2021a). Methodologie der Grenzforschung. In D. Gerst, M. Klessmann & H. Krämer (Eds.), *Grenzforschung. Handbuch für Wissenschaft und Studium* (pp. 121–140). Baden-Baden: Nomos.

Gerst, D. & Krämer, H. (2021b). The multiplication of border methodology. In C. Wille, D. Gerst & H. Krämer (Eds.), *Borders in Perspective—UniGR-CBS thematic issue. Identities and Methodologies of Border Studies: Recent Empirical and Conceptual Approaches* (Vol. 6, pp. 17–26).

Graham, S. & Healey, P. (1999). Relational concepts of space and place: Issues for planning theory and practice. *European Planning Studies, 7* (5), pp. 623–646.

Greider, T. & Garkovich, L. (1994). Landscapes: The Social Construction of Nature and the En vironment. *Rural Sociology, 59* (1), pp. 1–24.

Griffin, E. & Weeks, J. R. (2014). Peopling the Region. San Diego's Population Patterns. In P. R. Pryde (Ed.), *San Diego. An Introduction to the Region* (pp. 63–76). San Diego: Sunbelt Publications.

Häußermann, H. (2009). Der Suburbanisierung geht das Personal aus. Eine stadtsoziologische Zwischenbilanz. *Stadtbauwelt, 181* (12), pp. 52–57.

Hanlon, B. (2008). The decline of older, inner suburbs in metropolitan America. *Housing Policy Debate, 19* (3), pp. 423–456.

Hanlon, B. (2010). *Once the American Dream: Inner-ring suburbs of the metropolitan United States*. Philadelphia: Temple University Press.

Hanlon, B. & Airgood-Obrycki, W. (2018). Suburban revalorization: Residential infill and rehabilitation in Baltimore County's older suburbs. *Environment and Planning A: Economy and Space, 50* (4), pp. 895–921.

Hanlon, B., Short, J. R. & Vicino, T. (2009). *Cities and Suburbs. New Metropolitan Realities in the US*. London: Routledge.

Hanlon, B. & Vicino, T. (2007). The Fate of Inner Suburbs: Evidence from Metropolitan Baltimore. *Urban Geography, 28* (3), pp. 249–275.

Hanlon, B., Vicino, T. & Short, J. R. (2006). The New Metropolitan Reality in the US: Rethinking the Traditional Model. *Urban Studies, 43* (12), pp. 2129–2143.

Herfert, G. & Osterhage, F. (2012). Wohnen in der Stadt: Gibt es eine Trendwende zur Reurbanisierung? Ein quantitativ-analytischer Ansatz. In K. Brake & G. Herfert (Eds.), *Reurbanisierung* (pp. 86–112). Wiesbaden: Springer VS.

Herzog, L. A. (2010). *Return to the center: culture, public space, and city building in a global era.* Austin: University of Texas Press.

Herzog, L. A. (2015). *Global Suburbs: Urban Sprawl from the Rio Grande to Rio de Janeiro.* New York: Routledge.

Herzog, L. A. (2019). Suburbia/Postsuburbia. In A. M. Orum (Ed.), *The Wiley Blackwell Encyclopedia of Urban and Regional Studies* (pp. 1–7). Chichester: Wiley Blackwell.

Herzog, L. A. & Sohn, C. (2014). The Cross-Border Metropolis in a Global Age: A Conceptual Model and Empirical Evidence from the US–Mexico and European Border Regions. *Global Society, 28* (4), pp. 441–461.

Herzog, L. A. & Sohn, C. (2017). The co-mingling of bordering dynamics in the San Diego–Tijuana cross-border metropolis. *Territory, Politics, Governance, 7* (2), pp. 177–199.

Hesse, M. (2008a). Resilient Suburbs? Ungleiche Entwicklungsdynamiken suburbaner Räume in Nordamerika im Zeichen der Kreditkrise. *Geographische Zeitschrift, 96* (4), pp. 228–249.

Hesse, M. (2008b). Reurbanisierung? Urbane Diskurse, Deutungskonkurrenzen, konzeptionelle Konfusion. *Raumforschung und Raumordnung, 66* (5), pp. 415–428.

Hesse, M. (2010). Suburbs—the next slum? Explorations into the contested terrain of social construction and political discourse. *articulo. Journal of Urban Research, 43* (Special Issue 3).

Hillier, A. E. (2003). Redlining and the Home Owners' Loan Corporation. *Journal of Urban History, 29* (4), pp. 394–420.

Hitzler, R. & Eberle, T. S. (2004). Phenomenological Life-World Analysis. In U. Flick, E. von Kardoff & I. Steinke (Eds.), *A Companion to Qualitative Research* (pp. 67–71). London: Sage.

Hochstenbach, C. & Musterd, S. (2018). Gentrification and the suburbanization of poverty: Changing urban geographies through boom and bust periods. *Urban Geography, 39* (1), pp. 26–53.

Hollander, J. B., Polsky, C., Zinder, D. & Runfolda, D. (2018). Shrinking suburbs in a time of crisis. In B. Hanlon & T. Vicino (Eds.), *The Routledge Companion to the Suburbs* (pp. 1–14). London: Routledge.

Hooghe, L. & Marks, G. (2003). Unraveling the Central State, but How? Zypes of Multi-level Governance. *American Political Science Review, 97* (2), pp. 233–243.

Howard, P., Thompson, I., Waterton, E. & Atha, M. (2019). *The Routledge Companion to Landscape Studies* (2 ed.). London: Routledge.

Hudnut, W. H. (2003). *Halfway to everywhere: A portrait of America's first-tier suburbs.* Washington, D.C.: Urban Land Institute.

Husserl, E. (1970 [1954]). *The Crisis of European Sciences and Transcendental Philosophy. An Introduction to Phenomenological Philosophy.* Evanston: Northwestern University Press.

Iossifova, D. (2015). Borderland urbanism: seeing between enclaves. *Urban Geography, 36* (1), pp. 90–108.

Iossifova, D. (2019). Borderland. In A. M. Orum (Ed.), *The Wiley Blackwell Encyclopedia of Urban and Regional Studies* (pp. 1–4). Chichester: Wiley Blackwell.

Iossifova, D. (2020). Reading borders in the everyday: bordering as practice. In J. W. Scott (Ed.), *A Research Agend for Border Studies.* Cheltenham: Edward Elgar Publishing.

Jackson, K. T. (1985). *Crabgrass frontier: the suburbanization of the United States.* New York: Oxford University Press.

Joassart-Marcelli, P. (2021). *The $16 Taco: Contested Geographies of Food, Ethnicity, and Gentrification.* Seattle: University of Washington Press.

Joassart-Marcelli, P. & Bosco, F. J. (2020). The Taste of Gentrification. In A. H. Alkon, Y. Kato & J. Sbicca (Eds.), *A recipe for gentrification: Food, power, and resistance in the city* (pp. 31–53). New York: New York University Press.

Jones, M. (2009). Phase space: geography, relational thinking, and beyond. *Progress in Human geography, 33* (4), pp. 487–506.

Kayzar, B. (2006). *Analyzing Revitalization Outcomes in Downtown San Diego.* (PhD), University of California Santa Barbara and San Diego State University.

Kayzar, B. (2016). Ephemeral Conciliation: Community-Based Art and Redevelopment. *Yearbook of the Association of Pacific Coast Geographers, 78*, pp. 28–46.

Killory, C. (1993). Temporary Suburbs: The Lost Opportunity of San Diego's National Defense Housing Projects. *The Journal of San Diego History, 39* (1 & 2), p. n.pag.

Kruse, K. M. & Sugrue, T. J. (2006). *The New Suburban History.* Chicago: University of Chicago Press.

Kühne, O. (2012). *Stadt—Landschaft—Hybridität: Ästhetische Bezüge im postmodernen Los Angeles mit seinen modernen Persistenzen.* Wiesbaden: Springer VS.

Kühne, O. (2019). *Landscape Theories. A Brief Introduction.* Wiesbaden: Springer VS.

Kühne, O. & Schönwald, A. (2015a). *San Diego: Eigenlogiken, Widersprüche und Hybriditäten in und von ‚America´s finest city'.* Wiesbaden: Springer VS.

Kühne, O. & Schönwald, A. (2015b). San Diego: *Trouble in Paradise?* Zwischen Stadterneuerung, Reurbanisierung und restriktiver Steuerpolitik. *Geographische Rundschau, 67* (5), pp. 49–54.

Kühne, O. & Weber, F. (2019). *Hybrid California. Annäherungen an den Golden State, seine Entwicklungen, Ästhetisierungen und Inszenierungen.* Wiesbaden: Springer VS.

Lang, R. (2003). *Edgeless Cities: Exploring the Elusive Metropolis.* Washington, D.C.: Brookings Institution Press.

Lang, R., Sanchez, T. W. & Oner, A. C. (2009). Beyond Edge City: Office Geography in the New Metropolis. *Urban Geography, 30* (7), pp. 726–755.

Law, J. (2009). Actor Network Theory and Material Semiotics. In B. S. Turner (Ed.), *The New Blackwell Companion to Social Theory* (pp. 141–158). Oxford: Blackwell.

Le Texier, E. (2007). The Struggle against Gentrification in Barrio Logan. In R. Griswold del Castillo (Ed.), *Chicano San Diego: Cultural Space and the Struggle for Justice* (pp. 202–221). Tucson: University of Arizona Pr.

Lee, S. & Leigh, N. G. (2005). The Role of Inner Ring Suburbs in Metropolitan Smart Growth Strategies. *Journal of Planning Literature, 19* (3), pp. 330–346.

Lee, S. & Leigh, N. G. (2007). Intrametropolitan Spatial Differentiation and Decline of Inner-Ring Suburbs: A Comparison of Four US Metropolitan Areas. *Journal of Planning Education and Research, 27* (2), pp. 146–164.

Liao, K., Wehrhahn, R. & Breitung, W. (2019). Urban planners and the production of gated communities in China: A structure–agency approach. *Urban Studies, 56* (13), pp. 2635–2653.

Lucy, W. H. & Phillips, D. L. (2000). Suburban decline: The next urban crisis. *Issues in Science and Technology, XVII* (1), pp. 55–62.

Lynch, K. & Appleyard, D. (1974). *Temporary paradise: A look at the special landscape of the San Diego Region.* San Diego: San Diego City Planning Department.

Lyon, D. (2005). The border is everywhere: ID cards, surveillance and the other. In E. Zureik & M. B. Salter (Eds.), *Global Surveillance and Policing. Borders, security, identity* (pp. 66–82). Cullompton: Willan Publishing.

Masotti, L. H. & Hadden, J. K. (1973). *The urbanization of the suburbs.* Beverly Hills: SAGE Publications.

Masotti, L. H. & Hadden, J. K. (1974). *Suburbia in transition.* New York: New Viewpoints.

Matthes, G. (2014). Zur Quantifizierung von Reurbanisierungstendenzen. *Raumforschung und Raumordnung, 72* (4), pp. 323–336.

Mendoza, J. E. & Dupeyron, B. (2020). Economic Integration, Emerging Fields and Cross-border Governance: The Case of San Diego–Tijuana. *Journal of Borderlands Studies, 35* (4), pp. 55–74.

Menzl, M. (2014). Urbanisierungsprozesse in Suburbia? Überlegungen zur Ubiquität der urbanen Lebensweise. *Jahrbuch StadtRegion, 8* (1), pp. 43–60.

Merleau-Ponty, M. (1962). *Phenomenology of Perception.* London: Routledge.

Mirkowich, N. (1941). Urban Growth in the San Diego Region. *Economic Geography, 17* (3), pp. 308–310.

Mitchell, K. (1997). Different Diasporas and the Hype of Hybridity. *Environment and Planning D: Society and Space, 15* (5), pp. 533–553.

Murdoch, J. (2006). *Post-structuralist geography: a guide to relational space.* London: Sage.

Nienaber, B. & Wille, C. (2020). Cross-border cooperation in Europe: a relational perspective. *European Planning Studies, 28* (1), pp. 1–7.

Orfield, M. (2002). *American Metropolitics: The New Suburban Reality.* Washington, DC: Brookings Institution.

Pavlakovich-Kochi, V., Morehouse, B. J. & Wastl-Walter, D. (Eds.). (2004). *Challenged borderlands. Transcending political and cultural boundaries.* Aldershot: Ashgate.

Peña, S. (2007). Cross-Border Planning at the US-Mexico Border: An Institutional Approach. *Journal of Borderlands Studies, 22* (1), pp. 1–18.

Phelps, N. A. (2017). Introduction: Old Europe, New Suburbanization? In N. A. Phelps (Ed.), *Old Europe, New Suburbanization? Governance, Land, and Infrastructure in European Suburbanization* (pp. 3–17). Toronto: University of Toronto Press.

Phelps, N. A. (2018). In what sense a post-suburban era? In B. Hanlon & T. Vicino (Eds.), *The Routledge Companion to the Suburbs* (pp. 1–9). London: Routledge.

Phelps, N. A., Wood, A. M. & Valler, D. C. (2010). A Postsuburban World? An outline of a Research Agenda. *Environment and Planning A: Economy and Space, 42* (2), pp. 366–383.

Phelps, N. A. & Wu, F. (Eds.). (2011). *International Perspectives on Suburbanization. A Post-Suburban World?* London: Palgrave Macmillan.

Pryde, P. R. (2014). The Nature of the County. San Diego's Climate, Vegetation, and Wildlife. In P. R. Pryde (Ed.), *San Diego. An Introduction to the Region* (Fifth ed., pp. 29–45). San Diego: Sunbelt Publication.

Pryde, P. R. & Stutz, F. P. (2014). "America's Finest City". Communities of the City of San Diego. In P. R. Pryde (Ed.), *San Diego. An Introduction to the Region* (pp. 191–208). San Diego: Sunbelt Publications.

Puentes, R. & Orfield, M. (2002). *Valuing America's First Suburbs: A Policy Agenda for Older Suburbs in the Midwest.* Washington: The Brookings Institution Center on Urban and Metropolitan Policy.

Puentes, R. & Warren, D. (2006). *One-fifth of America: A comprehensive guide to America's first suburbs.* Washington: The Brookings Institution Metropolitan Policy Program.

Radeloff, V. C., Hammer, R. B., Stewart, S. I., Fried, J. S., Holcomb, S. S. & McKeefry, J. F. (2005). The wildland–urban interface in the United States. *Ecological applications, 15* (3), pp. 799–805.

Riehl, W. H. (1996). Das landschaftliche Auge. In G. Gröning & U. Herlyn (Eds.), *Landschaftswahrnehmung und Landschaftserfahrung* (pp. 28–68). Münster: LIT Verlag.

Roßmeier, A. (2019). *Quo vadis ,Temporary Paradise'?* Urfsurbanisierung und räumliche Konflikte im StadtLandHybriden San Diego. In K. Berr & C. Jenal (Eds.), *Landschaftskonflikte* (pp. 591–616). Wiesbaden: Springer VS.

Roßmeier, A. (2020a). COVID-19 in the borderland of San Diego and Tijuana: between re-bordering processes and contradictory realities. In C. Wille & R. Kanesu (Eds.), *Borders in Perspective UniGR-CBS Thematic Issue. Bordering in Pandemic Times: Insights into the COVID-19 Lockdown* (Vol. 4, pp. 57–61).

Roßmeier, A. (2020b). Hybrid Urban Borderlands. Von der Hybridität intra-urbaner Grenzen in der internationalen Metropolregion San Diego. In B. Caesar, J. Hollstegge, F. Weber & C. Wille (Eds.), *Geographien der Grenzen. Räume—Ordnungen—Verflechtungen* (pp. 357–389). Wiesbaden: Springer VS.

Roßmeier, A. & Weber, F. (2021). Hybrid Urban Borderlands—Redevelopment Efforts and Shifting Boundaries In and Around Downtown San Diego. *Journal of Borderlands Studies*, pp. 1–27.

Rumford, C. (2012). Towards a Multiperspectival Study of Borders. *Geopolitics, 17* (4), pp. 887–902.

Rumford, C. (2014). *Cosmopolitan Borders.* Hampshire: Palgrave Macmillan.

Rumpf, L. E. (2016). *The synthesis of social reality and the evolution of Barrio Logan: A critical view of the framing and representation of urban redevelopment.* Fielding Graduate University, Santa Barbara.

Sandul, P. J. P. (2014). *Calfiornia Dreaming. Boosterism, Memory, and Rural Suburbs in the Golden State.* Morgantown: West Virginia University Press.
Sarzynski, A. & Vicino, T. (2019). Shrinking Suburbs: Analyzing the Decline of American Suburban Spaces. *Sustainability, 11* (19), pp. 1–19.
Schnore, L. F. (1957). Metropolitan growth and decentralization. *American Journal of Sociology, 63* (2), pp. 171–180.
Schütz, A. (1970). *Collected Papers III. Studies in Phenomenological Philosophy.* Den Haag: Martinus Nijhoff.
Schütz, A. & Luckmann, T. (1973). *The Structures of the Life-World* (Vol. 1). Evanston, Illinois: Northwestern University Press.
Siedentop, S., Zakrzewski, P. & Stroms, P. (2018). A childless urban renaissance? Age-selective patterns of population change in North American and German Metropolitan areas. *Regional Studies, Regional Science, 5* (1), pp. 1–20.
Sieverts, T. (2003). *Cities without cities. An interpretation of the Zwischenstadt.* London: Routledge.
Sohn, C. & Lara-Valencia, F. (2013). Borders and cities: Perspectives from North America and Europe. *Journal of Borderlands Studies, 28* (2), pp. 181–190.
Soja, E. W. (1993). Los Angeles, eine nach außen gekehrte Stadt. Die Entwicklung der postmodernen in den USA. In V. Kreibich, B. Krilla & U. von Petz (Eds.), *Rom—Madrid—Athen. Die neue Rolle der städtischen Peripherie* (pp. 213–228). Dortmund: Selbstverlag.
Soja, E. W. (2014). *My Los Angeles. From Urban Restructuring to Regional Urbanization.* Berkley: University of California Press.
Spierings, B. (2012). Economic Flows, Spatial Folds and Intra-Urban Borders: Reflections on City Centre Redevelopment Plans From a European Border Studies Perspective. *Tijdschrift voor economische en sociale geografie, 103* (1), pp. 110–117.
Stilgoe, J. R. (1988). *Borderland: Origins of the American suburb, 1820–1939.* New Haven: Yale University Press.
Sweeney, G. & Hanlon, B. (2017). From old suburb to post-suburb: The politics of retrofit in the inner suburb of Upper Arlington, Ohio. *Journal of Urban Affairs, 39* (2), pp. 241–259.
Szytniewski, B., Spierings, B. & Van der Velde, M. (2020). Stretching the Border: Shopping, Petty Trade and Everyday Life Experiences in the Polish-Ukrainian Borderland. *International Journal of Urban and Regional Research, 44* (3), pp. 469–483.
Teaford, J. C. (2011). Suburbia and Post-suburbia: A Brief History. In N. A. Phelps & F. Wu (Eds.), *International Perspectives on Suburbanization* (pp. 15–34). London: Palgrave Macmillan.
Teaford, J. C. (2016 [1986]). *The twentieth-century American city: Problem, promise, and reality* (Third ed.). Baltimore: John Hopkins University Press.
Teaford, J. C. (2020 [1997]). *Post-Suburbia: Government and Politics in the Edge Cities* (Second ed.). Baltimore: Johns Hopkins University Press.
Tilley, C. (1994). *A phenomenology of landscape: places, paths, and monuments.* Oxford: Berg.
Tzaninis, Y. & Boterman, W. (2018). Beyond the urban–suburban dichotomy. Shifting mobilities and the transformation of suburbia. *City, 22* (1), pp. 43–62.
van Houtum, H. & van Naerssen, T. (2002). Bordering, Ordering and Othering. *Tijdschrift voor economische en sociale geografie, 93* (2), pp. 125–136.

Vicenzotti, V. & Qviström, M. (2018). *Zwischenstadt* as a travelling concept: towards a critical discussion of mobile ideas in transnational planning discourses on urban sprawl. *European Planning Studies, 26* (1), pp. 115–132.

Vicino, T. (2008a). The Quest to Confront Suburban Decline: Political Realities and Lessons. *Urban Affairs Review, 43* (4), pp. 553–581.

Vicino, T. (2008b). The spatial transformation of first-tier suburbs, 1970 to 2000: The case of metropolitan Baltimore. *Housing Policy Debate, 19* (3), pp. 479–518.

Villamizar-Santamaría, S. (2019). Gated Communities/Fortified Enclaves. In A. M. Orum (Ed.), *The Wiley Blackwell Encyclopedia of Urban and Regional Studies* (pp. 1–10). Chichester: Wiley Blackwell.

Waterton, E. (2013). Landscape and non-representational theories. In P. Howard, I. Thompson, E. Waterton & M. Atha (Eds.), *The Routledge Companion to Landscape Studies.* Abingdon: Routledge.

Waterton, E. (2019). More-than-representational landscapes. In P. Howard, I. Thompson, E. Waterton & M. Atha (Eds.), *The Routledge Companion to Landscape Studies* (2nd ed., pp. 91–101). London: Routledge.

Weber, F. (2019). Von divergierenden Grenzziehungen und Konflikten im StadtLandHybriden des Grand Paris. In K. Berr & C. Jenal (Eds.), *Landschaftskonflikte* (pp. 637–663). Wiesbaden: Springer VS.

Weber, F. (2022). Cross-border cooperation in the border region of Germany, France, and Luxembourg in times of Covid-19. *European Societies, 24* (3), pp. 1–28.

Weber, F. & Kühne, O. (2017). Hybrid suburbia: New research perspectives in France and Southern California. *Quaestiones Geographicae, 36* (4), pp. 17–28.

Weber, F. & Kühne, O. (2020). Hybride Grenzen. Stadt-Land-Entwicklungsprozesse im Großraum Paris aus der Perspektive der Border Studies. *Geographische Rundschau, 72* (1–2), pp. 48–53.

Weber, F. & Roßmeier, A. (2020). Les nouvelles marges urbaines au prisme des border studies: le cas de San Diego (Californie). *Revue de l'Institut de Sociologie de l'Université libre de Bruxelles, 90* (2020), pp. 61–79.

Weber, F., Wille, C., Caesar, B. & Hollstegge, J. (2020). Entwicklungslinien der Border Studies und Zugänge zu Geographien der Grenzen. In F. Weber, C. Wille, B. Caesar & J. Hollstegge (Eds.), *Geographien der Grenzen. Räume—Ordnungen—Verflechtungen* (pp. 3–22). Wiesbaden: Springer VS.

Webster, C., Glasze, G. & Frantz, K. (2002). The global spread of gated communities. *Environment and Planning B: Planning and Design, 29*, pp. 315–320.

Weinberg, D. (2009). Social Constructionism. In B. S. Turner (Ed.), *The New Blackwell Companion to Social Theory* (pp. 281–299). Oxford: Blackwell.

Whyte, W. H. (1958). Urban Sprawl. In W. H. Whyte (Ed.), *The Exploding Metropolis* (pp. 133–156). Berkeley: University of California Press.

Wylie, J. (2003). Landscape, performance and dwelling: a Glastonbury case study. In P. Cloke (Ed.), *Country Visions.* Harlow: Pearson Education Limited.

Wylie, J. (2019). Landscape and phenomenology. In P. Howard, I. Thompson, E. Waterton & M. Atha (Eds.), *The Routledge Companion to Landscape Studies* (2nd ed., pp. 127–138). London: Routledge.

Theoretical Perspectives, Operationalization, and the Conceptual Approach of Hybrid Urban Borderlands

2

The following chapter covers the theoretical and conceptual basis of the study. The theoretical foundation is presented in section 2.1: The phenomenological movement in philosophy and sociology in the tradition of Martin Heidegger (2005 [1994]), Edmund Husserl (Husserl 1913, 1970 [1954], 1973 [1929]), Maurice Merleau-Ponty (1962), and Alfred Schütz (1970; Schütz and Luckmann 1973), followed by the social constructivist perspective in the tradition of Peter L. Berger and Thomas Luckmann's (1966) sociology of knowledge. Subsequently, in section 2.2, landscape research is presented as one of the two fields in which the present analysis is embedded. In particular, the term 'landscape' and the social constructivist understanding of landscape are discussed (Gailing and Leibenath 2015; Greider and Garkovich 1994; P. J. Howard 2011; Kühne 2006, 2018a). Building on the social constructivist perspective and thus on the constructional character of landscape, the particular sensitivity of landscape research for aspects of hybridity will be outlined. It will be traced how the specific focus of landscape studies on *urban/rural hybridizations* emerged in response to social and cultural developments of the 20th and 21st centuries and how these hybridizations arise in a reciprocal relationship with the contingency of individual landscape conceptions (Kühne 2012a; Kühne et al. 2016; Kühne 2018a; Roßmeier 2020d).

In a next step, in section 2.3, the interdisciplinary research field of border studies will be introduced as the perspectival addition to the present landscape-theoretical foundation. Here, the genesis of the research field will be described on the basis of the common themes of border studies. It will be explained how the thematic expansion of the research object beyond the traditional scale of nation-state borders took place within border studies, and how urban borders and boundaries gained interest. In doing so, the notions of landscape and border will be operationalized in order to create an interface for the investigation of urban border and boundary situations, everyday (bordering) experiences, and individual

A. Rossmeier, *San Diego's Hybrid Urban Borderlands*,
Räume – Grenzen – Hybriditäten, https://doi.org/10.1007/978-3-658-42667-5_2

interpretations of inclusion/exclusion and own/alien in the context of urban redevelopment. Thus, in section 2.4 the conceptual-perspectival approach of hybrid urban borderlands will be elaborated. Subsequently, in a theoretical summary, the individual components of this chapter are brought together in order to derive related research questions (section 2.5). The chapter closes with critical reflections on the chosen theoretical and perspectival foundations and their implications for the present analysis of San Diego's inner-ring neighborhood redevelopment (section 2.6).

2.1 Methodological Foundation of the Analysis: Phenomenology and Social Constructivism

Social constructivist approaches are based on phenomenological considerations in particular and look back on a fruitful tradition in the social sciences in the course of the cultural and spatial turn in the second half of the 20th century (Burr 1995, 2005; Gergen 1994, 1999; Kühne 2018a, pp. 11–16; H. Scott 2004; Wylie 2003; 2019, p. 130). A fundamental difference of the social constructivist perspective and the phenomenological theory lies in their disciplinary anchoring. Accordingly, the two approaches also differ in interest and focus. While the social science theory of social constructivism stems from sociological considerations on social and individual knowledge production, phenomenology constitutes a philosophical and more fundamental way of thinking about the experiencing human or individual in the material world. Generally, phenomenology is "providing a philosophical foundation of the methodology of the social sciences" (Eberle 2014, p. 184) and thus of (social) constructivist thinking. Contrary to essentialist, positivist, and especially realist perspectives, social constructivist approaches stress the contingency and thus observer-dependency of reality as well as its constructive character (Berger and Luckmann 1966; Flick 2004; Schütz and Luckmann 1973; see Table 2.1). Phenomenology—as well as more recent approaches, such as 'more-than-representational' theories (Ingold 1993; Lorimer 2005; Thrift 2008; Waterton 2019) and the actor-network theory (Bosco 2015; Latour 1999; Law 2009) –, however, is focused on the intertwined relationship of subject and object (Chemero 2003; Moran 2000). Put more precisely, phenomenological research "aims to elucidate and express the meaning and nature of things in the world—of *phenomena*—through a focus upon human lived experience, perception, sensation and understanding" (Wylie 2019, p. 127 [original emphasis]; cf. Hitzler and Eberle 2004; Kühne 2019a, pp. 32–33; Soeffner 2004, pp. 97–98). In the following, the two theoretical approaches will be outlined and in a next step operationalized into a phenomenologically oriented social constructivist landscape and border research approach (section 2.4).

2.1.1 Phenomenological Philosophy and Proto-Sociology

Phenomenology is primarily a movement of continental philosophy, which developed into a theoretical or perspectival basis for sociological or "*proto*-sociological enterprise[s]" (Hitzler and Eberle 2004, p. 68 [original emphasis]) through the work of Alfred Schütz (1960 [1932]; Schütz and Luckmann 1973). Fundamentally, phenomenology "can be understood as the study and description of all phenomena" (Kühne 2019a, p. 31) in the world with a focus on the individual, on the human body, and on everyday lived or conscious experiences (Gergen 1999, p. 127; Merleau-Ponty 1962; Wylie 2007, p. 140). It is of interest how phenomena, describing the totality of "units, things, and events that present themselves to the world" (Kühne 2019a, p. 31) are experienced by a subject (Tilley 1994, p. 12). Thus, phenomenological considerations pursue the goal of analyzing and describing phenomena in the way "they *appear* to consciousness" (Moran 2000, p. 6 [original emphasis]; cf. Cresswell 2013, p. 112). Given its shared historical and philosophical roots with romanticism, phenomenology seeks to uncover invariant features and "deeper truths about humanity and nature" (Wylie 2019, p. 129) within a rather essentialist perspective. Within this understanding, however, subjectivity is not negated—the interest is built around "the nature of individual human subjectivity" (Wylie 2019, p. 129) and "of individual awareness" (Gergen 1999, p. 127), around the framework of perception and experience. Following Husserl (1970 [1954]), the framework and foundation of all knowledge, individual and everyday experiences, interpretations, and constructions is the life-world (see Figure 2.1). Husserl introduced the life-world concept into the discussion of the principles of philosophy as a critique against the objectivism of positivist sciences and their lack of reflection on truth and knowledge production. His undertaking is based on the premise that all socially constructed reality originates from a subject's orientation in the world, which constitutes the horizon of all phenomena (Hitzler and Honer 1984, p. 58). Thus, Husserl tried to establish his philosophy as broad foundation for other sciences, as a meta-science (Hitzler and Honer 1984, p. 58), which acknowledges the subjectivism of all constructions and interpretations, of all knowledge, scientific and prescientific (Husserl 1970 [1954], p. 121). Consequently, he laid the foundation for Peter L. Berger and Thomas Luckmann's (1966) sociology of knowledge and the scientific perspective of constructivism (cf. Kühne 2018a, pp. 11–16).

Building on Husserl's (1970 [1954]) preliminary work on transcendental phenomenology, Alfred Schütz (1970) conceptualized and operationalized their mundane phenomenology, a sociological approach of phenomenological life-world analysis (Flaherty 2009, pp. 221–226). In line with Husserl's critique of the objective and inductive approach of positivist science, phenomenological research in the tradition of Schütz is centered around subjective, individual experiences and deals with these insights in a reflexive manner (Hitzler and Eberle 2004, p. 67). In addition to Husserl's considerations on the life-world, Schütz' work is particularly indebted to Max Weber's (1981) sociology of *Verstehen* (engl.: understanding or interpreting; cf. Abel 1948) and his concept of 'subjectively intended meaning,' which an actor associates with a specific action. M. Weber defines acting as a meaningful, intentional behavior, an expression of active body control, the meaning of which is assigned from individual consciousness (Keller 2013, p. 27). In line with this, Schütz' approach within his mundane phenomenology focuses on the description of experiences of subjects and is dedicated to the process of understanding the subjectively intended meaning of their actions in the sense of M. Weber's interpretative sociology. His general aim and interest lies in the description of individual reality as it is constituted in subjective consciousness (Hitzler and Honer 1984, p. 59). In this way, Schütz is shifting the focus of sociological considerations to the individual subject and the epistemological reconstruction of the formal, invariant structures of the life-world (Kühne 2018a, p. 12). The life-world in this sense "is on the one hand a point of reference and on the other hand an implicit basis for research work in the social sciences" (Hitzler and Eberle 2004, p. 68; cf. Knoblauch 1995, p. 12). The associated life-world phenomenology is thus to be understood as the attempt to *understand understanding itself*, it is a process of "second degree" (Soeffner 2004, p. 97) understanding. Thus it becomes clear that phenomenology is "the foundation for the explanation of the social world, not the explanation itself"[1] (Hitzler and Honer 1984, p. 59). In this way, Schütz' approach is conceptualized as a "procedure of protosociological description"[2] (Hitzler and Honer 1984, p. 59), a hermeneutic undertaking that underlies sociological analysis and therefore allows a variety of possibilities for further theoretical and methodological development, as it will be shown in the following sections.

---

1 „das Fundament für das Erklären von Sozialwelt, nicht das Erklären selbst"

2 „protosoziologisches Deskriptionsverfahren"

Referring to the execution of social-scientific hermeneutics, Soeffner (2004, p. 95) describes the process of phenomenological interpreting as "other-understanding." In contrast to more or less unconscious processes of self-understanding, other-understanding means the process of interpreting and giving meaning to experiences, witnessed behaviors, and all phenomena in the world that have been expressed and thus already pre-interpreted by a subject. In line with this, phenomenological interpretations are to be seen as "constructs of constructs" (Soeffner 2004, p. 96), constructs of second degree, which are construed according to the observing subject's perspective and knowledge (Schütz 1972 [1962], p. 7). Thus, within phenomenological life-world analyses, understanding and interpreting—which constitute fundamental and unquestioned everyday activities—are performed as scientific operations within a specific methodological framework (Knoblauch 1995, p. 12). In doing so, the phenomenological observer takes on the role of a "scientific interpreter" (Soeffner 2004, p. 96), which operates, according to Schütz, in a "theoretical sub-world of meaning. This is a perspective of fundamental doubt in a taken for granted social reality" (Soeffner 2004, p. 97), which is taken in the attempt to reconstruct subjectively intended meaning, the constructs of first degree, as precisely as possible. This is due to the fact that other-understanding is only possible in fragments—constructs of first degree can only be reconstructed as approximation (Hitzler and Eberle 2004, pp. 68–69). Accordingly, "the semantic pre-constitution of the social world" needs to be taken into account methodologically, as Hitzler and Eberle (2004, p. 69) stress. In this sense, performed life-world analysis, such as the present study of everyday experiences in San Diego's central neighborhoods, "represents an attempt to describe the concrete forms of orientation, action and organization of individuals in and with their environment, and to interpret concrete action against this background" (Soeffner 2004, p. 98).

Within this quote, Soeffner is addressing a pivotal aspect and undertake of the agenda of phenomenology (Chemero 2003; Duineveld et al. 2017; Kühne 2019a, pp. 31–33; Wylie 2003, 2019): the role of environment or space within individual experiences and thus the intertwined relationship of object and subject, materiality and the subjective life-world (see Figure 2.1). The following sections on the operationalization of phenomenology and social constructivism will take up the aspect of the subject/object division and discuss their understanding of and relationship to environment and space, which is necessary for the elaboration of the theoretical-conceptual framework of the present research. In conclusion, it can be noted that phenomenological approaches in the tradition of Husserl and Schütz are an attempt to encounter subjective everyday experiences, the life-worldly backgrounds of certain behaviors, decisions, and perceptions in and in

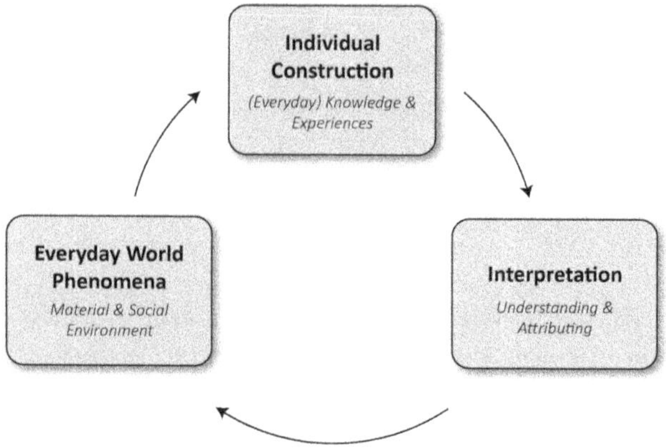

Figure 2.1 Life-World as the Horizon of Perception and Orientation: Interdependency of Everyday-World Phenomena, Individual Construction and Interpretation. (Source: Own representation based on Flick (2004) and Soeffner (2004))

relation with the material world. But how—and this is a fundamental question within Schütz' work—can "other human beings be understood if there is no direct access to their consciousness" (Hitzler and Eberle 2004, p. 69)? For this kind of approach to the social world, phenomenology is providing a theoretical foundation upon which the social constructivist perspective is built, as it will be outlined in the following sections.

2.1.2 Sociology of Knowledge and the Science Theoretical Position of Social Constructivism

Established in the German-language sociology in the 1920s by the early work of Max Scheler (1960 [1926]) and Karl Mannheim (1936) but also through English-language work by social scientists, such as Robert Merton (1957 [1949], 1973) and Louis Wirth, sociology of knowledge advanced to a substantial movement in sociology in the course of the 20th century (Kneer 2010). Particularly through the work of Peter L. Berger and Thomas Luckmann, sociology of knowledge received a thematic expansion and thus a significant development impulse in the 1960s and '70s. Berger and Luckmann's (1966) *The Social Construction of Reality* opened up the science theoretical position of social constructivism, drawing especially on phenomenological considerations, on Alfred Schütz' phenomenological sociology

(1970) and on the sociological concept of symbolic interactionism in the tradition of George H. Mead (2015 [1934]) and Herbert Blumer (1969). Thus, throughout the 1970s and '80s, the understanding of knowledge and reality in large parts of the social and cultural studies changed in a significant way, dividing these research fields continuingly from the realist positioned majority of the natural sciences in the course of the constructivist paradigm shift (Kneer 2009, p. 5; Kühne 2018b, p. 7). In the wake of this turn, different programmatic focuses within the constructivist position emerged, from moderate social constructivist approaches to radical constructivist concepts, each characterized through different premises and starting points for other science-theoretical positions. However, within all constructivist approaches, reality is attributed a subjective constructional character and thus lost its supposed immutability. Ultimately, reality was no longer understood as passive-objective dominant of society but as actively and subjectively generated outcome, which creates society in the first place (Berger and Luckmann 1966, pp. 27, 79). The following paragraphs trace the programmatic evolution of the phenomenological sociology of knowledge into the science theoretical position of social constructivism followed by a presentation of its central premises and a comparison with the positions of essentialism, positivism, and realism (see Table 2.1).

Influenced by various authors and different approaches to sociology of knowledge and its definition, Berger and Luckmann (1966) adapted their concept of knowledge and thus expanded the concern of the subdiscipline. Following the authors, sociology of knowledge pivotally aims to investigate how everyday knowledge coagulates to social reality. Within the goal to conceptually reorient the sociological consideration of knowledge and its task, Berger and Luckmann state: "a 'sociology of knowledge' will have to deal not only with the empirical variety of 'knowledge' in human societies but also with the processes by which any body of 'knowledge' comes to be socially established as 'reality'" (1966, p. 15). Unlike their scientific predecessors Scheler and Mannheim, who formed their sociologies of knowledge around the consideration of specialized knowledge, around theories and big ideologies, Berger and Luckmann devoted their attention to a much more shared and spread form: everyday knowledge (Kneer 2010, pp. 712–713; Weinberg 2009, p. 281). They argue that society is significantly shaped by everyday knowledge rather than highly theoretical interpretations and explanations of the world, which solely form a fragment of the social body of knowledge. Accordingly, "common-sense 'knowledge' rather than 'ideas' must be the central focus for the sociology of knowledge. It is precisely this 'knowledge' that constitutes the fabric of meanings without which no society could exist" (Berger and Luckmann 1966, p. 26). Thus, drawing especially on Schütz' (1972 [1962]) work on the life-world, Berger and Luckmann plea for an expansion of the object of investigation of sociology of knowledge. Ultimately, the 'everyday world' as described by Schütz and the everyday knowledge anchored in it move into the center of scientific interest of Berger and Luckmann's sociology of knowledge (1966, pp. 26–28). By posing the pivotal question of how it is "possible that subjective meanings *become* objective facticities" (1966, p. 30 [original emphasis]), Berger and Luckmann shifted the analytical focus toward the investigation of the "foundations of knowledge in everyday life" (Berger and Luckmann 1966, p. 28) and thus—in line with Schütz and Luckmann's preparatory thoughts for their work *The Structures of the Life-World* (1973)—toward the analysis of subjective experiences of everyday life in a descriptive, phenomenological approach. Building on this, social constructivist positioned research on the basis of empirical investigation is devoted to the question of which interpretations solidify into socially shared reality and which are negated or remain unshared (Kühne 2018a, p. 12), as it will be explained in the following section (2.2.2) on the operationalization of the research perspective in the direction of social constructivist landscape research.

First, however, Berger and Luckmann's conception of society and knowledge production is discussed. They conceive the development and structure of society and thus the construction of social reality as the interplay of three processes (see Figure 2.2): externalization, objectivation, and internalization (Berger and Luckmann 1966, pp. 78–79; cf. Kneer 2010, p. 713; Kühne 2018a, p. 14). These concurrent and dialectic processes describe how a social, everyday body of knowledge comes into being (as objective reality) and how it is adapted and transformed over and over again (through subjective reality). By means of the three moments of externalization, objectivation, and internalization, Berger and Luckmann outline the reciprocal relationship between objective-social and subjective-individual reality, which is based on the interplay of typification and interpretation (Gergen 1999, p. 128).

Firstly, the process of externalization describes the everyday expression of meaningful, intended behavior by subjects—in the sense of M. Weber (1981)—within social acts or interactions. Within this moment, subjects assign meaning to actions and objects. A special case of externalization, in recourse to the considerations of symbolic interactionism (Blumer 1969; Mead 2015 [1934]) is symbolic communication wherein signs are used to generate and transport meaning (Denzin 2004). Fundamentally, signs develop into symbols when the receiver of information assigns them the same meaning as the sender does (Kühne 2018a, p. 14), for instance in the case of traffic signs or fences. Thus, signs are objectified and integrated as symbols in the social body of knowledge. This is beneficial because

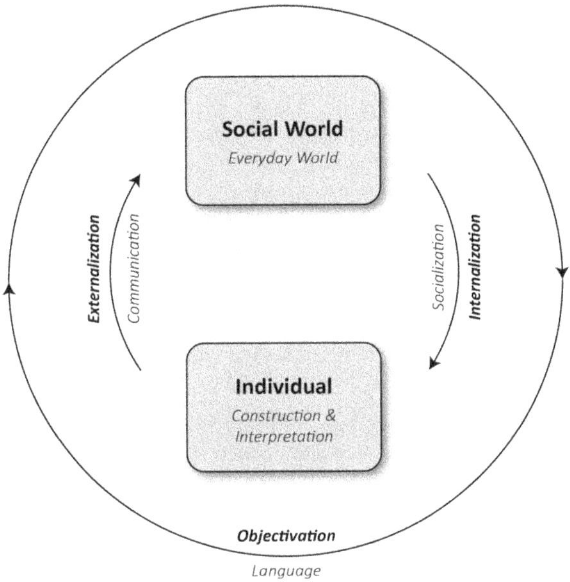

Figure 2.2 The Social Construction of Reality Through the Processes of Externalization (By Means of Communication), Internalization (By Means of Socialization), and Objectivation (By Means of Language). (Source: Own representation based on Berger and Luckmann (1966). The main processes of the social construction of reality are illustrated in bold, the means for these processes are shown in regular font-weight)

the social "stock of knowledge further supplies me with the typificatory schemes required for the major routines of everyday life" (Berger and Luckmann 1966, p. 58).

Accordingly, the process of objectivation describes how subjectively intended meaning assigned to actions, objects, or signs is becoming shared and socially accepted everyday reality through reciprocal typification (Berger and Luckmann 1966, pp. 43–48; cf. Schütz 1972 [1962], p. 6). Within the process of objectivation, a meaningful order of interpretations and classifications is established and institutionalized as everyday knowledge through repetitive habituation (Berger and Luckmann 1966, p. 56). In the tradition of Blumer (1969), it can be stated that the "meanings of things arise out of the process of social interaction" (Denzin 2004, p. 82; cf. Kühne 2018a, p. 14). Key instrument for such standardizations and normalizations is language (Burr 2005, pp. 7–8): language structures the everyday world by giving meaning, it "marks the coordinates of my life in society and fills that life with meaningful objects" (Berger and Luckmann 1966, p. 36), categories, and concepts. Thus, instead of conveying or depicting reality, language *produces* it, or—as Burr (2005, p. 11) puts it—"when people talk to each other, the world gets constructed." Therefore, language is seen as "a necessary pre-condition for thought" (Burr 2005, p. 8) rather than an expression of it. By means of language, an 'objective' world of shared interpretations, norms, and common-sense rules is developed and negotiated within everyday situations. Of particular importance hereby is the process of socialization, which describes the transfer of everyday knowledge from generation to generation (Berger and Luckmann 1966, pp. 139–174). Socialization means, on the one hand, the introduction of the subject into society with its interpretation and evaluation patterns and, on the other hand, the reproduction of society itself (cf. Kühne 2018a, pp. 79–80). Therefore, individuals are born into the social world of generated meaning and knowledge, which they eventually internalize (cf. Burr 2005, p. 7). The knowledge about nature reserves, for instance, is not innate, this specific land use concept is imparted to members of society by media, family and friends, at school or university and is thus learned and internalized within social acts (Schütz and Luckmann 1973).

In line with Berger and Luckmann (1966, p. 149), the process of internalization describes the beginning point of socialization. They understand internalization as "the immediate apprehension or interpretation of an objective event as expressing meaning, that is, as a manifestation of another's subjective processes which thereby becomes subjectively meaningful to myself" (Berger and Luckmann 1966, p. 149). In this sense, the subject absorbs and incorporates society through the ongoing process of socialization (Kneer 2010, p. 713). Thus, according to Berger and Luckmann, within the three moments of the social construction of reality, the process of internalization constitutes the stage that connects the perceived objectivity of the everyday world with subjective-individual processes of consciousness of the subject. The shared values, interpretations, and norms of action, in short, "the objectivated social world is retrojected into consciousness in the course of socialization" (Berger and Luckmann 1966, pp. 78–79). Conversely, within the process of externalization, subjective impulses are integrated into the objectified habitualizations, rules, and routines of society. Accordingly, Berger and Luckmann emphasize society both as objective and individual reality and thus as an ongoing dialectic process (see Figure 2.2). Precisely in this point lies the central premise of their sociology of knowledge, which advanced into the science theoretical position of social constructivism in the late 1960s and early '70s. Berger and Luckmann (1966, p. 79) summarize concisely: "Society is a human product. Society is an objective reality. Man is a social product." For no matter how rigid and 'real' social reality may seem, in line with social constructivism, it will at all times be the result of social negotiation processes of externalized, subjectively intended meaning.

Building on Berger and Luckmann's considerations on the construction of society and its reality, the central premises of the science-theoretical position of social constructivism are presented below. In their work *Collected Papers I. The Problem of Social Reality*, Schütz (1972 [1962], p. 5) already refers to constructs and their role in the process of perception of reality and thus provided the preliminary concept for subsequent constructivist approaches. Fundamentally, he comprehends constructs as "a set of abstractions, generalizations, formalizations and idealizations" (Schütz 1972 [1962], p. 5). In Schütz' understanding, constructs are the outcome of pre-conscious processes of abstraction and selection and thus build the foundation of all knowledge. As a consequence, within social constructivism, "there are no such things as facts pure and simple. All facts are from the outset selected from a universal context by the activities of our mind. There are, therefore, always interpreted facts" (Schütz 1972 [1962], p. 5; cf. Burr 2005, pp. 6, 9). Hence, social constructivism emphasizes that knowledge cannot be "a direct perception of reality" (Burr 2005, p. 9). However, it is not reality that is negated within social or radical constructivism (Flick 2004, p. 89) but the believe in absolute truth and the possibility to obtain absolute knowledge. Rather, it is assumed that an infinite number of multifarious and thus possibly contradictory realities exist side by side as subjective and "socially segregated sub-universes of meaning" (Berger and Luckmann 1966, p. 102). Thus, reality is comprehended as subjective abstraction. As a consequence, perception is seen less as a "passive-receptive process of representation but as an active-constructive process of production" (Flick 2004, p. 89), which draws on the combination of present and earlier perceptions (Schütz 1972 [1962], pp. 123–124). According to Gergen (1999, p. 49), the means of subjective production of reality and the understanding of the world are "social artifacts, products of historically and culturally situated interchanges among people."

In summary, it can be stated that social constructivist approaches emphasize the construction of reality through and within everyday social interactions and individual processes of perception. Within empirical, social constructivist research, "knowledge and the constructions it contains become the relevant means of access to the objects" (Flick 2004, p. 89) of investigation. A central premise hereby is the assumption that "knowledge grows and is communicated through social interactions" (Kühne 2018a, p. 14). Consequently, the social constructivist research interest lies in the study of social processes rather than 'fixed or given structures' (Burr 2005, p. 11), which traces back to the roots of constructivist thinking in symbolic interactionism (Blumer 1969; Denzin 2004; Mead 2015 [1934]). Generally, the central question of social constructivist approaches is *how* knowledge about certain topics and things is generated and negotiated within processes of social interaction. Accordingly, social constructivist inquiries with spatial reference, as adopted in the present study, focus upon the social processes by which perceptions, classifications, and interpretations of spaces acquire binding force. In order to outline social constructivist approaches more precisely, a comparison with the scientific-theoretical positions of essentialism, positivism, and realism is shown below (see Table 2.1). Following the comparison, the different perspectives and their specific premises will be explained briefly.

Table 2.1 Comparison of Science-Theoretical Positions

	Constructivism	Essentialism	Positivism	Realism
Word Origin	lat. *construere* (to construct)	lat. *essentia* (essence)	lat. *positivus* (positive)	lat. *realitas* (reality)
Premise	reality and knowledge as the result of social interaction and negotiation processes	entities own naturally inherent characteristics	production of knowledge based on actual, perceivable and verifiable findings	existence of material reality independent of human beings and their consciousness
Goals	revealing the constructive character of social reality	normative statements about the essence and characteristics of entities	objective and accurate descriptions or depictions	description of independently existing reality and external world
Interest/Focus	social negotiation processes and power structures	essential attributes of entities	sensually perceivable and empirically verifiable, 'positive' findings	condition of the existence of reality
Epistemological Principle	hermeneutic reconstruction	metaphysical observations	deduction and induction	deduction and induction
Dependency in Relation to the Observer	dependent	independent	independent	independent

(continued)

Table 2.1 (continued)

	Constructivism	**Essentialism**	**Positivism**	**Realism**
Role of the Observer in the Process of Perception	actively producing	passive-receptive	passive-receptive	passive-receptive
Perception and Concept of Space	space as process of social (re)production	territorial, space as real and existing material element	relational, space as a container	deterministic, space as real and existing material element

Source: Own composition based on Kneer (2009) and Kühne et al. (2018, p. 13; 2019, p. 123).

Essentialist (Lat. *essentia* = essence) perspectives in contrast to constructivist positions (Greider and Garkovich 1994, p. 2) emphasize that all things in the world—material and nonmaterial—naturally own inherent characteristics and thus identity. As one of the first science theoretical positions, essentialism stems from Platon's concept of ideas and was particularly influenced by Thomas von Aquin (1225–1274) in the direction of a religious and metaphysical approach (Kühne 2018b, p. 136). In line with these thoughts, essentialism conceives natural features, the 'true core' of objects in the world as readable and comprehendible by humans (Cresswell 2013, p. 177). This supposedly investigable core imparts an independent reality to objects and thus, in a pre-modern notion of science, value (Kühne 2018b, p. 137). The goal of essentialist approaches lies in the formulation of true statements about the inherent and observer-independent features of the object of investigation, which, in turn, allows and calls for a deduction of normative statements. These normative statements judge and advise according to a scale that is aligned to the essence of the considered object, for instance, whether a construction project corresponds to the nature of a specific landscape or not and thus to what extent it is beneficial or detrimental to the value of this landscape (cf. Kühne 2018a, p. 157; Linke 2019, p. 23). In this sense, essentialism typically forms its interest around the nature of things, around questions of *what* a specific object of investigation is.

Positivist (Lat. *positivus* = positive) approaches emerged from a modern notion of science in the 19th century and thus constitute a rejection of hitherto common metaphysical and essentialist interpretations of the world (Chilla et al. 2015, pp. 15–16). Within positivist thinking, the means of the production of knowledge are actual, perceivable, and verifiable findings of empirical research (Cresswell 2013, pp. 82–84). Objects of investigation (within geographical research particularly space or 'regions') are conceived as existent, countable, and distinguishable physical objects, not as social construction or as entities with a specific essence. Thus, positivist geographers are "seeking explanation rather than understanding" (Cresswell 2013, pp. 82–84). From a modern theory of science perspective, positivist approaches argue that knowledge about the world is produced and accumulated through the investigation of the relation between existing objects. In other words, positivist science "cannot understand the whole without testing the interconnections between the parts" (Greider and Garkovich 1994, p. 11). Ultimately, the goal of positivist positioned research is an accurate description and presentation of positive, observer-independent, and reproducible findings about these parts of the whole, i.e., of reality (Kühne 2018b, p. 137). Thus, rather than focusing on the essential whole, positivist research approaches the objects of investigation through its isolated phenomena, its layers and parts (Chilla et al. 2015, p. 16). In this sense, geographical spatial research is approaching space by means of the individual components it contains. Consequently, positivist studies do not make normative statements about the whole, such as in the case of essentialism, but about the examined individual layers regarding their relationship to one another (Chilla et al. 2015, p. 16).

Subsequently, realism is cited as a fourth science theoretical position next to social constructivism. Split into a multitude of different sub-positions, realist (Lat. *realitas* = reality) approaches generally stress the existence of objective reality and its independency of the observer and of social negotiation or construction processes (Kneer 2009, p. 11; cf. Burr 1998). The central premise of the variant of naïve realism lies in the "faith in reality of what we perceive" (Wetherell and Still 1998, p. 99) and is thus diametrically opposed to the social constructivist understanding of reality (Burr 2005, pp. 6–7). In line with their understanding of reality as objectively perceivable, naïve realist positions emphasize the possibility of approaching and revealing reality through empirical procedures of observation (Kühne 2018a, p. 11). In this sense, naïve realist perspectives comprehend perception as a passive process of reception rather than an active process of construction and interpretation.

However, in the course of the second half of the 20th century, not just within human geographical research, doubts increasingly arose about the conception of reality as observer independent as it is stressed within essentialist, positivist, and realist positions. Various paradigm shifts in this period of time led to extensive realignments of the geographical discipline, each shifting the scientific approach both perspectively and methodically (Kühne 2019a, pp. 69–72). In line with a contemporary, postmodern notion of science, current human geographical research as well as other social sciences are particularly characterized by social constructivist approaches and thus the "view that much that passes as 'reality' is produced by social forces" (Cresswell 2013, p. 250). Consequently, the following paragraphs will trace the evolution of social constructivist geographical research by outlining the major scientific turns of the 20th century. Hereby, the focus is set on the development of the concept of landscape and of social constructivist landscape research in which the present work is rooted.

2.2 Landscape Research: Historical and Conceptual Development, the 'Comeback' of Landscape as a Social Construct, and its Sensitivity to Aspects of Hybridity

In the following, *landscape* will be introduced and outlined as the first of two theoretical-conceptual key components of the present research work. Drawing on the semantically and theoretically ambiguous and open concept of landscape enables the present research to carve out aspects of (socio-)spatial hybridity and processuality in the context of neighborhood change and urban redevelopment. In the following section, first, the role of landscape in the historical development of the geographical discipline will be explained in detail. It will be traced how the concept of landscape in geography changed over time, developing from the discipline's central object of investigation to an endeavor understood as non-scientific, and ultimately to a product of social negotiation and power processes within its postmodern 'rediscovery.' Additionally, key aspects of the development of phenomenological considerations according to significant paradigm shifts in the 20th century will complement the explanations. In a next step, the social constructivist understanding of landscape will be outlined by presenting the central premises and approaches which fundamentally grasp landscape as "symbolic environments created by human acts of conferring meaning to nature and the environment" (Greider and Garkovich 1994, p. 1). In a post-structuralist way of thinking, landscape is attributed a subjective character and thus an infinite variety of possible interpretations and readings (cf. Denzer 2019; Wylie 2007, pp. 70–82). This opens up the possibility for landscape research to focus on aspects of hybridity and the combinations of traditional spatial concepts and interpretation categories, as it will be traced in the last subsection (2.2.4) on the hybridization of urban and suburban or rural phenomena.

2.2.1 The Historical Formation of Landscape Geography: Etymological, Art Historical, and Science-Theoretical Factors

The notion of landscape and the term itself have travelled and evolved through various epochs and paradigm shifts, look back on a diverse history of everyday use, and are key subjects of different disciplines, such as art history, geography, or philosophy (cf. Duineveld et al. 2017, p. 375). Thus, today, the concept of landscape comprises of a broad and ambiguous spectrum of meanings, which stems not only from different individual readings—in the sense of Berger and Luckmann (1966)—but also from diverging etymological or linguistic interpretations of various languages and cultures (Antrop 2019, p. 2; Kühne 2015; Schenk 2019). While the Old High German word '*lantscaf*,' which later became the German word '*Landschaft*,' carried a political-territorial meaning for various centuries (Berr and Schenk 2019, p. 26; Ipsen 2006, p. 73; Kühne 2018a, pp. 49–51; F. Weber 2018, pp. 46–48), the British meaning of 'landscape' rather referred to scenic interpretations and brings aesthetic aspects to the foreground. However, the US-American understanding of 'landscape'—at least among experts and scholars—was mainly shaped by German geographer Carl Otto Sauer and thus comprised of a rather territorial and materialized meaning with references to the homeland of a community (Antrop 2015, p. 55; Cosgrove 2004; Cresswell 2013, p. 139). The aesthetic dimension was first attributed to the German concept of 'Landschaft' within the time period of the Renaissance and its contemporary painting. In the following 16th and 17th century, landscape painting advanced into one of the most important art genres, which enhanced the aesthetic view of landscape substantially (Kühne 2018a, pp. 51–54). In the course of the epochs, landscape evolved more and more from the background of certain scenes to the main subject of paintings. What influences the everyday understanding of landscape persistently until today, however, is the scenic gaze depicted by the German romanticists of the 18th and 19th century, foremost Caspar David Friedrich, through whom landscape experienced a certain mystification and aesthetic valuation (see Figure 2.4). Today, in the German-speaking world, the aesthetic appropriation of the concept of landscape blends with ideological, homeland, political-territorial, religious, and scientific-analytical references (Kühne 2016, p. 15), which is also due to the major influence of geographical landscape research.

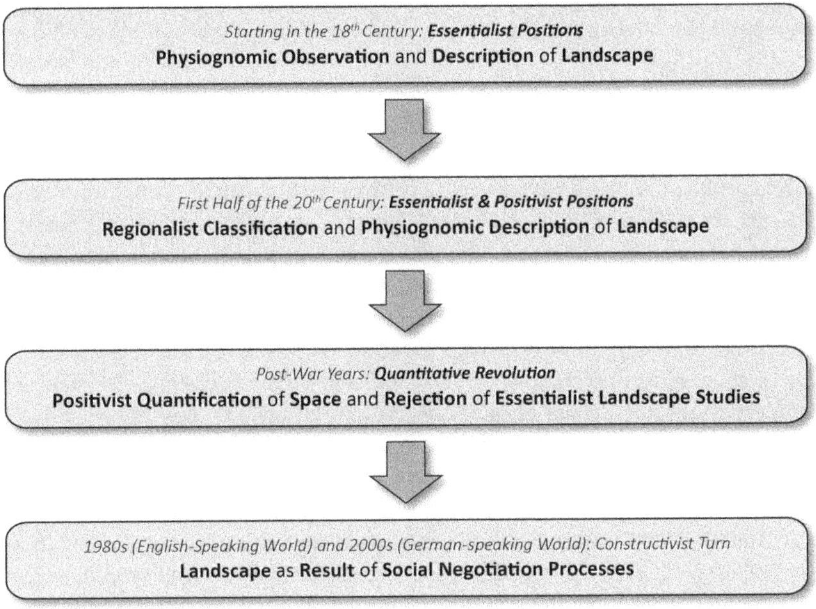

Figure 2.3 The Historical Development of Geographical Landscape Research. (Source: Own representation based on Kühne (2018a, pp. 49–78) and Kühne et al. (2018, p. 15))

The evolution and history of the geographical discipline (see Figure 2.3) is characterized by significant turns and shifts, particularly in respect to the scientific devotion to landscape and its conceptualization (Cresswell 2013). From the advent of geography as a university discipline in the 18th century, the observation of landscape and its physiognomy played a pivotal role (Schenk 2013). Especially from the 19th century until the post-war years, landscape enjoyed a certain "degree of prominence" (Cosgrove 1985, p. 45) and thus, landscape studies advanced to the key subject of geographical research (Hard 2002). Within a deterministic approach, early geographers committed themselves to the question of "the exact meaning of the word [landscape] and its 'scientific' definition" (Antrop 2019, p. 2). In the late 19th century, landscape studies in the German-speaking world were characterized by metaphysical-essentialist observations and interpretations and were aiming for the attribution of adjectives to the term. Thus, a multitude of different landscapes, such as cultural, natural, or urban landscapes, have been identified and put in relation to each another (Antrop 2015,

p. 56). Beginning in the 1920s, in the tradition of Sauer (1925), North American geographers devoted themselves to the then imposed natural/cultural landscape divide as well but within an emphasis on "the relationship between humans and nature" (Cresswell 2013, p. 79) or, in other words, on the influence of culture on nature. In Sauer's understanding, landscape constituted a material topography, something "out there" (cf. Wylie 2007, p. 20) and "a physical material reality we are immersed in with all our senses [...]. To put this another way, [for Sauer,] landscape [was] primarily factual and objective, it is an external, independent material field, a unified synthesis and arrangement of material forms and objects, and not a contrivance of our perception" (Wylie 2007, p. 20). In this sense, the geographical discipline of the 19th and the first half of the 20th century comprised of physiognomic descriptions or classifications of landscape and regions on the basis of essentialist observations (cf. Davoudi and Strange 2009b, p. 28; Berr and Kühne 2020, p. 50; Kühne et al. 2018, p. 15). Ultimately, this physiognomic-regionalist landscape approach was linked to the hope of bringing physical and human geography closer together (F. Weber and Kühne 2019, p. 62).

However, in the post-war years, the natural/cultural landscape differentiation in the tradition of Sauer and the concern with landscape in general "fell from favor" (Cosgrove 1985, p. 45). Early critics, such as Hartshorne (1939), voiced fundamental doubts about the scientific character of (essentialist) landscape geography (Antrop 2015, p. 56). In the 1950s and '60s, the scientific aversion against 'landscape' as a research topic culminated and ultimately called the discipline as a whole into question (Smith 1987). Geography "was labeled intellectually inadequate" (Cresswell 2013, p. 79) and landscape studies were declared socially irrelevant (Schenk 2013, p. 30). But it was not until 1969 that landscape was officially terminated from the human geographical agenda in the German-speaking research world. With the call for empirically verifiable and scientific findings, landscape was temporarily suppressed from the geographical discipline at the Geographer's Day in Kiel (see Kühne et al. 2019, pp. 119–120; F. Weber 2018, pp. 44–45). The so-called 'quantitative revolution' fostered the demand to replace essentialist descriptions of landscape in favor of quantifiable approaches to the concept of 'space.' Thus, the paradigm shift expresses the attempt to "transform the discipline into a scientific pursuit" (Kwan and Schwanen 2009, p. 283), by following the emergence of a modern notion of science and the concomitant rise of positivist positions (Davoudi and Strange 2009b, pp. 31–32).

In the following years, in response to the quantitative revolution, or rather as a counter-reaction to it, phenomenological approaches gained popularity. In the 1970s, human geographers raised fundamental doubts about the quantification of space, spatial science, and positivism in general (Cresswell 2013, p. 99). Drawing on the continental philosophies of meaning and on phenomenological thinking, humanist geographies strived to direct the scientific attention on the human, their bodies, and their experiences. The humanist subdiscipline assumed that subjectivity, emotions, and experiences are the means to explain the world, not positivist spatial science. Cresswell (2013, p. 104) formulates trenchantly: "The idea that there is a singular, measurable truth 'out there' waiting to be discovered and explained that can be accessed by adopting a lofty, God-like, objective 'view from nowhere' was dismissed. The notion that all-important things are quantifiable was questioned." In this sense, human beings were seen as intentional, and the research of their experiences was believed to require a different set of methods. Consequently, qualitative methods, such as interviews and participant observations, emerged within human geography, which continuously gained in popularity especially due to the change of perspective to constructivist approaches in the following decades.

Following the rise of postmodernism in the 1980s, a shift toward constructivist positions loomed in the English-speaking world of the geographical discipline. The 1980s and '90s were characterized by the onset of the 'new' cultural geography, which aimed for the 'comeback' of landscape into the field of human geography (Wylie 2003, pp. 138–139; 2007, pp. 91–93). In Britain, the work of Denis Cosgrove and Stephen Daniels played a crucial role in the development of a new landscape approach (Cosgrove 1985, 1998 [1984]; Cosgrove and Daniels 1988), which drew on the precursory work of Hoskins (1955). Newer North American landscape geography was influenced by James Duncan (1995) and particularly brought into relation with social constructivist theory through the work of Thomas Greider and Lorrain Garkovich (1994). A few years later, beginning in the 2000s, German-speaking geographies took up and incorporated constructivist landscape approaches as well (e.g., Crossey et al. 2020; Gailing and Leibenath 2015; Kost and Schönwald 2015; Kühne 2006; Linke 2019; for a detailed description of the paradigm shift see: Kühne et al. 2019, pp. 119–120; F. Weber 2018, pp. 44–45). Up until today, human geographical landscape research is largely based on the science theoretical position of social constructivism.

2.2.2 Social Constructivist Landscape Research: Central Premises and Conceptions

The 'anti-essentialist turn' (cf. Burr 2005, pp. 6–9) marks the entry of constructivist positions into the geographical discipline. The field of landscape studies benefited in particular from the conceptual and methodological turn as new research focusses became possible: Since the turn of the millenium, the social constructivist understanding of landscape formed the theoretical basis for a wide range of analyses in human geography and related spatial social sciences, for instance, for reaseach on tourism (Aschenbrand 2018; Metro-Roland and Soica 2019), industrial fallow land (Kühne 2016; Micheel 2012), suburbanization and urban/rural hybridization (Hofmeister and Kühne 2016; Kühne 2012a; Roßmeier 2020d; F. Weber and Kühne 2016), energy transition and the implementation of wind turbines (Leibenath and Otto 2012; Roßmeier and Weber 2021 [2018]; Roßmeier et al. 2018; F. Weber, Jenal, et al. 2017), conflict management (Berr and Jenal 2019; Berr et al. 2019; F. Weber et al. 2019), simulations, 3D representation, and virtual reality (Edler and Kühne 2019; Edler et al. 2018; Fontaine 2016; Vetter 2019), as well as other topics (cf. Kühne 2019c, p. 70). In the course of the constructivist turn, landscape was reintroduced into human geography by "American humanist geographers for the very reasons that their predecessors rejected it" (Cosgrove 1985, p. 45). The 'new' landscape or cultural geography aimed for "the incorporation of individual, imaginative and creative human experience into studies of the geographical environment, aspects which geographical science is claimed to have devalued at best and at worst, ignored" (Cosgrove 1985, p. 45).

Drawing on the phenomenological-sociological foundations of Schütz (1960 [1932]) and Berger and Luckmann (1966), social constructivist landscape theory negates conceptions of environment, landscape, and nature that derive from essentialist, positivist, or realist positions (see Table 2.1). Instead of conceiving landscape as an entity with inherent characteristics (essentialist), as quantifiable space (positivist), or as something real (realist), the social constructivist approach emphasizes landscape as actively and subjectively produced outcome of processes of social (power) negotiation (Gailing and Leibenath 2015). In other words, "nature and the environment are socially and culturally constructed [...] and become landscapes through social interaction and negotiation" (Greider and Garkovich 1994, p. 10). Thus, landscape is no longer understood as observer independent or as something that is 'seen,' rather it is conceptualized as a 'landscape eye' (Berr 2020, p. 190; in line with Riehl 1996) or 'a way of seeing' the world, as a "discursively constructed visual representation" (Wylie 2003, p. 141) of the observer (cf. Jauhiainen 2003, pp. 398–399). Accordingly, Cosgrove (1998 [1984]) states: "Landscape is not merely the world we see, it is a construction, a composition of that world." In this sense, Greider and Garkovich (1994), two early scholars in landscape studies drawing on the theoretical foundation of Berger and Luckmann (1966), formulate the central question of social constructivist landscape research not about *what* landscape is, but *how* landscape is socially charged with (symbolic) meaning. Consequently, new landscape studies are concerned with the exploration of "the cultural meanings of aspects of the physical environment and biophysical changes in this environment" (Greider and Garkovich 1994, p. 21), not the description of physical elements in space (cf. Kühne 2012b; F. Weber and Kühne 2019).

Thus, in addition to the conceptual reformulation of landscape in human geography, the adoption of the social constructivist theory "had a methodological correlate [...]. Instead of producing 'representations of landscape,' cultural geographies of landscape took these as their primary object of concern" (Wylie 2003, p. 139). Consequently, in the sense of hermeneutic interpretation (Soeffner 2004), social constructivist landscape research aims for understanding how individual and thus first-degree constructions of landscape—in line with Schütz (1972 [1962])—develop, change, and adapt over time, and how these constructions advance into shared, social stock of knowledge and common sense. Ultimately, this new research program draws on the dialectic interplay of the theoretical key processes of the sociology of knowledge of Berger and Luckmann (1966, pp. 78–79): Externalization, objectivation, and internalization (see Figure 2.2). With reference to the construction of landscape, the moment of externalization characterizes how meaning is conferred to physical objects of the environment through intentional, individual behavior—in the sense of M. Weber (1981)—or within social interaction, how material objects become landscape through individual thoughts and acts. Objectivation, then, means how subjective meaning ascribed to physical objects becomes part of the ideas of landscape that exist within society. The moment of internalization describes how individuals incorporate the social stock of knowledge about landscape into their consciousness, how they learn to see physical objects as landscape in the way their society does. Thus, internalization is the starting point for the process of 'landscape socialization' through which everyday knowledge about landscape is passed on from one generation to the next (cf. Kühne and Weber 2018, p. 531). Summing up, social constructivist landscape research is concerned with the reciprocal relationship between subjective-individual and objective-social perceptions, interpretations, and typifications of physical objects as landscape (cf. Gergen 1999, p. 128). Accordingly, within an anti-essentialist position, Greider and Garkovich (1994, p. 2) argument that "the processes of negotiation over landscapes, and subsequent social actions ought to be the foci of social science inquiry, because there are no natural meanings inherent in the world that is there."

Seeing landscapes as "sociocultural phenomena, not [as] physical phenomena" (Greider and Garkovich 1994, p. 1) implies a "hierarchical duality of meaning/materiality" (Wylie 2003, p. 141). While this hierarchical understanding that meaning is socially constructed and afterward applied on material elements of the environment opened up the way for a deeper consideration of text, discourse, and signs in the spatial sciences, phenomenological explorations of landscape receded during the 1980s and '90s (Kühne 2019a, p. 35). The material aspects of landscape were gradually pushed into the background of constructivist landscape research in favor of the "ideological, symbolic and discursive terms" (Wylie 2019, p. 130) of landscape (Duineveld et al. 2017, p. 375). However, in contrast to radical constructivist positions, social constructivist approaches allow for certain links to materiality and its consideration (Kühne 2018a, p. 15). As it will be shown in the following section, starting from a social constructivist understanding, various approaches to a combined examination of symbols, signs, *and* things developed since the 2000s (Kazig and Weichhart 2009).

2.2.3 Phenomenologically Oriented Variants of Constructivist (Landscape) Research

After the rise of humanist geographies in the 1970s and the following onset of constructivist approaches, phenomenological considerations of space and landscape have largely vanished from human geography (Wylie 2007, p. 140). However, since the last decade, the material and the relationship between society and materiality are receiving increasing attention in landscape geography—instead of focusing on one or the other, approaches considering both society *and* materiality are multiplying as variants of constructivist socio-spatial research (Chemero 2003; Duineveld et al. 2017; Kühne 2019a, pp. 31–33; Wylie 2003). In other words, phenomenological approaches of the material currently experience a 're-emergence' (Wylie 2019).

Precursory work already started in the 1980s with assemblage theory (De Landa 2006; Deleuze and Guattari 1988; Mattissek and Wiertz 2014) and in the 1990s with actor-network-theories (cf. Bosco 2015) developed by Michel Callon, Bruno Latour (1999), and John Law (2009). However, especially during the 2000s, human geography experienced a "recuperation of the material" (Whatmore 2006, p. 601) with the emergence of non- or more-than-representational approaches (Lorimer 2005, p. 85; cf. B. Anderson and Wylie 2009; Ingold 1993; Thrift 2008; Waterton 2019), which in part also focus on the human body and 'embodied experiences' (Turner 2009; Whatmore 2006, p. 602). Accordingly, Wylie (2019, p. 135) emphasizes: "To study landscape from a phenomenological perspective involves foregrounding lived, embodied experience and perception. In part, this is a 'practical' question of examining the varied practices and activities wherein people and landscape mutually interact."

Building on Husserl's (1913, 1970 [1954]) considerations, non- or more-than-representational approaches (Cresswell 2013, pp. 227–235) argue against the "conventional split between the conscious subject and the object of awareness" (Gergen 1999, p. 128), which characterizes representational theories based on positivist or constructivist positions (Kühne 2019a, p. 30). Landscape, in this sense, is more than just the representation of quantified elements (in positivist understanding) or individual or social interpretations (in social constructivist understanding), it is "more-than-visual and more-than-symbolic" (Wylie 2019, p. 132). For landscape or environment are not seen as "something alien to the spirit" (Soeffner 2004, p. 97) because in phenomenological thinking "subject and object—or self and other—are unified within experience" (Gergen 1999, p. 128). According to Husserl, experience is always relational, it is either directed "toward or absorbed by some pattern (object, person, etc.) in the external world" (Gergen 1999, p. 128; see generally on relational thinking M. Jones 2009; Murdoch 2006). Consequently, more-than-representational theories emphasize subject-object or "culture-nature relations" (Wylie 2019, p. 127; cf. Inglis and Bone 2006; Krauss 2019) in order to explore the importance of both the environment or the material *and* the social dimension of symbols and signs (Duineveld et al. 2017, p. 383)—"[t]he questions of landscape, in other words" (Wylie 2019, p. 128). Soeffner (2004, pp. 97–98 [original emphasis]) offers a detailed explanation for these complex relationships:

> "The human environment, or life-world, […] cannot be described either by a model of 'external/internal' or 'subject/object' or with the aid of spatial measurements and territorial demarcations. For us it is not something opposed, it is neither a cage nor an unlimited space but, rather, a horizon of perception, orientation and action. It moves with us, when we move, it changes us—and our action—when we change it. It does not exist without us, and we do not exist without it. But we *are* not our environment: we *have* it."

Sociological actor-network-theories (Cresswell 2013, pp. 250–254) take social constructivist thinking as a basis (Schulz-Schaeffer 2000) and thus share anti-essentialist, post-structuralist traits (Bosco 2015, p. 136). The aim of actor-network-theories is to integrate natural or non-human elements into observation that would be of interest in social constructivist research only if they play a role in the self-definition of individuals and cultural groups (Greider and Garkovich 1994, p. 21) or, in other words, if they receive aesthetic, emotional, or symbolic charge (Kühne 2019a, p. 34). Thus, it is possible—methodologically and empirically—to go beyond the scope of social constructivist landscape research and to uncover and trace "the many connections and relations among a variety of actors (human, non-human, material, discursive)" (Bosco 2015, p. 136). However, caution and attentiveness are needed to avoid a certain return to the traditional, essentialist search for 'true landscapes' and their supposed features. Accordingly, what is guiding (current) phenomenological approaches is not the search for meaning in the material world but the idea of expanding social constructivist research by examinations of the complex relations between object and subject, culture and nature, social and material dimension. Furthermore, this relational approach and the resulting sensitivity for spatial relations and their openness—in a post-structuralist sense—deepens the conceptual linkage of the landscape notion with border-theoretical questions on (dis)connection and (dis)continuity in space and society (Löw and Weidenhaus 2017), which constitutes an important link for the present work.

The previous sections (2.1–2.2.3) have set out the science-theoretical, perspectival foundation of the dissertation project. Subsequently, the next section will discuss the conceptual and methodological sensitivity of social constructivist landscape research to processes of socio-spatial hybridization. Thus, in a next step, the social constructivist landscape approach will be combined with border theory (presented in section 2.3) in order to develop and synthesize the key conceptualization of hybrid urban borderlands (section 2.4).

2.2.4 Hybrid Landscapes in the 21st Century: The Concept of Hybridity, the Socio-Spatial Phenomenon of Urban/Rural Hybridization, and Spatial Pastiches

Along with the advent of postmodernism, the spatialization and "co-existence of a multiplicity and variety of situation-dependent ways of life" (Burr 2005, p. 14) advanced. Fundamentally, postmodernity is "characterized by a rejection of meta-narratives, foundations, and essences" (Cresswell 2013, p. 196), hence it contrasts with the modern desire for purity and conformity and addresses ambiguities and contradictions instead (cf. Hoesterey 2001; Kühne 2016, p. 20). In the social sciences, the idea of postmodernism can be summarized briefly with social tendencies of flexibilization, individualization, and pluralization and is closely linked to social constructivist thinking (cf. Linke 2019, p. 5). In this sense, post-material western industrial societies underwent a value shift from modern, functional-cognitive interpretations of the world to a postmodern emphasis on emotional and aesthetic aspects (Kühne et al. 2019, p. 120). These social developments have increasingly been taken into account by means of the concept of hybridity.

Adopted from the field of biology, hybridity developed into an interpretation category for emerging social and cultural traits in the 1980s (Cresswell 2013, p. 171). First approaches for the adoption of the idea of hybridity into the consideration and interpretation of cultural contexts stem from Michail Bachtin (1895–1975) and his work on linguistic hybridization as well as from the urban sociologists of the Chicago school and their research on the hybrid 'marginal man' (cf. Ackermann 2004). Today, in the social sciences, the concept of hybridity is used as a label for cultural 'blending' and emphasizes the processuality of society and culture (Hein 2006; Kraidy 2005). Culture is understood as a constant process of negotiation and thus as a "complex, processual, and dynamic" (Kraidy 2002, p. 317) practice, not as a rigid, essential characteristic of society (cf. Kühne 2016, p. 18). Bhabha (2000, 2012), however, traces hybridity rather as a spatial phenomenon; they understand "hybridity as a form of in-between space, which they term the 'third space,' a space inherently critical of essentialism and conceptualizations of original or originary culture" (Mitchell 1997, p. 337). Bhabha's third space is characterized as the place where two moments commingle, which, however, does not induce homogenization or the creation of a third moment but "enables *other* positions to emerge" (Bhabha 1990, p. 211 [emphasis added]). Accordingly, in line with Bhabha (2000, p. 168), hybridity is not to be understood as a third term that emerges from the combination of two different moments but can be seen as an open discourse space that comprises unordered meanings and interpretations (cf. Struve 2013, p. 103). Hybridity, thus, stands for the emergence and persistence of plurality, difference, and in-betweenness, not for unambiguity or harmonization.

Since the 1990s and especially the 2000s, the concept of hybridity has been taken up in urban and landscape studies. When landscape is examined today in the social sciences, it is often with regard to the resolution of the differences between city and rural countryside, between urban and suburban spaces, focusing on the manifold hybrid forms and transitional areas that developed over the last century. Accordingly, a 'wide' landscape conception emerged, which enables urban/landscape research to take into account both undeveloped and developed spaces and everything in between (Kühne 2018b, p. 144). In their traditional meaning, the terms 'urban' and 'rural' describe two different societies, two different ways of life and "visions upon the environment" (Antrop 2015, p. 57). But "in the contemporary globalizing world there are no 'pure' rural or urban landscapes [...]. The dichotomy between urban and rural has been blurring already for decades" (Jauhiainen 2003, p. 396), resulting in ambiguous spaces and lifestyles that point in one direction and simultaneously in the other. "Landscape is no longer antipode, but part of the city and vice versa[3]" (Sieverts et al. 2005, p. 47). On the one hand, the 'city' penetrates into the 'landscape'—on the spatial-structural level as well as on the social, materially and immaterially—in the form of suburban development or the spread of high-speed internet connection and recent communication technology in increasingly remote areas through which diverse, cosmopolitan ways of life and work become possible. On the other hand, landscape blends progressively into urban contexts, e.g. in regard to the historically developed, romantic "notion of landscape as an essentially aesthetic way of looking at nature" (Kühne 2018a, p. 2; cf. Fischer 2011), which no longer applies only to 'wild' or supposedly 'untouched' nature but also to spaces equipped with buildings and other technological-structural elements (cf. Gailing 2015), as depicted in Figure 2.4 (on the social landscape perception of renewable energy technology see Roßmeier and Weber 2021 [2018]; Roßmeier et al. 2018; F. Weber, Jenal, et al. 2017; F. Weber, Roßmeier, et al. 2017). Hence, the postmodern settlement development of western societies "requires new interpretations" (Atha et al. 2019, p. xxvii; cf. Qviström 2019) and scientific readings that are no longer based on modern (Euclidian planning) logic (cf. Davoudi and Strange 2009a, 2009b; Davoudi 2012; Graham and Healey 1999; Kühne 2019a, p. 118).

A multitude of neologisms from social and spatial research and planning have been developed to take account of the complex hybrids of urban and rural spaces: The English-language world has spawned terms such as edge or edgeless city (Garreau 1992; Lang 2003), peri-urban (Allen 2003; Qviström 2019), urban sprawl (Duany et al. 2000; Herzog 2015; Whyte 1958), urfsurb (Kühne et al. 2016; F. Weber and Kühne 2017), or urban/rural hybrid (initially developed by Kühne 2012a; cf. Kühne 2019a) within the aim of making sense of the differentiated postmodern settlement structures. In German-language research and planning contexts, terms such as (Post-)Suburbium (Dangschat et al. 2001), Stadtland or StadtLandschaft (Hofmeister and Kühne 2016; Kropp 2015; Roßmeier 2016), and Zwischenstadt (Sieverts 2003, 2008; cf. Aring 1999; from a landscape perspective also Vicenzotti and Qviström 2018) have developed, each linking landscape and urban readings in different ways. According to Kühne (2016, pp. 13–14), interpretating urban contexts from a landscape(-theoretical) perspective seems beneficial not only in regard of the progressive everyday life interweaving of the dual modern categories of city and countryside but also because of the opportunity to incorporate the multilayered semantics of the landscape term (Hard 1969)

[3] „Landschaft ist nicht mehr Antipode, sondern Teil von Stadt und umgekehrt“

Figure 2.4 The Aesthetic Dimension of Landscape Inscribes Itself into the City: Scenic, Contemplative Gaze over the Smog Covered Urban/Rural Hybrid Los Angeles as a Contemporary Version of Romantic Landscape or Spatial Interpretation. (The installed park benches along the canyon trail facilitate aesthetic interpretations of the material surroundings. From a phenomenological sense, the hike up the rocky canyon can be interpreted as bodily (landscape) experience, which is unifying subject and object, social and material dimension. Furthermore, new (temporary) physical demarcations of everyday spaces take place in the course of the COVID-19 pandemic in the form of trail closures (fence and barrier tape on the left). Source: Picture Albert Roßmeier 2020, in reference to Kühne (2018a, p. 67) and Caspar David Friedrich's Wanderer above the Sea of Fog (1818))

into contemplation. Thus, drawing on its variety and openness of meaning, the 'wide' landscape concept can be used to frame the increasing complexification of urban, suburban, and rural spaces in an alternative and comprehensive way. Landscape, in the words of Vicenzotti (2019, p. 748), becomes a "theoretical lens for understanding the city[4]" (cf. Waldheim 2012, p. 15). One of the mentioned approaches that incorporates this landscape perspective is the concept of urban/rural hybrids (Kühne 2019a), which constitutes one of the two central pillars of the theoretical-conceptual operationalization of hybrid urban borderlands in the present work.

The concept of urban/rural hybridity highlights the fragmentation and pluralization of phenomena that are located between the poles of the urban and the rural, of nature and culture, of leisure and work, and describes processes that take place on the social as well as on the material level (Kühne 2019a, p. 124). In comparison to the Zwischenstadt idea (Sieverts 2003, 2008; Sieverts et al.

[4] „theoretische Linse zum Verstehen von Stadt"

2005), the concept of urban/rural hybridity traces blending and combination in various dimensions: Structural, functional, life-world, emotional, aesthetic, and cognitive aspects receive attention (Kühne and Weber 2019a, pp. 41–42). Furthermore, the socio-spatial phenomena of urban/rural hybrids are not characterized by a continuous density gradient from the urban inside to the suburban outside but by an unordered multitude of variously demarcated compartments with "varying degrees of urbanity and rurality" (Kühne 2019a, p. 124), density, and multifunctionality. In the course of the progressing urban/rural hybridization of western metropolitan regions, the postmodern urban grid or net structure and its sharply demarcated edges (Flusty 1994, 2001) are fading in favor of spatial pastiches, which are not characterized by clear delineation but by transitions and overlaps. Examples of compartments that have formed in urban/rural hybrids are edge cities or edgeless cities, which feature ambiguous boundaries and a high degree of processuality and disorder.

Spatial pastiches (on the integration of the concept of pastiche into spatial research see Aitken and Zonn 1994; Gottdiener 2000; Hetherington 1998) emerge as a result of land use changes in formerly monofunctional and separated modern spaces that have become postmodern compartments of urban/rural hybrids through the establishment of new or specialized uses, abandonment, or multifunctional reuse, such as in the example of urban redevelopment and reinterpretation of industrial structures into live/work lofts (cf. Vester 1993, pp. 27–32). Thus, the planned modern order dissolves and gives way for postmodern fragmentation and ambiguity or, in other words, spatial pastiches in the course of the postmodern multiplication of (aesthetic) interpretations (Dear 1986; cf. Kühne and Weber 2019b, p. 758). The development of pastiches is not to be understood as strict homogenization and dedifferentiation but as playful and sometimes unorganized *recombination* of differences, which creates new meanings, interpretations, and thus hybridizations (cf. Hoesterey 2001). Consequently, the conceptual approaches of urban/rural hybrids and spatial pastiches are not just about the resolution of the binary through mixing and commingling but also about recombination on the basis of new demarcations and exclusions. In particular, urban/rural hybridity addresses connection and dis-connection between the urban and the rural but also "between human and non-human; social and material; subjects and objects" (Whatmore 2002, p. 1), which entails a high compatibility to social constructivist as well as phenomenologically oriented theoretical approaches, such as actor-network-theories or the assemblage theory. In order to highlight not only aspects of hybridization in the present study but to also sharpen the focus on demarcations and exclusions, it is important to explore the links between the research fields of landscape and border studies. For, according to Ipsen (2006, p. 75), gaining a profound understanding of landscape is only possible by combining and relating different disciplines and approaches to each other.

2.3 The Interdisciplinary Field of Border Studies: Borders and Boundaries from a Social Constructivist Perspective and the Transposition on the Level of Urban Neighborhoods

In addition to the urban landscape notion and the focus on the processuality of spatial identity construction as well as spatial hybridization, the present analysis is grounded on a border-theoretical perspective, which will be presented below. The research field of border studies is based in different (social) sciences and academic fields, which offer diverse starting points for the study of—geographical and non-geographical—borders and boundaries. Ethnological, geographical, political science, and sociological research is devoted to theoretical questions of demarcations, mechanism of border enforcement and border crossing, or concrete border spaces and the complex interconnections of the two demarcated sides. In many cases, as it will be discussed, theory-driven border studies resort to a landscape perspective within the aim to conceptually do justice to the diversity as well as the spatiality of borders. This is where the present work starts, by further developing landscape approaches to border(ing) aspects and creating a uniting interface between these two theoretical-conceptual pillars that the author terms *hybrid urban borderlands* (section 2.4 and Figure 2.6).

First, however, the present understanding and use of the notions of borders and boundaries will be explained (section 2.3.1). In a next step, the historical and conceptual development of border studies will be outlined in order to explain the change of perspective the field underwent in the last decades: Border studies overcame the understanding of borders and boundaries as rigid diving lines between nations and developed a broad, relational conception of borders as practices and spaces where societies and cultures commingle and new hybrid forms emerge (section 2.3.2). Lastly, it will be demonstrated how the thematic opening of the research field enabled a transposition of theoretical-analytical border and boundary questions on the level of cities and communities and how the ideas of neighborhoods and demarcation are connected (section 2.3.3).

2.3.1 Borders and Boundaries: Present Use and Understanding of the Key Concepts of Border Studies

Within the aim to develop a concise conceptualization of hybrid urban borderlands, it seems necessary to outline how the specific terminology of border studies will be used in the following. For as different as the disciplinary approaches to the research field are, as ambiguous is the use of its central terms and concepts in literature, research, and everyday life. Unhelpfully, the terms 'boundary,' 'border,' and 'frontier' are often used synonymously and are not clearly distinguished from each other, although they historically and etymologically carry different meanings. As a consequence, a "definitional variety" persists (J. Anderson and O'Dowd 1999, p. 594; cf. Baud and van Schendel 1997; Haselsberger 2014; Newman 2003a).

In the present study, despite the prevalent polysemy, the boundary term will be used stringently to grasp linear, non-material demarcations of one particular aspect. It is comprehended as a "line that separates, encloses, and excludes, at a number of spatial and social scales" (Newman 2003a, p. 124). Although the demarcations understood here as boundaries do not represent physical-material structures in space, the focus lies on those that show spatial reference in terms of economic, cultural, social, or jurisdictional (dis)connection and (dis)continuity. This includes interpretative and symbolic-aesthetic boundary drawings as well as individual constructions and distinctions (equivalently conceptualized e.g. in Christmann and Weber 2020; Roßmeier 2020b; Roßmeier and Weber 2021; F. Weber and Roßmeier 2020; Wille and Weber 2020). Gender boundaries, however, will not be considered, even though these demarcations are as well reflected spatially.

The border concept as it is used in the following carries a stronger territorial meaning and refers especially to physical-material structures in space and lines on maps that define different nations, areas, cultures, and citizens (J. Anderson and O'Dowd 1999). Thus, borders are understood here as a *set* of boundaries, a bundle of overlapping non-material demarcations that materialize and consolidate into something thicker (Haselsberger 2014). Nevertheless, as it will be illustrated in the following sections, the present analysis does not resort to a purely territorial, positivist understanding of borders and boundaries as final products and fixed lines of order in a 'container space.' Instead, a relational perspective will be adopted (Graham and Healey 1999; M. Jones 2009; Massey 1995, 2005; Murdoch 2006; Nienaber and Wille 2020a), which conceives 'bordering' as open and ongoing negotiation process of social relations and interactions (Newman 2006a;

van Houtum and van Naerssen 2002). The frontier concept, in turn, will not be considered in the following. It commonly refers to unknown, empty, or uncultivated areas (at the edges) of national states and carries a certain spatiality in its meaning (Baud and van Schendel 1997, p. 213; Schetter and Müller-Koné 2021). However, as it will be presented below, the spatiality of borders and boundaries will be addressed with a different key concept of border studies: the *borderland* notion. Lastly, the frontier concept is mostly used in the context of (post-) colonialism, imperialism, and historical research on nation-state borders, which does not match the present research topic.

2.3.2 The Genesis of Border Studies: Conceptual Development and Common Themes

Along with the general science-theoretical development of the social sciences, work in the border studies or limology research field of the early 20th century has provided a rather descriptive approach to different kinds of national borders and political boundaries (Hartshorne 1936; S. B. Jones 1959; Lyde 1915). Guided by an essentialist-metaphysical worldview, political geographers have particularly carried out territorial analyses of spaces—understood as containers—and descriptions of the political, religious, or ethnic relations found 'within' them. In the course of the second half of the 20th century and the science-theoretical shift of the geographical discipline toward positivist thinking, border studies gradually moved their interest "away from the nature of the boundary's location and history, to its function" (Minghi 1963, p. 407; cf. Kolossov 2005). Correspondingly, generalizing approaches of model building and classification have been performed (Prescott 2015 [1987]; cf. generally Newman 2011; F. Weber et al. 2020), which drew on the understanding of borders as fixed lines and static "empirical entities which divide the global space into bounded units that change mainly as a consequence of conflicts" (Paasi 1998, p. 69).

During the 1980s and the '90s, following economic efforts of transnational trade organization (such as NAFTA), the global spread of the internet and new, digital communication technologies, as well as the fall of the 'Iron Curtain,' border studies have been guided particularly by the upcoming globalization discourse and the idea of a *deterritorialized* and borderless world (Newman 2006b; Ohmae 1990). However, the events of 9/11 and the emerging fear of global terrorism mark a turnaround—geopolitically and then also scientifically: The newly emerged securitization discourse has furthered a certain re-closure of national borders, which resonated in the border studies research field and brought up discussions about *reterritorialization* tendencies (Newman 2011, p. 34; Paasi 2011, p. 18; for a general discussion of the notion of territory see Paasi 2003) and the interplay of the diametral processes of *de-* and *rebordering* (e.g. Albert and Brock 1996). Furthermore, the incipient re-sealing of national borders does not just take place at the territorial edges of nation-states but materializes itself in manifold surveillance and security strategies throughout social space—in shopping centers, private urban parks and plazas, and especially at airports (Paasi 2011, p. 22)—or *vernacularizes* (Cooper et al. 2016 [2014]), meaning that border work is carried out by specific technology as well as agents recruited as border personnel, e.g. local law enforcement (Johnson and Jones 2016 [2014]). This diffusion and multiplication of borders challenges traditional boundary conceptions of inside/outside (Vaughan-Williams 2008). Hence, the notion of *borders everywhere* has emerged (Balibar 2004; cf. Lyon 2005; Rumford 2012, p. 894), expanding the research focus on "alternative border imaginaries 'beyond the line'" (Brambilla et al. 2015, p. 2).

At the end of the 20th century, the research field of border studies has experienced a substantial perspective shift. The upsurge of post-structuralist theories and relational thinking has facilitated the diversification and multiplication of border-theoretical concepts (among many Agnew 2008; Kolossov 2005; Massey 1995, p. 67; Paasi 2012; 2020, p. 2; Wastl-Walter 2011b). In parallel, the onset of postmodernity has led to an increasing interdisciplinarity in the social sciences, which connected border studies to diverse research fields and ideas beyond disciplinary boundaries (Amilhat Szary 2015; Brunet-Jailly 2005; Newman 2003a, 2006a). As an example, Paasi (2011, p. 27) refers to the idea of the other, which has been adopted from cultural studies and ultimately developed into the vital concept of *othering* (van Houtum and van Naerssen 2002) in border studies. In addition to the disciplinary opening, the science-theoretical shift of the research field toward social constructivist approaches—which is referred to as "processual shift" (Brambilla et al. 2015, p. 1)—has changed the idea of borders and boundaries fundamentally. Borders are now understood as socially constructed and changeable phenomena and thus constitute "mirror[s] of social relations" (Kolossov 2005, p. 610). Rather than "fixed point[s] in space or time" (van Houtum and van Naerssen 2002, p. 126), they represent complex "social practice[s] of spatial differentiation" (van Houtum and van Naerssen 2002, p. 126), which are always created anew. In this sense, Simmel (1997 [1903], p. 142) has argued that "the boundary is not a spatial fact with sociological consequences but a sociological fact that forms itself spatially." The object of investigation in the research field transformed from the territorial border line and its functions to diverse social, spatial, and institutional practices of *bordering* (Brambilla 2010, p. 82). The "overall idea of what borders are and what bordering means for social and human life has expanded dramatically" (Paasi 2020, p. 7).

Following this shift, research interest has moved to processes of social distinction, its reproduction and dissolution, and thus the scrutiny of "the changing relations between borders and society" (Rumford 2006, p. 155). The social constructivist approach to borders and boundaries allows border scholars of various disciplines to focus on the "*construction*, *organization* and *reproduction* of social life, territoriality and power" (Paasi 1998, p. 69 [original emphasis]) and to inquire into emerging "identities" (Newman 2003b, p. 15) and discourses found at, around, or on the subject borders. In summary: In the course of the 'disciplinary debordering' of border studies, borders and boundaries "have taken on a multi-dimensional meaning" (Newman 2011, p. 33; cf. Kolossov and Scott 2013; Laine 2016), although the focus of inquiry has been and continues to be the national level (O'Dowd 2010; Rumley and Minghi 2015 [1991], p. 2).

2.3.3 Changing the Scale: Urban Borders and Processes of Urban Bordering

As early as the 1960s, Minghi (1963, p. 428) pointed to the benefits of transposing the methodology that has been developed at the level of national border demarcations to other political levels. For borders are—according to (Balibar 2004, p. 1 [quoted after Cooper et al. 2016, p. 15])—found "wherever the movement of information, people and things is happening and is controlled." In recent decades, especially since the 2000s, a theoretical expansion has increasingly taken place, resulting in a fruitful scrutiny of borders and boundaries on different scales and of various forms (cf. Lazzarini 2015, p. 182; Newman 2003a, p. 129; 2003b, pp. 13–14). In contrast to previously held assumptions of a globalized—and hence borderless—world, border studies are focusing on various multi-layered demarcations today, which differ in terms of visibility, permeability, and materiality (Johnson and Jones 2016 [2014]; Rumford 2012, p. 894).

In this course, the focus of border studies expanded to urban space. At first, however, still within the traditional scale or view on the immediate vicinity of national borders. *Cross-border regions* and *transfrontier metropolises* have been identified and researched for their shared environmental resources, social, cultural, and economic relations, as well as governance structures across the 'stretched' border (among many Herzog 1991; Herzog and Sohn 2014, 2017;

Mendoza and Dupeyron 2020; Nienaber and Wille 2020b; Sohn 2014; Szytniewski and Spierings 2014; Wille 2016). A different approach to urban space has been initiated through the adoption of the notion of borders everywhere and the idea of border vernacularization. Following these thoughts, cities have been taken into account as spaces where border *work* is taking place (cf. Coleman 2007; Fauser 2017; Gilbert 2009; Lebuhn 2013; Varsanyi 2010). Common to these research approaches is the endeavor to explain the expansion of the national border beyond the actual border line within a focus on the deployment of border agents and emerging mechanisms of control and surveillance in cities and other places. Yet in these cases the extension of border-theoretical considerations to cities only takes place with reference to immigration and state border security issues, which rather illuminates national border policies than the everyday organization of urban space and life.

However, this is what is of particular interest in the present work: How small-scale borders and boundaries are (re)produced, adapt, and how they (re)structure everyday life in and around urbanizing neighborhoods. Corresponding approaches have increasingly taken place since the 2010s, starting from an emerging understanding of urban spaces as "social environments that are *created* by bordering processes" (J. W. Scott and Sohn 2018, p. 2 [emphasis added]). Lazzarini (2015, p. 177), for instance, traces the historic significance of borders and boundaries for urban settlements—from ancient Greek cities to postmodern metropolises—and states that the "idea of the city and the idea of the border are mutually related." Drawing on their research on everyday life in and around urban enclaves in China, Liao et al. (2018) and Breitung (2011, 2012) conclude that "urban space [...] is always made up of various boundaries created by administrative decisions, by processes of social or cultural segregation, by urban design, or by the mental representation of space as evident in imaginations, identities and human activity spaces" (Breitung 2011, p. 56). Similarly, focusing on place-making, urban infill, and redevelopment projects in Europe and North America, Spierings (2012, 2013) and Gibson and Canfield (2016, p. 74) conceptualize urban borders and boundaries "as either physical or non-physical divisions in a community, either internal or external, intended or unintended." The two scholars trace the effects of urban planning projects, which commonly "seek to create distinct [spatial] identities" (Gibson and Canfield 2016, p. 74), making a specific segment of the city "recognizable, visible and giving it form: it means making that space different, showing a discontinuity" (Lazzarini 2015, p. 177) and thus (re)creating, lifting, or shifting boundaries. Ultimately, these approaches show all together that urban borders and boundaries are complex phenomena, which exist not only physically-materially in the form of walls, fences, or large streets but also unfold everyday relevance as immaterial demarcations and symbolic disparities between dissimilar neighborhoods, places, and groups. Urban demarcations are mostly traversable, even invisible or irrelevant to certain parts of the population, and ultimately operate in the same way as the borders between nation-states.

Accordingly, the key concepts and ideas developed at the national level have also been applied at the micro level of (urban) neighborhoods. Urban border studies have successfully integrated and adapted the processual view on bordering (Lazzarini 2015; J. W. Scott and Sohn 2018) and the idea of debordering and rebordering (Liao et al. 2018). Generally, bordering is understood as "practice of creating, confirming and re-creating socio-spatial distinctions at the formal (e.g. political) as well as everyday level" (J. W. Scott and Sohn 2018, p. 5). In urban contexts, this approach applies in the same manner. For urban spaces are fundamentally characterized by the "interplay of closure, demarcation, and exclusion on the one hand and openness, connection, and participation on the other" (Roßmeier and Weber 2021, p. 6).Thus, urban bordering processes should be understood as everyday processes of coexistence, contestation, and identity building of the self in the world—or the neighborhood—of the other (Iossifova 2020). Lazzarini (2015, p. 182) puts it aptly: "urban life is involved in a ceaseless process of bordering."

In line with these thoughts, it becomes clear that urban bordering processes "constantly renegotiate and reshape the extent of spatial separation and integration" (Liao et al. 2018, p. 1108). This is connected to the notions of debordering and rebordering, which Liao et al. (2018) trace on the urban level by drawing on tendencies of physical, functional, and symbolic inclusion and exclusion. In their understanding, *physical debordering* on the neighborhood level "opens borders for people, either from both sides or from only one side, and creates spaces for common activities and exchanges between both sides" (Liao et al. 2018, p. 1095). In contrast, *physical rebordering* advances socio-spatial segregation tendencies and "restrictions to free movement" (Liao et al. 2018, p. 1095) through the (re)construction and reinforcement of walls or fences, highways, and other semi- or non-permeable barriers. Regarding the *symbolic* dimension of bordering, debordering of neighborhoods and communities promotes the creation of a shared sense of place, the "formation of common identities and weakening stigma" (Liao et al. 2018, p. 1096). *Symbolic rebordering*, in turn, means the amplification of ordering and othering processes (van Houtum and van Naerssen 2002) on the individual and social level and thus the intensification of "the fear of living with the 'other,' of wanting the self to be here and the other to be there, with a clear border separating the activity and interaction spaces of the two" (Newman 2006a, pp. 176–177).

In summary, urban borders and boundaries are largely involved in the organization of everyday life (relations) in and around neighborhoods. Furthermore, neighborhoods and communities cannot exist without boundaries and the constant distinction from everything around it (cf. Coulton et al. 2013; Mooney Melvin 1985; Suttles 1972; Taylor 2012). The 'here' needs to be distinguished from 'there' in order to become a *place* in space, whereby differentiation and demarcation can take place on multiple levels—in the physical, political, social, and symbolic dimension (see Figure 2.5). Thus, borders, boundaries, and bordering require a reference to the 'other side,' which could be a simple thought from the distant 'here' or social interaction in the physical space over 'there,' whereby the actual line between own/alien or self/other is either (re)constructed and (re)drawn or broken down and blurred in the sense of the notions of (symbolic) debordering and rebordering. Thus, from a post-structuralist understanding, it can be stated that bordering—on all levels of scale—should be understood as an ongoing and open power process over spatial and social relations, networks, and flows with parallel tendencies of border solidification and dissolution (cf. Paasi 1998, p. 80; 2011, p. 28; 2012, p. 2307). This duality points to the traditional function of borders as barriers, as much as it points to their "potential to constitute bridges and points of contact" (Newman 2006b, p. 143), as it will be explained in detail in the following.

2.4 Synthesis of Urban Landscape and Border Studies: *Hybrid Urban Borderlands* as Conceptual Basis of the Present Study

2.4.1 The Borderland Notion and the Relational Connection of Borders and Space

Following the globalization discourse and the rise of post-structuralist, relational thinking in the social and cultural sciences, the notion of *borders as bridges* has been put on the research agenda of border studies. This understanding emerges from the endeavor to conceptualize borders less as severing, static lines on maps but as facilities of connection, coexistence, and commingling (Newman 2011, pp. 37–38). For, on the one hand, even the areas around closed and rather unpassable borders share the same climate conditions and biophysical resources, which, however, may underlie very different strategies and policies on the two sides (Dedina 1995). On the other hand, open borders particularly promote movement, relations, and exchange across the line and thus the emergence of transition zones.

Within this view on "the mutually constitutive relationship between borders and connection" (Cooper and Rumford 2013, p. 110) the research focus shifts, inter alia, onto the "spatiality of borders" (Rumford 2006, p. 161), referencing—in the sense of Bhabha (2000, 2012)—a hybrid in-between space or 'third room' (cf. Babka et al. 2012; Easthope 1998, p. 341; Mitchell 1997, p. 260): the *borderland.*

The borderland notion constitutes an early conceptualization of the immediate area on both sides of the U.S.-Mexico border, which highlights cultural hybridity and everyday border(ing) experiences within the 'cross-border lifestyle' of the Mexican-American author Gloria Anzaldúa (2012 [1987]). The conceptualization of borderlands is regarded as an important impulse for North-American border studies in the 1980s and '90s and has since then influenced various border and boundary analyses on multiple scales—in the Americas and beyond (see generally Banerjee and Chen 2012; Boesen and Schnuer 2017; Fellner 2020, 2021; Iossifova 2019; Pavlakovich-Kochi et al. 2004). Generally, the borderland concept characterizes "an area through which a boundary line runs" (Morehouse 2004, p. 29) and—more importantly—"the spaces where the everyday reality of boundaries are played out" (Morehouse 2004, p. 29). More detailed and with reference to both large and small scale bordering experiences, Anzaldúa (2012 [1987], p. 25) states that "borderlands are physically present wherever two or more cultures edge each other, where people of different races occupy the same territory, where under, lower, middle and upper classes touch, where the space between two individuals shrinks with intimacy." Thus, the borderland constitutes a "transition zone between two distinct categories, rather than a clear cut-off line" (Newman 2011, p. 37), which comes into being through trans-boundary interaction and everyday bordering processes and experiences (Baud and van Schendel 1997, p. 221; Iossifova 2020) and not only through the fact of border proximity. They are the actively—but not necessarily consciously—produced socio-spatial outcome of the linkages across relatively open or undefined borders and boundaries, which form on all levels of scale—around (semi-)permeable national borders as well as around the contested diving lines of communities and neighborhoods.

In order to elaborate the multi-scalar formation of borderlands more clearly, the concepts of space and place need to be linked theoretically with the notions of borders and boundaries. The relational, post-structuralist view serves as an approach for this purpose, allowing the transdisciplinary connection between (urban) planning sciences and border studies to become more apparent. From a relational perspective neighborhoods or (urban) spaces in general are understood as the ever changing result of ambiguous and contingent social and physical processes, which ultimately "divide and connect things up into different kinds of collectives" (Thrift 2009, p. 95). In the social sciences this understanding has emerged in the course of the spatial turn, which initiated the extensive rejection of the notion of space as container or "*surface* upon which economic and social activity took place" (Graham and Healey 1999, p. 626 [original emphasis]; cf. M. Jones 2009; Murdoch 2006). In urban theory and planning sciences, too, this paradigm shift has triggered a departure from the Euclidian perspective on space and place as solely material conditions and "bounded entities […] defined with clear physical and perceptual boundaries" (Davoudi and Strange 2009b, p. 27), opening the way for a view of urban space as "essentially porous, open and interlinked" (Paasi 2014, p. 11; cf. Massey 1995, 2005). In this sense, the relational perspective moves away from the conception of unambiguously "bounded spaces" (Paasi 2014, p. 10) by foregrounding networks and flows and emphasizing aspects such as "fluidity, reflexivity, contingency, connectivity, multiplicity and polyvocality" (Davoudi and Strange 2009a, p. 5). Fundamentally, however, relational thinking is not arguing for a borderless world of infinitely an interlinked society and space but helps to further the conceptual multiplication and diversification of borders and boundaries throughout space and scale (cf. Paasi 2014; Wille 2021). For, in this tradition of thought, scale is itself understood as an "interconnected web [of relations] with contingent boundaries, constantly territorialized, and a site of political contestation" (Davoudi and Strange 2009b, p. 62).

Consequently, it becomes possible for post-structuralist, relational approaches to establish a theoretical bridge between "the concepts of (urban) space and borders, which are *both* fundamentally about (dis)connection and (dis)continuity of relations on multiple levels" (Roßmeier and Weber 2021, p. 6 [original emphasis]; cf. Haselsberger 2014). In line with this, borders and boundaries can be conceptualized "both [as] barriers and gateways" (Cooper and Rumford 2013, p. 110; cf. Nienaber and Wille 2020a) in everyday life. This functional duality of borders entails distinction, dissimilarity, and separation but may also lead to commingling, hybridization, and the reduction of difference across the diving line (Iossifova 2015, p. 104). In other words, borders and boundaries can create "interfaces of interaction and exchange [where] formal and informal boundary demarcations, delimitations, and definitions are [partly] undone and hybrid, urban borderlands emerge" (Iossifova 2019, p. 3; Newman 2011, p. 28). Ultimately, this perspective is shifting attention toward aspects of individuality, processuality, and hybridity and establishes a link to the social constructivist understanding of landscape as a *practice of creating (spatial) identity* and particularly to the idea of (socio-)spatial hybridization and pluralization, as it will be developed below.

2.4.2 Connective Approaches of Urban Landscape and Border Studies

In order to develop the present conceptual approach it seems necessary to trace the extent to which connections and references have already been established between the research fields of urban landscape and border studies. Urban and landscape studies have traditionally addressed demarcations and (dis)continuity in the planning context, however, rarely with reference to border theory or concepts. Correspondingly, Haselsberger (2014, p. 505) notes: "Over the past two decades a border studies literature has emerged, informed by a wide range of academic disciplines, such as geography, anthropology and political science, but not planning" (cf. Breitung 2011, p. 56). In their prominent work in the field of urban studies and design Kevin Lynch (1992), for instance, highlights demarcations for their role in mental maps and the visual structure of cities. By inquiring into the key design elements of urban space and the 'form of the city,' they identify paths, edges, nodes, districts, and landmarks. However, by taking a Euclidian planning perspective Lynch conceptualizes the focal elements of urban space rather as the outcome of planning processes than as the result of social and material relations or interdependencies. Thus, by highlighting edges, Lynch is referring to the visual aspect and the material dimension of borders but misses the political and particularly the social dimension of bordering (see Figure 2.5). Similarly, Vaughan et al. (2021) address community severance through large streets and traffic, however they do so within a Euclidean perspective as well and thus neglect individual and social aspects.

In postmodern urban theory borders have been addressed in more diverse ways—for instance, within a focus on the periphery of settlements or by looking at inner-city bordering processes. The first of the two approaches is showing particularly within the discussion of edgeless cities (Lang 2003). Although the shift of the boundaries of urban settlements into the periphery are generally addressed in the scientific revision of suburbanization processes, the concept of edgeless cities is explicitly referring to (non-existent) settlement boundaries. In their research on peripheral clusters of office buildings, Lang et al. (2009, p. 726) discuss the "'edgeless' nature" of settlements and thus the dissolution of their perceptible borders. Their findings identify edgeless cities as office clusters that extend across hundreds of square miles but "are *not* perceived as one place" (Lang et al. 2009, p. 732 [original emphasis]) due to their lack of meaningful identity and their unique boundaries that are "indeterminate [and] very difficult to define" (Lang et al. 2009, p. 733). The researchers of the *LA School of Urbanism*, in return, offer references to urban bordering processes 'on the inside.' In their postmodern adaption of Burgess' widely cited concentric ring model, Flusty (1994) identifies 'interdictory spaces' as the borders and boundaries between the "noncontiguous collage of parcelized, consumption-oriented landscapes" (Dear and Flusty 1998, p. 66). These spaces are by definition designed "to intercept or repel or filter would-be users" (Flusty 2001, p. 659) through the use of surveillance technology, guards, and other security-oriented methods. Border-theoretically framed Cooper et al. (2016 [2014]) call these phenomena "vernacularized borderwork," which turn border experiences from a rare encounter into "a daily occurrence" (Johnson and Jones 2016 [2014], p. 2). What is remarkable about Flusty's approach is the explicit reference to the notion of otherness and the process of redefining "the remainder [...] as 'Other'" (Flusty 2001, p. 659), which is already demonstrating the benefits of incorporating border theory into analytical urban landscape approaches.

Border Studies, in return, have so far drawn only little attention to the concept of landscape. One reason for this could be the fact that throughout the last half of the 20th century landscape research has commonly been associated with descriptive and classificatory approaches, and has thus been stigmatized as an 'unscientific endeavor' (Rumley and Minghi 2015 [1991], p. 3; cf. Cosgrove 1985, p. 45; Hard 2002, p. 173; Schenk 2006, p. 17; F. Weber 2016, p. 86). However, in the wake of the constructivist turn and a new, anti-essentialist devotion to landscape also border studies can benefit from the social constructivist landscape perspective and its sensitivity to hybridity. For in recent years urban landscape research has put "the increasing hybridization of the city and its surroundings" (Kühne 2018a, p. 98) on its research agenda and is thus shifting from mere descriptions of regions, bounded spaces, and 'true landscapes' to the inquiry of individual interpretations of today's interlocked urban/rural life-worlds. In the early 1990s, embracing the perspective reorientation of landscape research and aiming for the liberation of landscape "from the conceptual closet to which it has been confined", Rumley and Minghi (2015 [1991], p. 1) have already been looking for ways to combine the notions of landscape and border by means of the term *border landscapes*. However, in their introduction to *The Geography of Border Landscapes*, they are not differentiating between the terms border landscape and borderland and use them synonymously.

The *borderscapes* concept and the notion of *borderscaping*, in turn, constitute a more recent approach toward a conceptual linking of the two fields (Brambilla 2015; Brambilla and Jones 2020; Dell'Agnese and Amilhat Szary 2015). Through the attachment of the suffix '-scape' from the word 'landscape,' Brambilla (2010, p. 83) is explicitly drawing on the social constructivist understanding of landscape within the goal to emphasize aspects of subjectivity. Borderscaping is, in this sense, referencing the etymological meaning of scaping as "shaping and carving" (Brambilla and Jones 2020, p. 289) and emphasizes the "political project of 'making'" (Brambilla and Jones 2020, p. 289). Thus, the borderscapes/borderscaping notion is primarily concerned with the critical discussion of negotiation processes in the "normative dimension" (Brambilla et al. 2015, p. 2), which, however, lies not within the focus of the current research project. Furthermore, Brambilla (2015, p. 22, cf. p. 33) argues for an adoption of the suffix '-scape' instead of the prefix 'land-' in order to "liberate the conceptual potential of new terms […] from restrictions imposed by the etymological ambivalence that characterises the term landscape" (Brambilla 2015, p. 22). However, the criticized ambiguity and contingency of landscapes or, in the sense of Hard (1969, p. 3), the large 'semantic court' of the concept of landscape is of particular interest in the present study because it allows to shift the focus on the plurality and openness (of meaning) of demarcations and urban space in general. Furthermore, also aesthetic, cognitive, and emotional aspects are addressed by the established concepts of landscape studies and other terms that embrace the semantically indistinct prefix 'land-' (Kühne 2016, pp. 13–14). Ultimately, the borderscapes concept has undoubtedly contributed to the linking of landscape and border and has enriched the discussion of recent years in regard to the complexification of borders. What is integrated from the derivation of borderscapes into the present conceptualization is the understanding of landscape as a practice. However, the notion of borderlands and the prefix 'land-' seem more suitable for bringing attention to (the blurring of) urban borders and boundaries and spatial hybridities in general. Consequently, the present research builds upon the borderland notion by Anzaldúa (2012 [1987]) and the urban borderland adaption as developed initially by Iossifova (2013, 2015, 2019) and discussed further by Liao et al. (2018).

2.4.3 Hybrid Urban Borderlands—Ambiguous Spaces and Four-dimensional Practices

The borderland concept and its analytical focus are adapted here by a breakdown into four bordering dimensions (see Figure 2.5) and by carving out in greater detail its potential for highlighting spatial compartmentalization, hybridization, and pastiches by means of the urban landscape perspective. Anzaldúa (2012 [1987], p. 25) is already including aspects of cultural hybridity and blending in her conceptualization by means of the mestizo/mestiza term and is referring to the U.S.-Mexico borderland as "a vague and undetermined […] place of contradictions." Iossifova (2019, p. 1) is highlighting aspects of urban coexistence and interaction, thus elaborating socio-spatial hybridities on the urban scale. Liao et al. (2018, p. 1093), in turn, are building on Iossifova's concept of urban borderlands, however, within a focus on (simultaneous) rebordering and debordering tendencies. Here, in order to enlarge upon the manifold hybridizations between the neighborhoods of San Diego's inner ring, urban borders and boundaries are considered within four bordering dimensions.

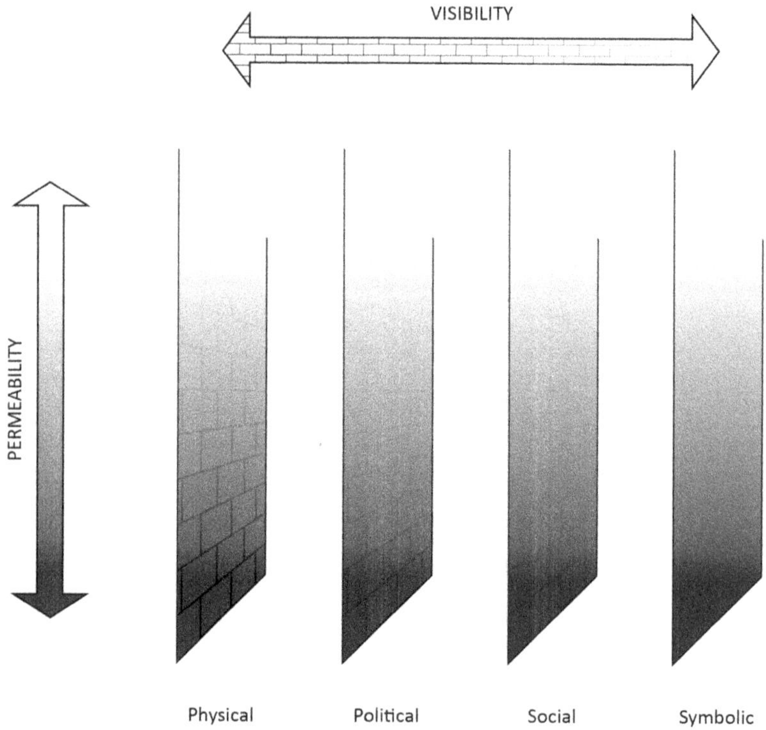

Figure 2.5 Bordering Dimensions. (Source: Own representation)

Everyday life bordering can be understood as a set of complex processes, which unfold—in various degrees—simultaneously in the interconnected *physical, political, social*, and *symbolic dimension.* Taking the context of neighborhood (re)construction as an example, borders and boundaries adapt, are (re)drawn, or lifted in the familiar everyday urban landscape within these four dimensions: In the *physical dimension*, which contains the strongest visual components of bordering, new physical differentiations become obvious in the form of new walls, prohibition signs, or surveillance technology or, in the sense of debordering, through a dissolution of the architectural differentiation line of areas with old, small housing structures on the one side and large, new buildings on the other. Furthermore, neighborhood development could mean the drawing of new *political boundaries* in the form of redistributing planning responsibilities through a shift of the official district boundary, the incorporation of a city, or simply the implementation of neighborhood signs and other branding efforts. (De)bordering in the *social dimension* in the case of neighborhood restructuring could mean enhanced accessibility and increased appropriation or participation of a formerly excluded group but at the same time also the strengthening of aspects such as exclusiveness, otherness, and segregation. Lastly, *symbolic de- and rebordering* becomes conceivable in the sense of changes in reputation, atmospheres, aesthetics, or common use of space creating an either inviting or appealing image of a place or leading to aversion and withdrawal (cf. Berglund 2019; Berglund and Gregory 2019; Gregory 2019). Symbolic border phenomena are commonly less visible but might still develop into boundaries that "can be as sealed as the strongest of inter-state borders separating two belligerent countries from each other" (Newman 2006a, p. 177). Summing up, all kinds of bordering and debordering processes take place in different, yet interconnected dimensions and become particularly effective on the individual level. Thus, just like landscape or space in general (cf. Berr 2020; Duttmann et al. 2020; Riehl 1996), borders and boundaries are to be understood as practices, as individual construction and interpretation processes that are learned—or internalized—through biographical socialization and everyday experiences, constantly updated, and *becoming*.

An additional aspect that needs further attention in the conceptualization of hybrid urban borderlands is that bordering processes do not always take place precisely. The oscillation between openness and closedness and the varying distinctness or visibility of borders and boundaries entail ambiguous situations of simultaneity and hybridity, which break with simple dichotomies (Paasi 2012, p. 2307) and create *third* spatial categories in the sense of Bhabha (2000; 2012; cf. Easthope 1998; Mitchell 1997). The circumstance that bordering takes place in different dimensions adds to the possibility of unclear and blurred delimitation because borders and boundaries do not always lie exactly on top of each other in space and time but can exist next to each other in an unordered way, creating "mobile, fluid and relational" (Lazzarini 2015, p. 177) in-between spaces or transition zones, which are "continually redefining identity and otherness, dynamics of inclusion and exclusion, citizenship and participation forms" (Lazzarini

2015, p. 177). Thus, urban borders and boundaries obtain a certain spatiality, which goes beyond traditional concepts of the "bounded spaces of a formal urban geography" (Iossifova 2015, p. 104) and Euclidian urban planning theory. What comes into being are dynamic and open spatial alternatives in the sense of a postmodern-relational (cf. Graham and Healey 1999; Löw and Weidenhaus 2017; Murdoch 2006) and hybrid geography (cf. Whatmore 2002), which bring aspects of contingency, fluidity, and mobility into focus.

The connection between urban border theory and urban landscape studies lies particularly in this ambiguous spatiality and the dual interplay of dissolution and containment, which resembles the development of urban/rural hybrids and spatial pastiches. For urban/rural hybridization occurs, on the one hand, through the dissolution of the counterpart and the binary or, in other words, of boundaries between two compartments, resulting in more or less differentiated edges and a tangled web of different uses and areas (Kühne 2017, p. 21; cf. Jauhiainen 2003, pp. 396–397). These hybridizations and spatial pastiches emerge as a side effect of functional changes in the postmodern urban landscape, which promote socio-spatial reformation, reinterpretation, and renegotiation (see generally Aitken and Zonn 1994; Hoesterey 2001; Kühne and Weber 2019b; Vester 1993). On the other hand, however, the overall process of urban/rural hybridization simultaneously entails new forms of inclusion and exclusion in the attempts "to define edges and limits and to construct a degree of order within [these] 'unordered' situations of change" (J. W. Scott and Sohn 2018, p. 13). Political, social, and symbolic boundaries in the postmodern urban fabric may 'harden' into physical borders, while physical borders may dissolve and leave behind "fuzzy and indeterminate and thus contentious" (J. W. Scott and Sohn 2018, pp. 13–14) boundaries of socio-spatial distinction and attribution (Bourne 1996, p. 165). Thus, urban/rural hybridization furthers social and spatial compartmentalization and the reformation of categories. This, however, might lead to the social contestation of demarcations, dichotomy, and otherness. For as Newman (2011, p. 35) states, the "social construction of compartments and their borders is necessary for the ordering of society, but will always be contested, because there can rarely be a single border which is totally congruous with the absolute category." In line with these thoughts, hybrid urban borderlands are conceptualized as the contingently interpreted and fluid transition zones "in-between sociospatially dissimilar, adjacent parts of cities" (Iossifova 2019, p. 1), the spaces where modern categorizations of duality and opposition break and hybrid situations of connection and ambiguity emerge (Roßmeier 2020b; Roßmeier and Weber 2021).

It is important to note, however, that the aim of the present conceptualization is not to provide a fixed and all-encompassing border theory. For, in line with the argumentation of Wastl-Walter (2011a) and Paasi (2011), borders and boundaries should be approached *in context* rather than from a distant position and through a universal theory. Consequently, the developed concept of hybrid urban borderlands constitutes a versatile perspective for urban borderland research, which is meant to be "used and re-conceptualized further in various empirical settings" (Paasi 2011, p. 19). In this respect, the aim of the present study is not merely a wider understanding of the neighborhood changes in and around downtown San Diego but also to explore the potential for differentiation within border studies by paying special attention to aspects of emerging and transitional hybridity in the context of urban and rural developments. This becomes possible, on the one hand, by drawing on the common themes of the border studies research field and, on the other hand, by contextual theorization of San Diego's inner-city borders and boundaries (see Chapter 6). Finally, this approach follows Johnson and Jones (2016 [2014], 5), who argue that "the geographies of borders are more expansive, and as a result the conceptual tools we use to understand them must be expanded" (cf. Cooper et al. 2016 [2014], p. 16) as well.

2.5 Deduction of the Central Research Questions: Theoretical Summary and Operationalization

Following the presentation of the science-theoretical and conceptual foundations of this work, a theoretical summary and operationalization is given. This serves the purpose of demonstrating how the research topic of manifold urbanization processes in San Diego's inner-ring neighborhoods can be approached within the underlying positions and perspectives and which questions and focal points become possible. Fundamentally, the aim of the present work is to elaborate a broader conceptualization of (urban) landscape research by means of a border-theoretical approach in the direction of hybrid urban borderlands (see Figure 2.6). On top of the given presentation of the historical development of landscape and border studies, it was crucial for the conceptualization of hybrid urban borderlands to create a theoretical framework that defines the perspective and methodological trajectories of the present analysis. As it will be explained in the following, the social constructivist position serves this purpose in supplementation with aspects of non- or more-than-representational theories (cf. B. Anderson and Wylie 2009; Ingold 1993; Lorimer 2005; Thrift 2008; Waterton 2019).

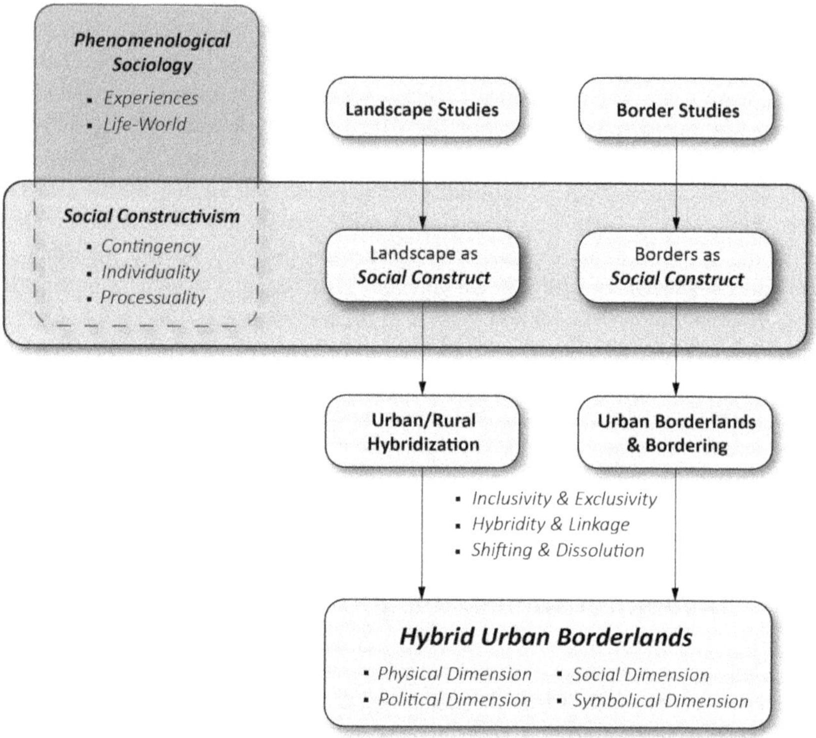

Figure 2.6 Theoretical-Conceptual Synthesis of Hybrid Urban Borderlands. (Source: Own representation)

The social constructivist approach in the tradition of Berger and Luckmann (1966) is interested in the constructional character of phenomena and the social mechanisms by which these are charged with (symbolic) meaning (Burr 1995; Gergen 1999). Drawing on Schütz' work (1972 [1962]) on the life-world, Berger and Luckmann shifted the focus of their sociology of knowledge on the everyday world and the everyday knowledge it contains—in other words, the infinite multiplicity of individual realities. In the present research context, social constructivist thinking implies an understanding of urban landscape, borders, and boundaries as socially constructed phenomena, shifting the analytical focus on individual processes of construction and perception, (landscape or border) socialization, and subsequently the social contingency and processuality of landscape, demarcations, and space in general (cf. Cosgrove 1998 [1984]; Greider and Garkovich 1994; Kühne 2019a, p. 33; Riehl 1996; Schenk 2019). Generally, in opposition to the science-theoretical position of positivism, social constructivism aims for *understanding* and *interpretation* rather than explanation; in contrast to essentialist approaches, social constructivism asks *how* a certain body of knowledge comes into being instead of what a certain object is (cf. Burr 2005, p. 11). Accordingly, the social constructivist approach seeks to *understand how* urban landscape, neighborhoods, or borderlands emerge as (active) individual achievements of subjective construction and perception, rather than describing these phenomena from a single point of view or tracing their supposed nature. For no matter how rigid and 'real' social reality and all appearing phenomena may seem, following social constructivism, they will at all times be the result of social negotiation processes of externalized, subjectively intended meaning.

Phenomenological sociology in the tradition of Husserl (1913, 1970 [1954]) and Schütz (1970) is in a similar manner concerned with 'feeling' or understanding individual reality, experiences, and actions (Hitzler and Eberle 2004, p. 59; Soeffner 2004, p. 97). In the present study, drawing on phenomenological thinking enables the focus to shift on practices and the subjectively intended meanings of actions (in the sense of M. Weber 1981), on experiences and perceptions in and of the life-world—the redeveloping neighborhoods—within a particular view on material aspects of urban landscape and bordering (cf. Flaherty 2009; Tilley 1994; Wylie 2019). For phenomenological approaches seek to explore the intertwined relations of subject and object, society and materiality, or, in the context of phenomenological landscape research, "the varied practices and activities wherein people and [the physical foundations of] landscape mutually interact" (Wylie 2019, p. 135; cf. Chemero 2003; Moran 2000). However, the efforts of non- or more-than-representational research in general and the present study in particular are not aiming for finding meaning in the material world but are driven by the idea of expanding the theoretical focus of social constructivism through investigations of the complex interactions of the social and the material dimension. This is achieved by giving attention to *subjective everyday life experiences*, such as multi-dimensional bordering experiences (see Figure 2.5), through which subject and object are inseparably connected. Because in the understanding of Husserl, experience is a relational practice that is either directed "toward or absorbed by some pattern (object, person, etc.) in the external world" (Gergen 1999, p. 128). Relational thinking is in this sense generally concerned with the dissolution of "the boundaries between objects and space" (M. Jones 2009, p. 491). Consequently, not only in the present analysis, researching everyday life experiences means to take the interactions between and the relations of subject-object and human-landscape into focus.

In summary, the different classification or valuation of materiality within the two approaches of social constructivism and phenomenology can be explained by means of the theoretical systematization of Berger and Luckmann's sociology of knowledge (see Figure 2.2): While social constructivism pays attention to material aspects solely within the process of externalization (cf. Kühne 2019a, p. 22), more-than-representational theories go beyond this approach and value material aspects both in the process of externalization *and* internalization. For landscape, borders, or any materialized phenomenon is from a non- or more-than-representational understanding—as their titling indicates—*more* than a simple representation of symbolic charges and "cultural identities" (Greider and Garkovich 1994, p. 2) because of the *experiential relationship* a material element forms with a subject, which inevitably implements this element into the individual's subjective life-world.

In line with these thoughts, it seems beneficial for the present research to triangulate the two approaches into a phenomenologically oriented social constructivist perspective that seeks to *understand* urban/rural hybridization, multi-dimensional bordering processes, and the hybrid urban borderlands they form. In concrete terms, the present study aims to trace and illustrate the individual interpretations of the redevelopment processes in San Diego's inner-ring suburbs and how the socio-spatial changes, but especially the subjective attributions, lead to the (re)drawing and adaption of borders and boundaries and the creation of hybrid, fuzzy situations of co-existence, interaction, and commingling. In line with the argumentation of Ashley (1988; as described by Paasi 1998, pp. 75–76), the focus is directed less on the description of exact border and boundary locations in San Diego but especially on the social processes by which the multi-dimensional demarcations are created, maintained, and reproduced in everyday life situations of social and material interaction. Accordingly, the interest forms around social ordering and othering processes and individual experiences in the course of neighborhood change, urban infill development, gentrification tendencies, and the socially driven development trend of hybrid 'urban village' living. With regard to these processes, it is important to take a close look at the changing "relationships between (re)investment and disinvestment, affluence and poverty, and accessibility and isolation across metropolitan space" (Markley 2018, p. 625) and the boundaries created by this, which "may be as invisible as they are tangible and, equally, as perceived as they are real" (Newman 2006a, p. 177). Therefore, the present analysis is directed toward the study of everyday life "border realities" (Brambilla 2010, p. 81), experiences, and interpretations of the hybrid processes of inner-city urbanization and redevelopment in San Diego through a multi-method and "multi-perspective approach" (Rumford 2012, p. 893) that is becoming increasingly known as 'neopragmatic' research approach (Chilla et al. 2015; Kühne 2019b; Kühne and Jenal 2020a, 2021, 2020b).

2.6 Critical Reflections on the Present Theoretical Foundation

After the presentation of the potentials and focusses of the theoretical foundations of this work, the following paragraphs deal with limitations and common points of criticism.

Phenomenological philosophy and sociology face a large number of criticisms and are subject of heated discussions concerning their focus, application or operationalization, and the depth of their insights they may or may not reach (detailed in Seamon 2019). A common point of criticism is the supposed vagueness of phenomenology itself and the alleged, strong subjectivity and thus arbitrariness of phenomenological findings. Drawing on the philosopher Herbert Spiegelberg, Seamon (2019, p. 38) states "that there are as many phenomenologies as there are phenomenologists." Within the aim to emphasize the breadth of common interpretations of phenomenology, Seamon (2019, pp. 46–48) lists 23 definitions of phenomenological philosophy and eponymous research, which unveil diverse methodological approaches (cf. Finlay 2009). In this sense, Finlay (2011, p. 19) describes phenomenology as a "continuum with pure rigorous, scientific description on one end and fluidly poetic interpretation on the other." Furthermore, phenomenological findings are accused of strong subjectivity—not only when the descriptions of observed phenomena fall in the more artistic end of the spectrum. According to the criticism of Hard (1995, p. 133), the best possible extent of intersubjectivity of phenomenological descriptions lies in the reader's approval, which is based on their similar experiences. This is also due to the fact that phenomenological approaches refrain "from any causal or genetic hypotheses, as well as from assertions about the ontological status of the phenomena analyzed" (Berger and Luckmann 1966, p. 34), which, in turn, allows for the triangulation with social constructivist positions. However, caution is required when turning to materiality in order to not discard the achievements of post-structuralism and constructivism again, as Duineveld et al. (2017, p. 382) argue. Accordingly, what is guiding the present phenomenologically oriented approach is not the search for essence in the material world but the idea of expanding social constructivist research by examinations of the complex relations between object and subject, social and material dimension.

Another important component of the present research approach, the concept of hybridity, has been subject to various points of criticisms in the course of its synthesis, which need to be addressed. Kühne and Weber (2019a, pp. 31–32) discuss a list of common criticisms, which will guide the following brief presentation. Drawing on Ackermann (2004, p. 152), they point out that the concept's discussion of differences might lead away from its main concern, the aspects of connection and commingling. Ha (2015), for example, is demoting the use of the concept hybridity as 'postmodern, late-capitalist hype,' which suffers from its supposed modernist perspective on nations and nation-state structured societies. Their argumentation builds on a more profound point of criticism, which is the accusation of bearing hidden essentialisms in the conceptualization on the basis of assumed previous 'purity' of the different hybridized compartments. In reaction to this, approaches drawing on hybridity, such as the present one, negate their essentialist foundation, however, in a next step, experience a consequential criticism. If the considered hybridized compartments have not been of 'pure' condition, then the concept of hybridity is nothing more than a playful tautology, testifying another 'stage' of commingling of always already blended, 'impure' elements (Ackermann 2004, p. 153). In this sense, hybridity is accused of lacking theoretical-conceptual depth and constituting nothing more than a 'loose' concept, therefore being applicable to an infinite variety of research subjects. Consequently, discussing aspects of hybridity might seem unnecessary and insufficiently illuminating. However, the potentials and benefits of the concept lie particularly in the discussion of a *new* stage of commingling and thus of processuality, which can be used to emphasize and shed new light on various change processes. Following this argumentation, the present approach incorporates the concept of hybridity because of its ability to work out the processual aspect of neighborhood change and individual adaptions and (re)interpretations. Finally, therein lies the connection to the social-constructivist position and the post-structuralist perspective, which are both concerned with aspects of processual change and development.

Retrospectively, it can be concluded that within this chapter connections have been established between all the individual conceptual and perspective components and science-theoretical foundations of the present work. This underlines the theoretical stringency of the developed conceptualization of hybrid urban borderlands, which serves as perspectival-methodological guidance of this study. On the basis of the connected foundations of the hybrid urban borderlands notion, the methodic components of the present analysis will be derived and outlined in detail in the following chapter.

References

Abel, T. (1948). The Operation Called Verstehen. *American Journal of Sociology, 54* (3), pp. 211–218.

Ackermann, A. (2004). Das Eigene und das Fremde: Hybridität, Vielfalt und Kulturtransfers. In F. Jäger & J. Rüsen (Eds.), *Handbuch der Kulturwissenschaften. Band III: Themen und Tendenzen* (pp. 139–154). Stuttgart: J. B. Metzler Verlag.

Agnew, J. (2008). Borders on the mind: re-framing border thinking. *Ethics & Global Politics, 1* (4), pp. 175–191.

Aitken, S. C. & Zonn, L. (1994). Re-presenting the place pastiche. In S. C. Aitken & L. Zonn (Eds.), *Place, power, situation, and spectacle. A geography of film* (pp. 3–25). Lanham: Rowman & Littlefield.

Albert, M. & Brock, L. (1996). Debordering the world of states: New spaces in international relations. *New Political Science, 18* (1), pp. 69–106.

Allen, A. (2003). Environmental planning and management of the peri-urban interface: perspectives on an emerging field. *Environment and Urbanization, 15* (1), pp. 135–148.

Amilhat Szary, A.-L. (2015). Boundaries and Borders. In J. Agnew, V. Mamadouh, A. J. Secor & J. Sharp (Eds.), *The Wiley-Blackwell Companion to Political Geography* (pp. 13–25). Oxford: Wiley Blackwell.

Anderson, B. & Wylie, J. (2009). On Geography and Materiality. *Environment and Planning A: Economy and Space, 41* (2), pp. 318–335.

Anderson, J. & O'Dowd, L. (1999). Borders, Border Regions and Territoriality: Contradictory Meanings, Changing Significance. *Regional Studies, 33* (7), pp. 593–604.

Antrop, M. (2015). Interacting Cultural, Psychological and Geographical Factors of Landscape Preference. In D. Bruns, O. Kühne, A. Schönwald & S. Theile (Eds.), *Landscape Culture—Culturing Landscapes. The Differentiated Construction of Landscape*. Wiesbaden: Springer VS.

Antrop, M. (2019). A brief history of landscape research. In P. Howard, I. Thompson, E. Waterton & M. Atha (Eds.), *The Routledge Companion to Landscape Studies* (2nd ed., pp. 1–15). London: Routledge.

Anzaldúa, G. (2012 [1987]). *Borderlands/La frontera. The New Mestiza*. San Francisco: Aunt Lute Books.

Aring, J. (1999). *Suburbia—Postsuburbia—Zwischenstadt. Die jüngere Wohnsiedlungsentwicklung im Umland der großen Städte Westdeutschlands und Folgerungen für die regionale Planung und Steuerung*. Hannover: Verl. der ARL.

Aschenbrand, E. (2018). 'A landscape like a painting?'—the staging of landscape in hiking and cycling tourism. *Zeitschrift für Tourismuswissenschaft, 10* (1), pp. 121–141.

Ashley, R. K. (1988). Untying the Sovereign State: A Double Reading of the Anarchy Problematique *Millennium, 17* (2), pp. 227–262.

Atha, M., Howard, P., Thompson, I. & Waterton, E. (2019). Introduction. Ways of knowing and being with landscapes: a beginning. In P. Howard, I. Thompson, E. Waterton & M. Atha (Eds.), *The Routledge Companion to Landscape Studies* (2nd ed., pp. xix-xxviii). London: Routledge.

Babka, A., Malle, J. & Schmidt, M. (2012). Einleitung. In A. Babka & J. Malle (Eds.), *Dritte Räume. Homi K. Bhabhas Kulturtheorie. Anwendungen. Kritik. Reflexionen* (pp. 9–26). Wien: Turia + Kant.

Balibar, É. (2004). *We, the people of Europe. Reflections on transnational citizenship.* Princeton: Princeton University Press.

Banerjee, P. & Chen, X. (2012). Living in in-between spaces: A structure-agency analysis of the India-China and India-Bangladesh borderlands. *Cities, 34*, pp. 18–29.

Baud, M. & van Schendel, W. (1997). Toward a Comparative History of Borderlands. *Journal of World History, 8* (2, Fall 1997), pp. 211–242.

Berger, P. L. & Luckmann, T. (1966). *The social construction of reality*. New York: Anchor.

Berglund, L. (2019). Excluded by design: informality versus tactical urbanism in the redevelopment of Detroit neighborhoods. *Journal of Cultural Geography, 36* (2), pp. 144–181.

Berglund, L. & Gregory, S. (2019). Introduction: the aesthetics of neighborhood change. *Journal of Cultural Geography, 36* (2), pp. 117–120.

Berr, K. (2020). Visuality, Aesthetics, and Landscape: For the Clarification and Self-Awareness of Constructivist Landscape Research. In D. Edler, C. Jenal & O. Kühne (Eds.), *Modern Approaches to the Visualization of Landscapes* (pp. 189–215). Wiesbaden: Springer VS.

Berr, K. & Jenal, C. (Eds.). (2019). *Landschaftskonflikte*. Wiesbaden: Springer VS.

Berr, K., Jenal, C. & Kindler, H. (2019). Landschaftskonflikte. In O. Kühne, F. Weber, K. Berr & C. Jenal (Eds.), *Handbuch Landschaft* (pp. 367–382). Wiesbaden: Springer VS.

Berr, K. & Kühne, O. (2020). *"Und das ungeheure Bild der Landschaft..." The Genesis of Landscape Understanding in the German-speaking Regions.* Wiesbaden: Springer VS.

Berr, K. & Schenk, W. (2019). Begriffsgeschichte. In O. Kühne, F. Weber, K. Berr & C. Jenal (Eds.), *Handbuch Landschaft* (pp. 23–38). Wiesbaden: Springer VS.

Bhabha, H. K. (1990). *Nation and Narration.* New York: Routledge.

Bhabha, H. K. (2000). *Die Verortung der Kultur*. Tübingen: Stauffenburg.

Bhabha, H. K. (2012). Über kulturelle Hybridität: Tradition und Übersetzung. In A. Babka & G. Posselt (Eds.), *Über kulturelle Hybridität. Tradition und Übersetzung* (pp. 17–57). Wien: Turia + Kant.

Blumer, H. (1969). Symbolic Interactionsm. London: University of California Press.

Boesen, E. & Schnuer, G. (Eds.). (2017). *European Borderlands. Living with Barriers and Bridges*. Abingdon: Routledge.

Bosco, F. J. (2015). Actor-network theory, networks, and relational geographies. In S. C. Aitken & G. Valentine (Eds.), *Approaches to human geography. Philosophies, theories, people and practices* (2 ed., pp. 150–162). Los Angeles: Sage.

Bourne, L. S. (1996). Reinventing the Suburbs: Old Myths and New Realities. *Progress in Planning, 46* (3), pp. 163–184.

Brambilla, C. (2010). Borders Still Exist! What are Borders? In B. Riccio & C. Brambilla (Eds.), *Transnational Migration, Cosmopolitanism and Dis-located Borders* (pp. 73–85). Rimini: Guaraldi.

Brambilla, C. (2015). Exploring the critical potential of the borderscapes concept. *Geopolitics, 20* (1), pp. 14–34.

Brambilla, C. & Jones, R. (2020). Rethinking borders, violence, and conflict: From sovereign power to borderscapes as sites of struggles. *Environment and Planning D: Society and Space, 38* (2), pp. 287–305.

Brambilla, C., Laine, J., Scott, J. W. & Bocchi, G. (2015). Introduction: Thinking, mapping, acting and living borders under contemporary globalisation. In C. Brambilla, J. Laine, J. W. Scott & G. Bocchi (Eds.), *Borderscaping: Imaginations and practices of border making* (pp. 1–9). Burlington: Ashgate.

Breitung, W. (2011). Borders and the City: Intra-Urban Boundaries in Guangzhou (China). *Quaestiones Geographicae, 30* (4), pp. 55–61.

Breitung, W. (2012). Enclave Urbanism in China: Attitudes Towards Gated Communities in Guangzhou. *Urban Geography, 33* (2), pp. 278–294.

Brunet-Jailly, E. (2005). Theorizing borders: An interdisciplinary perspective. *Geopolitics, 10* (4), pp. 633–649.

Burr, V. (1995). *An Introdcution to Social Constructionism.* New York: Routledge.

Burr, V. (1998). Overview: Realism, Relativism, Social Constructionism and Discourse. In I. Parker (Ed.), *Social Constructionism, Discourse and Realism* (pp. 13–26). London: Sage.

Burr, V. (2005). *Social constructionism.* London: Routledge.

Chemero, A. (2003). An Outline of a Theory of Affordances. *Ecological Psychology, 15* (2), pp. 181–195.

Chilla, T., Kühne, O., Weber, F. & Weber, F. (2015). ‚Neopragmatische' Argumente zur Vereinbarkeit von konzeptioneller Diskussion und Praxis der Regionalentwicklung. In O. Kühne & F. Weber (Eds.), *Bausteine der Regionalentwicklung* (pp. 13–24). Wiesbaden: Springer VS.

Christmann, N. & Weber, F. (2020). Stadtlandgrenzräume *UniGR-Border Glossary.* http://cbs.unigr.eu/de/ressourcen/digitales-glossar: forthcoming.

Coleman, M. (2007). Immigration Geopolitics Beyond the Mexico–US Border. *Antipode, 39* (1), pp. 54–76.

Cooper, A., Perkins, C. & Rumford, C. (2016 [2014]). The Vernacularization of Borders. In R. Jones & C. Johnson (Eds.), *Placing the Border in Everyday Life* (pp. 15–32). London: Routledge.

Cooper, A. & Rumford, C. (2013). Monumentalising the Border: Bordering Through Connectivity. *Mobilities, 8* (1), pp. 107–124.

Cosgrove, D. (1985). Prospect, perspective and the evolution of the landscape idea. *Transactions of the Institute of British Geographers, 10* (1), pp. 45–62.

Cosgrove, D. (1998 [1984]). *Social Formation and Symbolic Landscape.* Madison: The University of Wisconsin Press.

Cosgrove, D. (2004). *Lanscape and Landschaft.* Paper presented at the "Spatial Turn in History Symposium German Historical Institute", GHI Bulletin.

Cosgrove, D. & Daniels, S. (1988). *The iconography of landscape. Essays on the symbolic representation, design and use of past environments* (Cambridge Studies in Historical Geography, Vol. 9). Cambridge: Cambridge University Press.

Coulton, C., Jennings, M. Z. & Chan, T. (2013). How Big is My Neighborhood? Individual and Contextual Effects on Perceptions of Neighborhood Scale. *American Journal of Community Psychology, 51* (1), pp. 140–150.

Cresswell, T. (2013). *Geographic Thought: A Critical Introduction.* Chichester: Wiley-Blackwell.

Crossey, N., Roßmeier, A. & Weber, F. (2020). Natura 2000 in Bayern. Über die weitere Etablierung des EU-weiten Schutzgebietsnetzes. *Naturschutz und Landschaftsplanung, 52* (01), pp. 10–15.

Dangschat, J., Brake, K. & Herfert, G. (Eds.). (2001). *Suburbanisierung in Deutschland: aktuelle Tendenzen*. Opladen: Leske+ Budrich.

Davoudi, S. (2012). The Legacy of Positivism and the Emergence of Interpretive Tradition in Spatial Planning. *Regional Studies, 46* (4), pp. 429–441.

Davoudi, S. & Strange, I. (2009a). Introduction. In S. Davoudi & I. Strange (Eds.), *Conceptions of Space and Place in Strategic Spatial Planning* (pp. 1–6). London: Routledge.

Davoudi, S. & Strange, I. (2009b). Space and place in the twentieth century planning: an analytical framework and a historical review. In S. Davoudi & I. Strange (Eds.), *Conceptions of Space and Place in Strategic Spatial Planning* (pp. 7–42). London: Routledge.

De Landa, M. (2006). *A New Philosophy of Society: Assemblage Theory and Social Complexity*. London: Continuum.

Dear, M. (1986). Postmodernism and Planning. *Environment and Planning D: Society and Space, 4* (3), pp. 367–384.

Dear, M. & Flusty, S. (1998). Postmodern Urbanism. *Annals of the Association of American Geographers, 88* (1), pp. 50–72.

Dedina, S. (1995). The political ecology of transboundary development: Land use, flood control and politics in the Tijuana river valley. *Journal of Borderlands Studies, 10* (1), pp. 89–110.

Deleuze, G. & Guattari, F. (1988). *A thousand plateaus: Capitalism and Schizophrenia.* London: Bloomsbury Publishing.

Dell'Agnese, E. & Amilhat Szary, A.-L. (2015). Borderscapes: From Border Landscapes to Border Aesthetics. *Geopolitics, 20* (1), pp. 4–13.

Denzer, V. (2019). Landschaft als Text. In O. Kühne, F. Weber, K. Berr & C. Jenal (Eds.), *Handbuch Landschaft* (pp. 81–89). Wiesbaden: Springer VS.

Denzin, N. K. (2004). Symbolic Interactionism. In U. Flick, E. von Kardoff & I. Steinke (Eds.), *A Companion to Qualitative Research* (pp. 81–87). London: Sage.

Duany, A., Plater-Zyberk, E. & Speck, J. (2000). *Suburban Nation: The Rise of Sprawl and the Decline of the American Dream.* New York: North Press Point.

Duineveld, M., Van Assche, K. & Beunen, R. (2017). Re-Conceptualising political landscapes after the material turn: a typology of material events. *Landscape Research, 42* (4), pp. 375–384.

Duncan, J. (1995). Landscape Geography, 1993–94. *Progress in Human geography, 19* (3), pp. 414–422.

Duttmann, R., Kühne, O. & Weber, F. (Eds.). (2020). *Landschaft als Prozess*. Wiesbaden: Springer VS.

Easthope, A. (1998). Bhabha, hybridity and identiy. *Textual Practice, 12* (2), pp. 341–348.

Eberle, T. S. (2014). Phenomenology as a Research Method. In U. Flick (Ed.), *The SAGE Handbook of Qualitative Data Analysis* (pp. 184–202). Los Angeles: SAGE.

Edler, D. & Kühne, O. (2019). Nicht-visuelle Landschaften. In O. Kühne, F. Weber, K. Berr & C. Jenal (Eds.), *Handbuch Landschaft* (pp. 599–612). Wiesbaden: Springer VS.

Edler, D., Kühne, O., Jenal, C., Vetter, M. & Dickmann, F. (2018). Potenziale der Raumvisualisierung in Virtual Reality (VR) für die sozialkonstruktivistische Landschaftsforschung. *KN—Journal of Cartography and Geographic Information, 68* (5), pp. 245–254.

Fauser, M. (2017). The Emergence of Urban Border Spaces in Europe. *Journal of Borderlands Studies, 34* (4), pp. 605–622.

Fellner, A. M. (2020). Counter-Mapping Corporeal Borderlands: Border Imaginaries in the Americas. In F. Weber, C. Wille, B. Caesar & J. Hollstegge (Eds.), *Geographien der Grenzen. Räume—Ordnungen—Verflechtungen* (pp. 287–300). Wiesbaden: Springer VS.

Fellner, A. M. (2021). Grenze und Ästhetik: Repräsentationen von Grenzen in den kulturwissenschaftlichen *Border Studies*. In D. Gerst, M. Klessmann & H. Krämer (Eds.), *Grenzforschung. Handbuch für Wissenschaft und Studium* (pp. 436–456). Baden-Baden: Nomos.

Finlay, L. (2009). Debating Phenomenological Research Methods. *Phenomenology & Practice, 3* (1), pp. 6–25.

Finlay, L. (2011). *Phenomenology for therapists: Researching the lived world*. Chichester: Wiley-Blackwell.

Fischer, L. (2011). *Landschaft und der doppelte Effekt der Arbeit. Vortrag auf der Konferenz: Konstituierung von Kulturlandschaft: Wie wird Landschaft gemacht?* Given lecture on 12th May 2011 in Hannover.

Flaherty, M. H. (2009). Phenomenology. In B. S. Turner (Ed.), *The New Blackwell Companion to Social Theory* (pp. 218–234). Oxford: Blackwell.

Flick, U. (2004). Constructivism. In U. Flick, E. von Kardoff & I. Steinke (Eds.), *A Companion to Qualitative Research* (pp. 88–94). London: SAGE.

Flusty, S. (1994). *Building Paranoia: The Proliferation of Interdictory Space and the Erosion of Spatial Justice*. West Hollywood: Los Angeles Forum for Architecture and Urban Design.

Flusty, S. (2001). The Banality of Interdiction: Surveillance, Control and the Displacement of Diversity. *International Journal of Urban and Regional Research, 25* (3), pp. 658–664.

Fontaine, D. (2016). *Simulierte Landschaften in der Postmoderne: Reflexionen und Befunde zu Disneyland, Wolfersheim und GTA V*. Wiesbaden: Springer VS.

Gailing, L. (2015). Die Transformation suburbaner Räume in westlichen Gesellschaften und die Perspektive der sozialwissenschaftlichen Landschaftsforschung. In O. Kühne, K. Gawroński & J. Hernik (Eds.), *Transformation und Landschaft: Die Folgen sozialer Wandlungsprozesse auf Landschaft* (pp. 83–93). Wiesbaden: Springer VS.

Gailing, L. & Leibenath, M. (2015). The Social Construction of Landscapes: Two Theoretical Lenses and Their Empirical Applications. *Landscape Research, 40* (2), pp. 123–138.

Garreau, J. (1992). *Edge city: Life on the New Frontier*. New York: Anchor.

Gergen, K. (1994). *Realities and relationships: Soundings in social construction*. Cambridge: Harvard University Press.

Gergen, K. (1999). *An Invitation to Social Construction*. Thousank Oaks: Sage.

Gibson, H. J. & Canfield, J. L. (2016). The Non-gated Gated Community of Stapleton. *Journal of Borderlands Studies, 31* (1), pp. 73–89.

Gilbert, L. (2009). Immigration as Local Politics: Re-Bordering Immigration and Multiculturalism through Deterrence and Incapacitation. *International Journal of Urban and Regional Research, 33* (1), pp. 26–42.

Gottdiener, M. (2000). The consumption of space and the spaces of consumption. In M. Gottdiener (Ed.), *New forms of consumption. Consumers, culture, and commodification* (pp. 265–284). Lanham: Rowman & Littlefield.

Graham, S. & Healey, P. (1999). Relational concepts of space and place: Issues for planning theory and practice. *European Planning Studies, 7* (5), pp. 623–646.

Gregory, S. (2019). Authenticity and luxury branding in a renewing Detroit landscape. *Journal of Cultural Geography, 36* (2), pp. 182–210.
Greider, T. & Garkovich, L. (1994). Landscapes: The Social Construction of Nature and the En vironment. *Rural Sociology, 59* (1), pp. 1–24.
Ha, K. N. (2015). *Hype um Hybridität: kultureller Differenzkonsum und postmoderne Verwertungstechniken im Spätkapitalismus.* Bielefeld: transcript Verlag.
Hard, G. (1969). Das Wort Landschaft und sein semantischer Hof. Zur Methode und Ergebnis eines linguistischen Tests. *Wirkendes Wort, 9*, pp. 3–14.
Hard, G. (1995). *Spuren und Spurenleser. Zur Theorie und Ästhetik des Spurenlesens in der Vegetation und anderswo: Vol.16. Osnabrücker Studien zur Geographie.* Osnabrück: Universitätsverlag Rasch.
Hard, G. (2002). Zu Begriff und Geschichte von „Natur" und „Landschaft" in der Geogra phie des 19. und 20. Jahrhunderts [first published in 1983]. In G. Hard (Ed.), *Landschaft und Raum. Aufsätze zur Theorie der Geographie* (Osnabrücker Studien zur Geographie, Vol. 22, pp. 171–210). Osnabrück: Universitätsverlag Rasch.
Hartshorne, R. (1936). Suggestions on the terminology of political boundaries. *Annals of the Association of American Geographers, 26* (1), pp. 56–57.
Haselsberger, B. (2014). Decoding borders. Appreciating border impacts on space and people. *Planning Theory & Pratice, 15* (4), pp. 505–526.
Heidegger, M. (2005 [1994]). *Introduction to Phenomenological Research.* Bloomington: Indiana University Press.
Hein, K. (2006). *Hybride Identitäten. Bastelbiografien im Spannungsverhältnis zwischen Lateinamerika und Europa.* Bielefeld: transcript Verlag.
Herzog, L. A. (1991). Cross-national Urban Structure in the Era of Global Cities: The US-Mexico Transfrontier Metropolis. *Urban Studies, 28* (4), pp. 519–533.
Herzog, L. A. (2015). *Global Suburbs: Urban Sprawl from the Rio Grande to Rio de Janeiro.* New York: Routledge.
Herzog, L. A. & Sohn, C. (2014). The Cross-Border Metropolis in a Global Age: A Conceptual Model and Empirical Evidence from the US–Mexico and European Border Regions. *Global Society, 28* (4), pp. 441–461.
Herzog, L. A. & Sohn, C. (2017). The co-mingling of bordering dynamics in the San Diego–Tijuana cross-border metropolis. *Territory, Politics, Governance, 7* (2), pp. 177–199.
Hetherington, K. (1998). *Expressions of identity: Space, performance, politics.* London: Sage.
Hitzler, R. & Eberle, T. S. (2004). Phenomenological Life-World Analysis. In U. Flick, E. von Kardoff & I. Steinke (Eds.), *A Companion to Qualitative Research* (pp. 67–71). London: Sage.
Hitzler, R. & Honer, A. (1984). Lebenswelt—Milieu—Situation: terminologische Vorschläge zur theoretischen Verständigung. *Kölner Zeitschrift für Soziologie und Sozialpsychologie, 36* (1), pp. 56–74.
Hoesterey, I. (2001). *Pastiche: Cultural memory in art, film, literature.* Bloomington: Indiana University Press.
Hofmeister, S. & Kühne, O. (Eds.). (2016). *StadtLandschaften. Die neue Hybridität von Stadt und Land.* Wiesbaden: Springer VS.
Hoskins, W. G. (1955). *The Making of the English Landscape.* London: Penguin Books.
Howard, P. J. (2011). *An Introduction to Landscape.* London: Routledge.

Husserl, E. (1913). *Ideen zu einer reinen Phänomenologie und phänomenologischen Philosophie. Erstes Buch: Allgemeine Einführung in die Phänomenologie.* Halle (Saale): Niemeyer.

Husserl, E. (1970 [1954]). *The Crisis of European Sciences and Transcendental Philosophy. An Introduction to Phenomenological Philosophy.* Evanston: Northwestern University Press.

Husserl, E. (1973 [1929]). *Cartesianische Meditationen und Pariser Vorträge.* Den Haag: Martinus Nijhoff.

Inglis, D. & Bone, J. (2006). Boundary Maintenance, Border Crossing and the Nature/Culture Divide. *European Journal of Social Theory, 9* (2), pp. 272–287.

Ingold, T. (1993). The Temporality of the Landscape. *World archaeology, 25* (2), pp. 152–174.

Iossifova, D. (2013). Searching for common ground. Urban borderlands in a world of borders and boundaries. *Cities, 34*, pp. 1–5.

Iossifova, D. (2015). Borderland urbanism: seeing between enclaves. *Urban Geography, 36* (1), pp. 90–108.

Iossifova, D. (2019). Borderland. In A. M. Orum (Ed.), *The Wiley Blackwell Encyclopedia of Urban and Regional Studies* (pp. 1–4). Chichester: Wiley Blackwell.

Iossifova, D. (2020). Reading borders in the everyday: bordering as practice. In J. W. Scott (Ed.), *A Research Agend for Border Studies.* Cheltenham: Edward Elgar Publishing.

Ipsen, D. (2006). *Ort und Landschaft.* Wiesbaden: Springer VS.

Jauhiainen, J. S. (2003). Learning from Tartu — Towards Post-Postmodern Landscapes. In H. Palang & G. Fry (Eds.), *Landscape Interfaces* (Landscape Series, Vol. 1). Dordrecht: Springer.

Johnson, C. & Jones, R. (2016 [2014]). Where is the Border? In R. Jones & C. Johnson (Eds.), *Placing the Border in Everyday Life* (pp. 1–11). London: Routledge.

Jones, M. (2009). Phase space: geography, relational thinking, and beyond. *Progress in Human geography, 33* (4), pp. 487–506.

Jones, S. B. (1959). Boundary concepts in the setting of place and time. *Annals of the Association of American Geographers, 49* (3), pp. 241–255.

Kazig, R. & Weichhart, P. (2009). Die Neuthematisierung der materiellen Welt in der Humangeographie. *Berichte zur deutschen Landeskunde, 83* (2), pp. 109–128.

Keller, R. (2013). Das Wissen der Wörter und Diskurse. Über Sprache und Wissen in der Wissenssoziologischen Diskursanalyse. In W. Viehöfer, R. Keller & W. Schneider (Eds.), *Diskurs—Sprache—Wissen. Interdisziplinäre Beiträge zum Verhältnis von Sprache und Wissen in der Diksursforschung* (pp. 21–50). Wiesbaden: Springer VS.

Kneer, G. (2009). Jenseits von Realismus und Antirealismus. Eine Verteidigung des Sozialkonstruktivismus gegenüber seinen postkonstruktivistischen Kritikern / Beyond Realism and Anti-Realism: A Defense of Social Constructivism Against Its Post-Constructivist Critics. *Zeitschrift für Soziologie, 38* (1), pp. 5–25.

Kneer, G. (2010). Wissenssoziologie. In G. Kneer & M. Schroer (Eds.), *Handbuch Spezielle Soziologien* (pp. 707–723). Wiesbaden: Springer VS.

Knoblauch, H. (1995). *Kommunikationskultur: die kommunikative Konstruktion kultureller Kontexte.* Berlin: de Gruyter.

Kolossov, V. (2005). Border studies: Changing Perspectives and Theoretical Approaches. *Geopolitics, 10* (4), pp. 606–632.

Kolossov, V. & Scott, J. W. (2013). Selected conceptual issues in border studies. *Belgeo, 1*, p. 41 paragraphs.

Kost, S. & Schönwald, A. (Eds.). (2015). *Landschaftswandel–Wandel von Machtstrukturen*. Wiesbaden: Springer VS.

Kraidy, M. M. (2002). Hybridity in Cultural Globalization. *Communication Theory, 12* (3), pp. 316–339.

Kraidy, M. M. (2005). *Hybridity, or the Cultural Logic of Globalization*. Philadelphia: Temple University Press.

Krauss, W. (2019). Postenvironmental landscapes in the Anthropocene. In P. Howard, I. Thompson, E. Waterton & M. Atha (Eds.), *The Routledge Companion to Landscape Studies* (2nd ed., pp. 62–73). London: Routledge.

Kropp, C. (2015). Regionale StadtLandschaften–Muster der lebensweltlichen Erfahrung postindustrieller Raumproduktion zwischen Homogenisierung und Fragmentierung. *Raumforschung und Raumordnung, 73* (2), pp. 91–106.

Kühne, O. (2006). *Landschaft in der Postmoderne. Das Beispiel des Saarlandes*. Wiesbaden: Deutscher Universitäts-Verlag.

Kühne, O. (2012a). *Stadt—Landschaft—Hybridität: Ästhetische Bezüge im postmodernen Los Angeles mit seinen modernen Persistenzen*. Wiesbaden: Springer VS.

Kühne, O. (2012b). Urban Nature Between Modern and Postmodern Aesthetics: Reflections Based on the Social Constructivist Approach. *Quaestiones Geographicae, 31* (2), pp. 61–70.

Kühne, O. (2015). Historical Developments. The Evolution of the Concept of Landscape in German Linguistic Areas. In D. Bruns, O. Kühne, A. Schönwald & S. Theile (Eds.), *Landscape Culture—Culturing Landscapes. The Differentiated Construction of Landscapes* (pp. 43–52). Wiesbaden: Springer VS.

Kühne, O. (2016). Transformation, Hybridisierung, Streben nach Eindeutigkeit und Urbanizing former Suburbs (URFSURBS): Entwicklungen postmoderner Stadtlandhybride in Süd kalifornien und in Altindustrieräumen Mitteleuropas—Beobachtungen aus der Perspektive sozialkonstruktivistischer Landschaftsforschung. In S. Hofmeister & O. Kühne (Eds.), *StadtLandschaften. Die neue Hybridität von Stadt und Land* (pp. 13–36). Wiesbaden: Springer VS.

Kühne, O. (2017). Hybridisierungstendenzen, Raumpastiches und URFSURBs in Südkalifornien als Herausforderung für die Planung. In K. Berr (Ed.), *Architektur- und Planungsethik* (pp. 15–32). Wiesbaden: Springer VS.

Kühne, O. (2018a). *Landscape and Power in Space as a Geographical Socio-Aesthetical Construct*. Dordrecht: Springer International.

Kühne, O. (2018b). *Landschaftstheorie und Landschaftspraxis. Eine Einführung aus sozialkonstruktivistischer Perspektive* (2 ed.). Wiesbaden: Springer VS.

Kühne, O. (2019a). *Landscape Theories. A Brief Introduction*. Wiesbaden: Springer VS.

Kühne, O. (2019b). Sich abzeichnende theoretische Perspektiven für die Landschaftsforschung: Neopragmatismus, Akteur-Netzwerk-Theorie und Assemblage-Theorie. In O. Kühne, F. Weber, K. Berr & C. Jenal (Eds.), *Handbuch Landschaft* (pp. 153–162). Wiesbaden: Springer VS.

Kühne, O. (2019c). Sozialkonstruktivistische Landschaftstheorie. In O. Kühne, F. Weber, K. Berr & C. Jenal (Eds.), *Handbuch Landschaft* (pp. 69–79). Wiesbaden: Springer VS.

Kühne, O. & Jenal, C. (2020a). Baton Rouge (Louisiana): On the Importance of Thematic Cartography for 'Neopragmatic Horizontal Geography'. *KN—Journal of Cartography and Geographic Information*, pp. 1–9.

Kühne, O. & Jenal, C. (2020b). *Baton Rouge–the Multivillage Metropolis. A Neopragmatic Landscape Biographical Approach on Spatial Pastiches, Hybridization, and Differentiation.* Wiesbaden: Springer VS.

Kühne, O. & Jenal, C. (2021). Baton Rouge–A Neopragmatic Regional Geographical Approach. *urban science, 5* (17), pp. 1–23.

Kühne, O., Schönwald, A. & Weber, F. (2016). Urban/rural hybrids: The urbanisation of former suburbs (URFSURBS). *Quaestiones Geographicae, 35* (4), pp. 23–34.

Kühne, O. & Weber, F. (2018). Conflicts and negotiation processes in the course of power grid extension in Germany. *Landscape Research, 43* (4), pp. 529–541.

Kühne, O. & Weber, F. (2019a). *Hybrid California. Annäherungen an den Golden State, seine Entwicklungen, Ästhetisierungen und Inszenierungen.* Wiesbaden: Springer VS.

Kühne, O. & Weber, F. (2019b). Postmoderne Zugriffe und Differenzierungen von Stadt und Land(schaft): Stadtlandhybride, räumliche Pastiches und URFSURBS. In O. Kühne, F. Weber, K. Berr & C. Jenal (Eds.), *Handbuch Landschaft* (pp. 755–770). Wiesbaden: Springer VS.

Kühne, O., Weber, F. & Jenal, C. (2018). *Neue Landschaftsgeographie. Ein Überblick.* Wiesbaden: Springer VS.

Kühne, O., Weber, F. & Jenal, C. (2019). Neue Landschaftsgeographie. In O. Kühne, F. Weber, K. Berr & C. Jenal (Eds.), *Handbuch Landschaft* (pp. 119–134). Wiesbaden: Springer VS.

Kwan, M.-P. & Schwanen, T. (2009). Quantitative Revolution 2: The Critical (Re)Turn. *The Professional Geographer, 61* (3), pp. 281–291.

Laine, J. (2016). The Multiscalar Production of Borders. *Geopolitics, 21* (3), pp. 465–482.

Lang, R. (2003). *Edgeless Cities: Exploring the Elusive Metropolis.* Washington, D.C.: Brookings Institution Press.

Lang, R., Sanchez, T. W. & Oner, A. C. (2009). Beyond Edge City: Office Geography in the New Metropolis. *Urban Geography, 30* (7), pp. 726–755.

Latour, B. (1999). *Pandora's hope: Essays on the reality of science studies.* London: Harvard University Press.

Law, J. (2009). Actor Network Theory and Material Semiotics. In B. S. Turner (Ed.), *The New Blackwell Companion to Social Theory* (pp. 141–158). Oxford: Blackwell.

Lazzarini, A. (2015). Metamorphosis of City Borders. In C. Brambilla, J. Laine, J. W. Scott & G. Bocchi (Eds.), *Borderscaping: Imaginations and Practices of Border Making* (pp. 177–185). Burlington: Ashgate.

Lebuhn, H. (2013). Local border practices and urban citizenship in Europe. Exploring urban borderlands. *City: Analysis of Urban Trends, Culture, Theory, Policy, Action, 17* (1), pp. 37–51.

Leibenath, M. & Otto, A. (2012). Diskursive Konstituierung von Kulturlandschaft am Beispiel politischer Windenergiediskurse in Deutschland. *Raumforschung und Raumordnung, 70* (2), pp. 119–131.

Liao, K., Breitung, W. & Wehrhahn, R. (2018). Debordering and rebordering in the residential borderlands of suburban Guangzhou. *Urban Geography, 39* (7), pp. 1092–1112.

Linke, S. (2019). *Die Ästhetik medialer Landschaftskonstrukte. Theoretische Reflexionen und empirische Befunde.* Wiesbaden: Springer VS.
Löw, M. & Weidenhaus, G. (2017). Borders that relate: Conceptualizing boundaries in relational space. *Current Sociology Monograph, 65* (4), pp. 553–570.
Lorimer, H. (2005). Cultural geography: the busyness of being 'more-than-representational'. *Progress in Human geography, 29* (1), pp. 83–94.
Lyde, L. W. (1915). *Some frontiers of tomorrow: An aspiration for Europe.* London: Black.
Lynch, K. (1992). *The image of the city* (21th ed.). Cambridge: MIT Press.
Lyon, D. (2005). The border is everywhere: ID cards, surveillance and the other. In E. Zureik & M. B. Salter (Eds.), *Global Surveillance and Policing. Borders, security, identity* (pp. 66–82). Cullompton: Willan Publishing.
Mannheim, K. (1936). *Ideology und Utopia. An Introduction to the Sociology of Knowledge.* London: Routledge.
Markley, S. (2018). Suburban gentrification? Examining the geographies of New Urbanism in Atlanta's inner suburbs. *Urban Geography, 39* (4), pp. 606–630.
Massey, D. (1995). The Conceptualization of Space. In D. Massey & P. M. Jess (Eds.), *A Place in the World* (pp. 45–85). Oxford: Oxford University Press.
Massey, D. (2005). *For Space.* London: Sage.
Mattissek, A. & Wiertz, T. (2014). Materialität und Macht im Spiegel der Assemblage-Theorie: Erkundungen am Beispiel der Waldpolitik in Thailand. *Geographica Helvetica, 69* (3), pp. 157–169.
Mead, G. H. (2015 [1934]). *Mind, Self, and Society: The Definitive Edition.* Chicago: The University of Chicago Press.
Mendoza, J. E. & Dupeyron, B. (2020). Economic Integration, Emerging Fields and Cross-border Governance: The Case of San Diego–Tijuana. *Journal of Borderlands Studies, 35* (4), pp. 55–74.
Merleau-Ponty, M. (1962). *Phenomenology of Perception.* London: Routledge.
Merton, R. K. (1957 [1949]). *Social theory and social structure.* Glencoe: The Free Press.
Merton, R. K. (1973). *The Sociology of Science: Theoretical and Empirical Investigations.* Chicago: The University of Chicago Press.
Metro-Roland, M. M. & Soica, S. (2019). Landscape and tourism. In P. Howard, I. Thompson, E. Waterton & M. Atha (Eds.), *The Routledge Companion to Landscape Studies* (2nd ed., pp. 354–365). London: Routledge.
Micheel, M. (2012). Alltagsweltliche Konstruktionen von Kulturlandschaft. *Raumforschung und Raumordnung Spatial Research and Planning, 70* (2), pp. 107–117.
Minghi, J. V. (1963). Boundary studies in political geography. *Annals of the Association of American Geographers, 53* (3), pp. 407–428.
Mitchell, K. (1997). Different Diasporas and the Hype of Hybridity. *Environment and Planning D: Society and Space, 15* (5), pp. 533–553.
Mooney Melvin, P. (1985). Changing Contexts: Neighborhood Definition and Urban Organization. *American Quaterly, 37* (3), pp. 357–367.
Moran, D. (2000). *Introduction to Phenomenology.* London: Routledge.
Morehouse, B. J. (2004). Theoretical Approaches to Border Spaces and Identities. In V. Pavlakovich-Kochi, B. J. Morehouse & D. Wastl-Walter (Eds.), *Challenged Borderlands. Transcending Political and Cultural Boundaries* (pp. 19–39). Aldershot: Ashgate.
Murdoch, J. (2006). *Post-structuralist geography: a guide to relational space.* London: Sage.

Newman, D. (2003a). Boundaries. In J. Agnew, K. Mitchell & G. Toal (Eds.), *A Companion to Political Geography* (pp. 123–131). Oxford: Blackwell.

Newman, D. (2003b). On borders and power: A theoretical framework. *Journal of Borderlands Studies, 18* (1), pp. 13–25.

Newman, D. (2006a). Borders and Bordering: Towards an Interdisciplinary Dialogue. *European Journal of Social Theory, 9* (2), pp. 171–186.

Newman, D. (2006b). The lines that continue to separate us: borders in our 'borderless' world. *Progress in Human geography, 30* (2), pp. 143–161.

Newman, D. (2011). Contemporary Research Agendas in Border Studies: An Overview. In D. Wastl-Walter (Ed.), *The Ashgate Research Companion to Border Studies* (pp. 33–47). Farnham: Ashgate.

Nienaber, B. & Wille, C. (2020a). Cross-border cooperation in Europe: a relational perspective. *European Planning Studies, 28* (1), pp. 1–7.

Nienaber, B. & Wille, C. (2020b). Cross-Border Cooperation in Europe. Networks—Governance—Territorialisation. *European Planning Studies, 28* (1; Special Issue).

O'Dowd, L. (2010). From a 'borderless world' to a 'world of borders': 'bringing history back in'. *Environment and Planning D: Society and Space, 28* (6), pp. 1031–1050.

Ohmae, K. (1990). *Borderless world*. London: HarperCollins.

Paasi, A. (1998). Boundaries as social processes: Territoriality in the world of flows. *Geopolitics, 3* (1), pp. 69–88.

Paasi, A. (2003). Territory. In J. Agnew, K. Mitchell & G. Toal (Eds.), *A Companion to Political Geography* (pp. 109–122). Oxford: Blackwell.

Paasi, A. (2011). A *Border Theory*: An Unattainable Dream or a Realistic Aim for Border Scholars? In D. Wastl-Walter (Ed.), *The Ashgate Research Companion to Border Studies* (pp. 11–31). Farnham: Ashgate.

Paasi, A. (2012). Border Studies Reanimated: Going beyond the Territorial/Relational Devide. *Environment and Planning A: Economy and Space, 44* (10), pp. 2303–2309.

Paasi, A. (2014). The shifting landscape of border studies and the challenge of relational thinking. In M. Bufon, J. Minghi & A. Paasi (Eds.), *The New European Frontiers: Social and Spatial (Re)Integration Issues in Multicultural and Border Regions* (pp. 361–379). Newcastle: Cambridge Scholars Publishing.

Paasi, A. (2020). Problematizing 'Bordering, Ordering and Othering' as Manifestations of Socio-Spatial Fetishism. *Tijdschrift voor economische en sociale geografie* (1), pp. 18–25.

Pavlakovich-Kochi, V., Morehouse, B. J. & Wastl-Walter, D. (Eds.). (2004). *Challenged borderlands. Transcending political and cultural boundaries*. Aldershot: Ashgate.

Prescott, J. R. V. (2015 [1987]). *Political Frontiers and Boundaries*. London: Routledge.

Qviström, M. (2019). Peri-urban landscape studies. In P. Howard, I. Thompson, E. Waterton & M. Atha (Eds.), *The Routledge Companion to Landscape Studies* (2nd ed., pp. 523–533). London: Routledge.

Riehl, W. H. (1996). Das landschaftliche Auge. In G. Gröning & U. Herlyn (Eds.), *Landschaftswahrnehmung und Landschaftserfahrung* (pp. 28–68). Münster: LIT Verlag.

Roßmeier, A. (2016). Die Amerikanisierung der europäischen Stadtlandschaft—Freising als Edge City? *morphé. rural—suburban—urban* (2), pp. 61–80.

Roßmeier, A. (2020b). Urban/Rural Hybridity in Pictures. The Creation of Neighborhood Images Using the Example of San Diego's Urbanizing Inner-Ring Suburbs East Village and Barrio Logan. In D. Edler, C. Jenal & O. Kühne (Eds.), *Modern Approaches to the Visualization of Landscapes* (pp. 477–496). Wiesbaden: Springer VS.

Roßmeier, A. (2020d). COVID-19 in the borderland of San Diego and Tijuana: between re-bordering processes and contradictory realities. In C. Wille & R. Kanesu (Eds.), *Borders in Perspective UniGR-CBS Thematic Issue. Bordering in Pandemic Times: Insights into the COVID-19 Lockdown* (4), pp. 57–61).

Roßmeier, A. & Weber, F. (2021). Hybrid Urban Borderlands—Redevelopment Efforts and Shifting Boundaries In and Around Downtown San Diego. *Journal of Borderlands Studies*, pp. 1–27.

Roßmeier, A. & Weber, F. (2021 [2018]). Stormy times. Civic engagement in wind power development: between support and rejection. In A. Stefansky & A. Göb (Eds.), *'All change please!'—challenges and opportunities of the energy transition* (pp. 52–79). Hannover: Academy for Territorial Development in the Leibniz Association.

Roßmeier, A., Weber, F. & Kühne, O. (2018). Wandel und gesellschaftliche Resonanz–Diskurse um Landschaft und Partizipation beim Windkraftausbau. In O. Kühne & F. Weber (Eds.), *Bausteine der Energiewende* (pp. 653–679). Wiesbaden: Springer VS.

Rumford, C. (2006). Introduction: Theorizing Borders. *European Journal of Social Theory, 9* (2), pp. 155–169.

Rumford, C. (2012). Towards a Multiperspectival Study of Borders. *Geopolitics, 17* (4), pp. 887–902.

Rumley, D. & Minghi, J. V. (2015 [1991]). Introduction: The border landscape concept. In D. Rumley & J. V. Minghi (Eds.), *The Geography of Border Landscapes* (pp. 1–14). London: Routledge.

Sauer, C. O. (1925). The Morphology of Landscape. In J. Leighly (Ed.), *Land and Life. A Selection from the Writings of Carl Ortwin Sauer* (pp. 315–350). Berkley: University of California Press.

Scheler, M. (1960 [1926]). *Die Wissensformen und die Gesellschaft*. Bern: Francke Verlag.

Schenk, W. (2006). Der Terminus „geschwachsene Kulturlandschaft" im Kontext öffentlicher und raumwissenschaftlicher Diskurse zu „Landschaft" und „Kulturlandschaft". In U. Matthiesen, R. Danielzyk, S. Heiland & S. Tzschaschel (Eds.), *Kulturlandschaften als Herausforderung für die Raumplanung: Verständnisse—Erfahrungen—Perspektiven* (pp. 9–21). Hannover: Verlag der ARL.

Schenk, W. (2013). Landschaft als zweifache sekundäre Bildung—historische Aspekte im aktuellen Gebrauch von Landschaft im deutschsprachigen Raum, namentlich in der Geographie. In D. Bruns & O. Kühne (Eds.), *Landschaften: Theorie, Praxis und internationale Bezüge* (pp. 23–36). Schwerin: Oceano Verlag.

Schenk, W. (2019). Landscape. In L. Kühnhardt & T. Mayer (Eds.), *The Bonn Handbook of Globality* (pp. 621–633). Cham: Springer.

Schetter, C. & Müller-Koné, M. (2021). *Frontier*—ein Gegenbegriff zur Grenze? In D. Gerst, M. Klessmann & H. Krämer (Eds.), *Grenzforschung. Handbuch für Wissenschaft und Studium* (pp. 240–253). Baden-Baden: Nomos.

Schütz, A. (1960 [1932]). *Der sinnhafte Aufbau der sozialen Welt. Eine Einleitung in die Verstehende Soziologie* (2 ed.). Wien: Julius Springer.

Schütz, A. (1970). *Collected Papers III. Studies in Phenomenological Philosophy*. Den Haag: Martinus Nijhoff.

Schütz, A. (1972 [1962]). *Collected Papers I. The Problem of Social Reality*. Den Haag: Martinus Nijhoff.
Schütz, A. & Luckmann, T. (1973). *The Structures of the Life-World* (Vol. 1). Evanston, Illinois: Northwestern University Press.
Schulz-Schaeffer, I. (2000). Akteur-Netzwerk-Theorie: Zur Koevolution von Gesellschaft, Natur und Technik. In J. Weyer (Ed.), *Soziale Netzwerke: Konzepte und Methoden der sozialwissenschaftlichen Netzwerkforschung*. München: Oldenbourg.
Scott, H. (2004). Cultural Turns. In J. Duncan, N. C. Johnson & R. H. Schein (Eds.), *A Companion to Cultural Geography* (pp. 24–37). Malden: Blackwell.
Scott, J. W. & Sohn, C. (2018). Place-making and the bordering of urban space: Interpreting the emergence of new neighbourhoods in Berlin and Budapest. *European Urban and Regional Studies, 1* (4), pp. 1–17.
Seamon, D. (2019). Whither phenomenology? Thrity years of *EAP*. *Environmental & Architectural Phenomenology, 30* (2; 1990–2019: Special 30th-anniversary issue!), pp. 37–44.
Sieverts, T. (2003). *Cities without cities. An interpretation of the Zwischenstadt*. London: Routledge.
Sieverts, T. (2008). *Zwischenstadt: zwischen Ort und Welt, Raum und Zeit, Stadt und Land* (3rd ed.). Berlin: Birkhäuser.
Sieverts, T., Koch, M., Stein, U. & Steinbusch, M. (2005). *Zwischenstadt—Inzwischen Stadt? Entdecken, Begreifen, Verändern*. Wuppertal: Müller + Busmann.
Simmel, G. (1997 [1903]). The Sociology of Space. In D. Frisby & M. Featherstone (Eds.), *Simmel on Culture: Selected Writings* (pp. 137–170). London: Sage.
Smith, N. (1987). "Academic War Over the Field of Geography": The Elimination of Geography at Harvard, 1947–1951. *Annals of the Association of American Geographers, 77* (2), pp. 155–172.
Soeffner, H.-G. (2004). Social Scientific Hermeneutics. In U. Flick, E. von Kardoff & I. Steinke (Eds.), *A Companion to Qualitative Research* (pp. 95–100). London: Sage.
Sohn, C. (2014). The Border as a Resource in the Global Urban Space: A Contribution to the Cross-Border Metropolis Hypothesis. *International Journal of Urban and Regional Research, 38* (5), pp. 1697–1711.
Spierings, B. (2012). Economic Flows, Spatial Folds and Intra-Urban Borders: Reflections on City Centre Redevelopment Plans From a European Border Studies Perspective. *Tijdschrift voor economische en sociale geografie, 103* (1), pp. 110–117.
Spierings, B. (2013). Fixing missing links in shopping routes: Reflections on intra-urban borders and city centre redevelopment in Nijmegen, The Netherlands. *Cities, 34*, pp. 44–51.
Struve, K. (2013). *Zur Aktualität von Homi K. Bhabha. Einleitung in sein Werk*. Wiesbaden: Springer VS.
Suttles, G. D. (1972). *The social construction of communities*. Chicago: University of Chicago Press.
Szytniewski, B. & Spierings, B. (2014). Encounters with Otherness: Implications of (Un)familiarity for Daily Life in Borderlands. *Journal of Borderlands Studies, 29* (3), pp. 339–351.
Taylor, R. B. (2012). Defining Neighborhoods in Space and Time. *Cityscape: A Journal of Policy Development and Research, 14* (2), pp. 225–230.
Thrift, N. (2008). *Non-representational theory: Space, politics, affect*. London: Routledge.

Thrift, N. (2009). Space: the Fundamental Stuff of Geography. In N. J. Clifford, S. L. Holloway, S. P. Rice & G. Valentine (Eds.), *Key Concepts in Geographie* (2nd ed., pp. 97–118). London: Sage.

Tilley, C. (1994). *A phenomenology of landscape: places, paths, and monuments*. Oxford: Berg.

Turner, B. S. (2009). The Sociology of the Body. In B. S. Turner (Ed.), *The New Blackwell Companion to Social Theory* (pp. 513–532). Oxford: Blackwell.

van Houtum, H. & van Naerssen, T. (2002). Bordering, Ordering and Othering. *Tijdschrift voor economische en sociale geografie, 93* (2), pp. 125–136.

Varsanyi, M. (2010). *Taking local control: Immigration policy activism in US cities and states.* Stanford: Stanford University Press.

Vaughan, L., Anciaes, P. R. & Mindell, J. S. (2021). Cars, conflict and community severance. In B. Appleyard & D. Appleyard (Eds.), *Livable Streets 2.0*. Amsterdam: Elsevier.

Vaughan-Williams, N. (2008). Borderwork beyond Inside/Outside? Frontex, the Citizen–Detective and the War on Terror. *Space and Polity, 12* (1), pp. 63–79.

Vester, H.-G. (1993). *Soziologie der Postmoderne*. München: Quintessenz.

Vetter, M. (2019). 3D-Visualisierung von Landschaft—Ein Ausblick auf zukünftige Entwicklungen. In O. Kühne, F. Weber, K. Berr & C. Jenal (Eds.), *Handbuch Landschaft* (pp. 559–573). Wiesbaden: Springer VS.

Vicenzotti, V. (2019). Die Landschaft der Zwischenstadt. In O. Kühne, F. Weber, K. Berr & C. Jenal (Eds.), *Handbuch Landschaft* (pp. 743–753). Wiesbaden: Springer VS.

Vicenzotti, V. & Qviström, M. (2018). *Zwischenstadt* as a travelling concept: towards a critical discussion of mobile ideas in transnational planning discourses on urban sprawl. *European Planning Studies, 26* (1), pp. 115–132.

Waldheim, C. (2012). Introduction. In C. Waldheim (Ed.), *The Landscape Urbanism Reader* (pp. 13–19). New York: Princeton Architectural Press.

Wastl-Walter, D. (2011a). Introduction. In D. Wastl-Walter (Ed.), *The Ashgate Research Companion to Border Studies* (pp. 1–8). Farnham: Ashgate.

Wastl-Walter, D. (Ed.) (2011b). *The Ashgate Research Companion to Border Studies.* Farnham: Ashgate.

Waterton, E. (2019). More-than-representational landscapes. In P. Howard, I. Thompson, E. Waterton & M. Atha (Eds.), *The Routledge Companion to Landscape Studies* (2nd ed., pp. 91–101). London: Routledge.

Weber, F. (2016). The Potential of Discourse Theory for Landscape Research. *Prace Komisji Krajobrazu Kulturowego. Dissertations of Cultural Landscape Commission, 31*, pp. 85–102.

Weber, F. (2018). *Konflikte um die Energiewende. Vom Diskurs zur Praxis.* Wiesbaden: Springer VS.

Weber, F., Crossey, N., Roßmeier, A. & Kühne, O. (2019). Natura 2000 in Bayern–Konflikte, Kompetenzen und Kommunikation. *ANLiegen Natur, 41* (1), pp. 199–204.

Weber, F., Jenal, C., Roßmeier, A. & Kühne, O. (2017). Conflicts around Germany's Energiewende: Discourse patterns of citizens' initiatives. *Quaestiones Geographicae, 36* (4), pp. 117–130.

Weber, F. & Kühne, O. (2016). *Fraktale Metropolen: Stadtentwicklung zwischen Devianz, Polarisierung und Hybridisierung.* Wiesbaden: Springer VS.

Weber, F. & Kühne, O. (2017). Hybrid suburbia: New research perspectives in France and Southern California. *Quaestiones Geographicae, 36* (4), pp. 17–28.

Weber, F. & Kühne, O. (2019). Essentialistische Landschafts- und positivistische Raumforschung. In O. Kühne, F. Weber, K. Berr & C. Jenal (Eds.), *Handbuch Landschaft* (pp. 57–68). Wiesbaden: Springer VS.

Weber, F. & Roßmeier, A. (2020). Les nouvelles marges urbaines au prisme des border studies: le cas de San Diego (Californie). *Revue de l'Institut de Sociologie de l'Université libre de Bruxelles, 90* (2020), pp. 61–79.

Weber, F., Wille, C., Caesar, B. & Hollstegge, J. (2020). Entwicklungslinien der Border Studies und Zugänge zu Geographien der Grenzen. In F. Weber, C. Wille, B. Caesar & J. Hollstegge (Eds.), *Geographien der Grenzen. Räume—Ordnungen—Verflechtungen* (pp. 3–22). Wiesbaden: Springer VS.

Weber, M. (1981). Some Categories of Interpretive Sociology. *The Sociological Quaterly, 22* (2), pp. 151–180.

Weinberg, D. (2009). Social Constructionism. In B. S. Turner (Ed.), *The New Blackwell Companion to Social Theory* (pp. 281–299). Oxford: Blackwell.

Wetherell, M. & Still, A. (1998). Realism and relativism. In R. Sapsford, A. Still, M. Wetherell, D. Miell & R. Stevens (Eds.), *Theory and Social Psychology* (pp. 99–114). London: Sage.

Whatmore, S. (2002). *Hybrid Geographies. Natures cultures spaces.* London: Sage.

Whatmore, S. (2006). Materialist returns: practising cultural geography in and for a more-than-human world. *cultural geographies, 2006* (13), pp. 600–609.

Whyte, W. H. (1958). Urban Sprawl. In W. H. Whyte (Ed.), *The Exploding Metropolis* (pp. 133–156). Berkeley: University of California Press.

Wille, C. (2016). Grenzüberschreitender Arbeitsmarkt in der Großregion SaarLorLux: Politische Visionen und empirische Wirklichkeiten. In W. H. Lorig, S. Regolot & S. Henn (Eds.), *Die Großregion SaarLorLux. Politischer Anspruch, Wirklichkeiten, Perspektiven* (pp. 115–143). Wiesbaden: Springer VS.

Wille, C. (2021). Vom processual shift zum complexity shift: Aktuelle analytische Trends der Grenzforschung. In D. Gerst, M. Klessmann & H. Krämer (Eds.), *Grenzforschung. Handbuch für Wissenschaft und Studium* (pp. 106–120). Baden-Baden: Nomos.

Wille, C. & Weber, F. (2020). Analyzing border geographies in times of COVID-19. In G. Mein & J. Pause (Eds.), *Self and Society in the Corona Crisis. Perspectives from the Humanities and Social Sciences* (Vol. 2, The Ends of Humanities, pp. 1–17). Esch-sur-Alzette: Melusina Press.

Wylie, J. (2003). Landscape, performance and dwelling: a Glastonbury case study. In P. Cloke (Ed.), *Country Visions.* Harlow: Pearson Education Limited.

Wylie, J. (2007). *Landscape.* London, New York: Routledge.

Wylie, J. (2019). Landscape and phenomenology. In P. Howard, I. Thompson, E. Waterton & M. Atha (Eds.), *The Routledge Companion to Landscape Studies* (2nd ed., pp. 127–138). London: Routledge.

Methodological Considerations and Methodical Approach

3

The chapter proceeds as follows. First, fundamental considerations on neighborhood as a socio-spatial research category will be presented (section 3.1). In this section, methodological implications regarding the social and individual construction of neighborhoods and their multi-dimensional demarcation will be discussed. In a next step, the methodical research approach will be outlined (section 3.2), which follows the triangulated theoretical-perspectival foundation of social constructivist and phenomenological thinking (outlined in chapter 2). Aiming for a poly-vocal research of various subjective interpretations and everyday experiences, a multi-method approach will be developed, which follows the principles of neopragmatic research triangulation (cf. Kühne and Jenal 2020a, 2021; Kühne and Weber 2019) as it will be outlined in detail. The chapter continues with a discussion of the individual research components (section 3.3): guided and unconstrained interviews supplemented with participant observation, cartographic and photographic representations, and census data to underline the qualitative findings. Finally, the interpretative evaluation process will be outlined, which follows a qualitative content analysis in the tradition of Mayring (2004, 2019).

3.1 Defining and Bounding Neighborhoods: Difficulties and Present Social Constructivist Approach

In addition to the presented theoretical-perspectival foundation of the present study, the methodological derivation must also take general implications of the broad concept of *neighborhood* into account. Generally, throughout literature and research, neighborhood and other close terms, such as community, are discussed as complex and "context dependent" (Sperling 2012, p. 222) conceptualizations

A. Rossmeier, *San Diego's Hybrid Urban Borderlands*,
Räume – Grenzen – Hybriditäten, https://doi.org/10.1007/978-3-658-42667-5_3

that constitute "some of the most notoriously slippery [in] social science" (Taylor 2012, p. 225; cf. Mooney Melvin 1985). Accordingly, qualitative as well as quantitative-statistical considerations of neighborhoods draw on a wide range of definitions and methodical research approaches (Galster 2001; Nicotera 2007). The neighborhood term is characterized by a "definitional ambiguity" (Coulton 2012, p. 232), which is leaving researchers with the fundamental question: "What exactly comprises the neighborhood" (Lohmann and McMurran 2009, p. 67)?

In consequence of this conceptual diversity, neighborhood research in urban social sciences is—according to Coulton et al. (2013, p. 140)—traditionally suffering from "simplifying assumptions about boundaries, often relying on census geography to operationalize the neighborhood units" the studies focus on (cf. Coulton et al. 2001; Sperling 2012). These approaches stand in particular contrast to social constructivist positions according to which neighborhoods are not merely something territorial and uniformly bounded but are strongly shaped by social and individual and thus non-material aspects. In this sense, B. A. Lee et al. (1994, p. 252) argue that "unlike cities or counties, neighborhoods typically are neither formal governmental jurisdictions nor clearly demarcated territorial units. Rather, they are social constructions named and bounded differently by different individuals" (cf. Suttles 1972). According to this conceptualization, in this study neighborhoods are understood as individually *perceived* and *demarcated* places, processually and contingently charged with meaning. What is integrated into and what is excluded from individual neighborhood constructions is from this perspective the result of subjective perception, identity building, and ordering processes rather than of administrative (neighborhood designation) or statistical categorization (census tracts[1]).

Arguing from a social constructivist urban landscape perspective and its focus "on the transformation of the physical environment into [urban] landscapes through cultural symbols" (Greider and Garkovich 1994, p. 1), neighborhoods come into being equivalently—through the transformation of subjectively selected elements of the physical environment into familiar life-world environments of hometowns by means of individual perception and social or cultural processes

[1] For the periodically executed censuses in the United States, the *U.S. Census Bureau* defines different socio-spatial, territorial categories. One category, which is listed in the official glossary, is the census tract, "a small, relatively permanent statistical subdivision of a county delineated by a local committee of census data users for the purpose of presenting data" (U.S. Census Bureau 2021, p. n.pag.), which then allows for socio-spatial comparison. Generally, census tracts offer a detailed, wide range of statistical data collected decennially and are thus used extensively in neighborhood research as well as for planning and policy making purposes.

that ultimately create meaning. In line with this, neighborhoods are understood as the socio-spatial outcome of the complex relations between physical elements, the individual, and society and are in this understanding particularly conveying the post-structuralist idea that "space is made not of structures but of relations" (Davoudi and Strange 2009, p. 56). However, in consequence, the resulting individuality and contingency of neighborhoods is adding to the complexity of the task to define neighborhoods and is posing the subsequent questions of how to draw common and uniform boundaries of neighborhoods and if this endeavor is even possible. For what has been agreed on in constructivist and post-structuralist positioned neighborhood research is that "neighborhood boundaries are not static but often dynamic and contested" (Coulton 2012, p. 232; cf. Curto 2019).

In contrast, guided by Euclidean logics of space and scale, urban policymakers and planners often draw on fixed spatial units (Davoudi 2012) and neglect the "lack of congruence among local actors' perceptions of boundaries" (Galster 2001, p. 2112; cf. Rumford 2012, p. 899). Countering the positivist nature of standardized planning and surveying approaches, social constructivist positioned neighborhood research is drawing on "resident-informed methods that allow variations in perception to be investigated" (Coulton 2012, p. 231) and multiple, possibly contradictive visible and non-visible boundaries to emerge. For as Newman (2011, p. 33) states: "There is no single border situation"—neither on the national level nor on the level of neighborhoods. For this reason, Graham and Healey (1999, p. 629) argue that *simple* plans and other monological planning discourses or instruments are not able to do justice to the complex diversity of place perceptions, experiences, and constructions—regardless of whether the respective space is subjectively composed as landscape, neighborhood, or any other individual conception. Accordingly, "to attempt to capture the multiple, dynamic, and contingent, lived worlds of places or cities, a growing number of urban commentators now stress the need for maintaining *multiple* perspectives of the city simultaneously" (Graham and Healey 1999, p. 629 [original emphasis]). Equivalently, in the field of border studies, Newman (2006) and Rumford (2012) as well outline the necessity of *multi-perspectival* research approaches and the focus on "individual border narratives and experiences" (Newman 2006, p. 143). Drawing on the notion of *borders everywhere* in the tradition of Balibar (2004), Rumford (2012, p. 893) claims that "contemporary transformations cannot be properly understood from a single privileged vantage point" for the simple reason that borders "mean different things to different people, and work differently on different groups" (Rumford 2012, p. 894).

Consequently, the present study aims to explore how the recent neighborhood changes in and around downtown San Diego are "experienced, lived and interpreted by those who inhabit" (Brambilla 2015, p. 27) the changing neighborhoods, the "ordinary people" (Rumford 2012, p. 898). This approach follows the idea of studying border issues *from the inside* of the border rather than observing them *from the two sides* it separates—a relational approach that Rumford (2014) calls 'seeing like a border' and that is in the field of phenomenological landscape studies equivalently described as "in-the-landscape observation" (Kühne 2019, p. 33; cf. Wylie 2007). Because where borders lie, "what forms they take, what purpose they serve and who is involved in maintaining them" (Rumford 2014, pp. 16–17) can be more comprehensively understood when we *see like* a border or a boundary and thus focus on the accompanying manifold everyday perspectives rather than by *looking on* the border from an essentialist and positivist understanding of borders as territorial, given facts (Gerst and Krämer 2021, p. 131). This complex endeavor, however, requires a suitable methodical approach that is capable of capturing multiple voices and border perceptions.

Generally, and thus also in the context of neighborhood research, qualitative, ethnographic research methods, such as interviews, participant observations, and neighborhood walks, are particularly suitable for capturing individual interpretations, experiences, and spatial behaviors or routines. However, quantitative data sets as well prove profitable in the fields of urban and border research: for instance, census data can reveal socio-spatial differences, which can then be contextualized by means of qualitative analyses. In line with this logic, the present research project is drawing on a mixed method approach (see Figure 3.1). For methodical triangulation offers the prospect of a more comprehensive approach to the research object (cf. Cresswell 2013, pp. 196–197). Furthermore, combining different methods and science-theoretical perspectives becomes inevitable within the aim of tracing and understanding diverse bordering processes in the different conceptualized dimensions (processes concerning physical borders, political-administrative demarcations, individual or social, and symbolic-aesthetic boundaries; see Figure 2.5) since these are—according to the argumentation of Gerst and Krämer (2021, p. 134)—each connected to varying *methodological border perspectives*. Thus, the compiled methodical design for the present research project must not only be capable of performing multi-perspectival research in the sense of a poly-vocal survey but also in terms of the combination of different science-theoretical, methodological perspectives. The following section on neo-pragmatic research design (Chilla et al. 2015; Kühne and Jenal 2020a; 2020b, pp. 10–20; 2021; Kühne and Weber 2019, pp. 47–50) discusses how the combination of different methods of various science-theoretical and methodological origin can succeed and form a coherent and logical research approach to the multi-dimensional demarcation of neighborhoods.

3.2 Neopragmatic Approach: Achieving Multi-Perspectival Research Through Methodological and Methodical Triangulation

The present neopragmatic approach to methodological and methodical triangulation constitutes a postmodern translation of the key paradigms of the philosophical tradition of pragmatism for spatial-scientific analyses, here in the context of urban landscape and border issues. The philosophical roots of pragmatism trace back to the 16th century and the English philosopher Francis Bacon as well as to the 18th century German philosopher Immanuel Kant. However, it was not until the late 19th century that the philosophical program of pragmatism had been developed in detail in North America, first by Charles S. Peirce and later by John Dewey and William James. In their work, they traced pragmatism as a programmatic way of finding and (temporarily) assigning truth to selected knowledge that has been declared useful 'by all who investigate,' as Peirce writes in 1878 (cited after Barnes 2008, p. 1545). In other words, in line with James' interpretation, truth was assigned "to beliefs that allowed us to achieve what we wanted to accomplish" (Barnes 2008, p. 1544). Thus, philosophic pragmatism is ultimately "concerned with improving the conditions that enable individuals to thrive in their everyday lives" (Rumens and Kelemen 2016, p. 3). According to this primary aim, pragmatism constitutes a program that moves theoretical considerations more in the background by foregrounding practical criteria instead—in pragmatic logic "practical consequences and effects should determine actions, not moral principles or higher-level theories[2]" (Kühne and Weber 2019, p. 48; initially developed into a 'horizontal geographical' approach by Kühne 2018). Thus, ideas become, in the sense of Dewey's instrumentalism, "instruments that should be picked up, dropped, and modified at will to meet the exigencies of our changing life condition" (Barnes 2008, p. 1544). It is particularly this interpretation the present neopragmatic triangulation approach builds on.

Generally, neopragmatism is associated with thinkers such as Richard Rorty (1979) and Hilary Putnam (1995) and emerged in the 1980 s as a postmodern adaption of the early 20th century American pragmatism. Neopragmatism, however, is not to be understood as a new form of theory or philosophical movement, it is rather a conceptual approach for framing and bundling different theories and thus, according to Kühne and Weber (2019, p. 48), a 'pragmatic meta-theory' (cf. Rumens and Kelemen 2016). In this sense, the neopragmatic approach follows

[2] „die praktischen Konsequenzen und Wirkungen sollten das Handeln bestimmen, nicht jedoch moralische Grundsätze oder übergeordnete Theorien"

the pragmatic program in its conceptualization as "an idea about ideas" (Menand 2001, p. xi) rather than being a superordinate idea or theory itself. Where the two programs differ, however, is in their perspective level: While neopragmatism has developed a meta-perspective, which enables a stronger focus on the level of ideas, theories, and concepts, pragmatism operates at the object level. In other words, postmodern neopragmatism is characterized by its transposition of pragmatic concerns of usefulness and consequences onto the level of theories and concepts (Kühne and Weber 2019, p. 49).

Consequently, according to Barnes (2008, p. 1551), making use of neopragmatism in geographical research means to determine and question the usefulness of chosen concepts and methods for the desired research purpose. In short, if a concept doesn't "accomplish its purpose [...] it needs to be redone or scrapped" (Barnes 2008, p. 1552). Particularly connectable to this approach is the poststructuralist view of theory as "a kind of toolbox where particular sets of ideas allow us to understand particular bits of living in the world" (Cresswell 2013, pp. 196–197). In line with this understanding, neopragmatic research approaches allow the triangulation of different science-theoretical perspectives, their methodologies, and resulting methods—even if they underlie contradicting logics—as long as the combination promises to paint a larger and more differentiated picture (Chilla et al. 2015; Kühne and Jenal 2020b, p. 12) and thus, in the words of Denzin (2017, p. 304), to "maximize the validity of field efforts." In this sense, drawing on the neopragmatic program enables the present research approach to combine social constructivist framed, 'distant' research methods (guided and unconstrained interviews) with phenomenologically framed 'in-the-landscape' observations (neighborhood walks) as well as positivist framed methods (cartographic and photographic representation) into a multi-methodical triangulation of different methodologies and perspectives (see Figure 3.1; the individual methods will be outlined in section 3.3). Further, also the inclusion of quantitative survey data (census data) in the sense of 'data triangulation' becomes possible (cf. Flick 2004, p. 178).

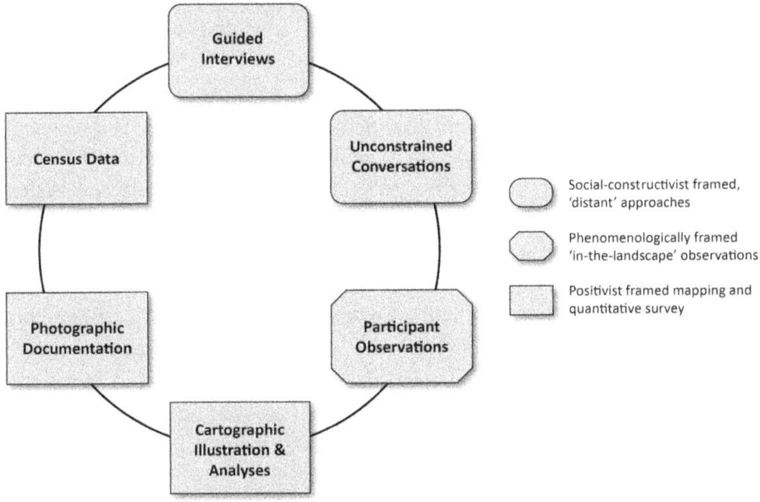

Figure 3.1 Neopragmatic Research Design—Social Constructivist Framed Methodological-Perspectival, Methodical, and Data Triangulation. (Source: Own representation)

As outlined in detail in chapter 2, social constructivism constitutes the science-theoretical framework of the present study, which means—in neopragmatic logic—that all triangulated methodical parts (see Figure 3.1) and the individual results they yield will be interpreted against this perspectival background, even if they origin from different perspective positions. The rather positivist, representational cartographic and photographic approaches as well as the census data are, in this sense, not used as tools to capture and depict the singular, objective 'truth out there' but are applied as additional means to approach and understand the complex social realities of San Diego's central neighborhoods (cf. Kühne and Weber 2020, p. 75). In the same manner, the goal of the present integration of the phenomenologically framed neighborhood walks is not to search for meaning in the material and thus to contest the social constructivist understanding of a 'meaningless world' (Greider and Garkovich 1994, p. 2) but to extend the social constructivist approach by examining the complex relationships between the material and the social. As outlined in detail in section 2.5, making use of phenomenological and non- or more-than-representational thinking (cf. B. Anderson and Wylie 2009; Ingold 1993; Lorimer 2005; Thrift 2008; Waterton 2019) allows the present research, on the one hand, to give special attention to the physical foundations of urban borderlands, landscape, and neighborhoods and, on the other hand, to expand the analytical focus to subjective everyday life *experiences*. Because in order to make sense of the physical, political, social/individual, and symbolic b/ordering processes (see Figure 2.5) that take place in the inner-ring neighborhoods of San Diego, it is, firstly, in the words of Brambilla (2010, p. 82), necessary "to recover the relationships between social and spatial aspects that are played out at [these complex] border" and boundary situations (cf. Iossifova 2015, p. 101). Secondly, in the words of Newman and Paasi (1998, p. 197), everyday life experiences ought to be brought into focus because it is this particular sphere "where the meanings of (state) boundaries are ultimately reproduced and contested" (cf. Cooper et al. 2016 [2014], p. 23). Ultimately, in addition to the social constructivist search for individual constructions and interpretations in the present study, in a phenomenological sense, subjective *border and boundaries experiences* (cf. Mein and Pause 2020; Wille and Weber 2020) become relevant.

3.3 Present Research Triangulation: Presentation of the Individual Methods and Their Execution

In the following, it will be outlined how the individual methods of this study have been carried out in the research process. Additionally, the discussion of the individual methods serves a further justification of the method selection for the present triangulation.

3.3.1 Semi-Structured and Unstructured Interviews: Focusing on Individual Perception and Interpretation

Following the aim of tracing individual constructions and interpretations of the (re-)development processes, hybridizations, and neighborhood changes in and around downtown San Diego, the present study is drawing especially on qualitative interviews. During the research process, two kinds of interviews have been executed: semi-structured guided interviews and 'ero-epic conversations' (Girtler 2001), which are unstructured, informal interviews. Generally, guided interviews offer a thematic frame through introductory comments and prepared questions, thus defining a direction in which the interviews are meant to flow while still leaving enough flexibility for deviating, individual impulses. While the interviewer uses the guideline to ensure that all of the important topics and questions are addressed, the interviewee is able to express individual thoughts and interpretations throughout the openly designed questions. Thus, the course of the interview can be re-focused according to the individual situation and in reaction to the shared subjective insights.

In addition to the semi-structured interviews, unstructured interviews constitute useful tools for the acquisition of individual constructions and experiences. The fundamental difference between the two interview types lies in the way in which the interviewer and the interviewee interact: While semi-structured interviews put the interviewee on the spot, informal and unstructured interviews, such as ero-epic conversations, are characterized by the principle of equality (Girtler 2001, p. 147). This means, that the interview situation is designed as an open, informal conversation in which the interviewer and the interviewee are meant to express themselves freely and ask each other questions. Thus, it is not only the interviewee but also the interviewer who shares individual perspectives and assessments. In this sense, ero-epic conversations are conversational situations that are neither introduced with a prepared guideline nor contain designed questions but are performed on a par with the interviewee, thus aiming for insights into the individual, everyday world context of the shared subjective constructions, interpretations, and experiences.

In the time frame from May 2019 to April 2020, the author has executed 39 semi-structured and unstructured interviews, each lasting between 5 and 120 minutes (see Table 3.1). The majority of the interviews took place during a research visit in the San Diego/Tijuana area in 2019, three additional phone interviews have been added in the spring of 2020. Within the aim of covering diverse, broad insights into the social perception and interpretation of the redevelopment and b/ordering processes of San Diego's central neighborhoods, the interviews are divided roughly equally between three perspective groups (which cover various sexes and different ages): 'residents' (14 interviews), 'economy/development & planning' (13 interviews), and 'academia & research' (12 interviews). By means of this perspective classification and diversification both individual everyday life assessments as well as expert views have been captured and combined. However, it is important to mention that the interview guideline and the questions have not been adapted to the specific groups. This is due to the reason that the overall topics of urban redevelopment and neighborhood change have a high level of everyday relevance, thus offering manifold points of contact for San Diego's residents. Furthermore, the present research topics are particularly relevant in the everyday business of developers, planners, as well as large groups of social scientists and can thus also be discussed easily and in depth from different expert perspectives. For this purpose, some of the interviewees have been specifically recruited for their position as city officials, developers, or local scientists, while others—such as residents and users of certain spaces—were contacted through a snowball procedure or simply addressed in the field. Accordingly, the recruitment was mostly done via email or phone but also personally at the end of different community and planning meetings or during participant observations and other activities in the field. Generally, in this project, interviews were performed until the insights reached 'thematic saturation' and no further relevant or new information was gained during the interviews. Some of the interviews have been recorded, while unrecorded interviews have been transcribed verbatim or noted down as detailed as possible after the interviews concluded. The evaluation process will be discussed in detail in section 3.3.4.

Table 3.1 List of Interviews and Their Perspective Classification

Role	Date	Approximate Duration	Situation	Identifier
Resident	2019/05/18	45 Minutes	Face-to-face	R-01
Resident	2019/05/20	10 Minutes	Face-to-face	R-02
Resident	2019/05/21	25 Minutes	Face-to-face	R-03
Resident	2019/05/23	20 Minutes	Face-to-face	R-04
Resident	2019/05/24	15 Minutes	Face-to-face	R-05
Resident	2019/05/25	10 Minutes	Face-to-face	R-06
Resident	2019/05/28	35 Minutes	Phone	R-07
Former Resident	2019/06/06	15 Minutes	Face-to-face	R-08
Resident	2019/09/06	30 Minutes	Face-to-face	R-09
Resident	2019/09/10	70 Minutes	Face-to-face	R-10
Resident	2019/09/10	60 Minutes	Phone	R-11
Resident	2019/09/21	60 Minutes	Phone	R-12
Resident	2020/02/17	60 Minutes	Phone	R-13
Resident (Tijuana)	2019/07/10	120 Minutes	Face-to-face	R-TJ
Business Owner	2019/05/18	10 Minutes	Face-to-face	EP-01
Development	2019/05/21	35 Minutes	Phone	EP-02
Development	2019/05/23	25 Minutes	Phone	EP-03
Development	2019/05/24	40 Minutes	Face-to-face	EP-04
Development	2019/05/24	45 Minutes	Face-to-face	EP-05
Development	2019/05/28	40 Minutes	Face-to-face	EP-06

(continued)

Table 3.1 (continued)

Role	Date	Approximate Duration	Situation	Identifier
Development	2019/06/05	40 Minutes	Phone	EP-07
Development	2019/06/20	50 Minutes	Phone	EP-08
Planning	2019/07/09	35 Minutes	Face-to-face	EP-09
Development	2019/08/08	15 Minutes	Face-to-face	EP-10
Planning	2019/08/27	45 Minutes	Phone	EP-11
Planning	2019/09/05	45 Minutes	Phone	EP-12
Planning (Tijuana)	2020/04/30	60 Minutes	Phone	EP-TJ
Academia & Research	2019/05/22	30 Minutes	Phone	AR-01
Academia & Research	2019/06/11	45 Minutes	Phone	AR-02
Academia & Research	2019/06/12	50 Minutes	Phone	AR-03
Academia & Research	2019/06/14	25 Minutes	Phone	AR-04
Academia & Research	2019/06/25	50 Minutes	Phone	AR-05
Academia & Research	2019/06/27	50 Minutes	Phone	AR-06
Graduate Student	2019/07/11	60 Minutes	Face-to-face	AR-07
Academia & Research	2019/08/15	80 Minutes	Phone	AR-08
Academia & Research	2019/09/05	45 Minutes	Phone	AR-09
Academia & Research	2019/09/06	70 Minutes	Phone	AR-10
Academia & Research	2019/10/10	90 Minutes	Phone	AR-11
Academia & Research (Tijuana)	2020/04/30	50 Minutes	Phone	AR-TJ

Source: Own representation and listing.

3.3.2 Participant Observations and Phenomenological Neighborhood Walks: Inquiring into Practices, Experiences, and the Interrelations of the Social and the Material

In order to play out the developed theoretical-perspectival combination of social constructivism with phenomenological and non or more-than-representational thinking, the present methodical approach is triangulating the outlined interview types with the phenomenological-ethnographic methods of participant field observations and neighborhood walks (cf. Ingold and Vergunst 2008; Lorimer

2005; Lüders 2004; Sidaway 2009; Thrift 2008; Wylie 2005). While the interviews focus on individual interpretations and thus solely on social aspects, the observations and walks are dedicated to the inquiry of the interrelations between subject and object, which are—according to Husserl—unified within experience (Gergen 1999, p. 128). Consequently, by bringing the two perspectival research approaches of social constructivism and phenomenology together, the neighborhood changes and hybridizations in San Diego can be analyzed with regard to individual and social construction processes, (bordering) experiences, and (ordering and othering) practices—including aspects of the social as well as the physical, political, and symbolic-aesthetic dimension of bordering. The aim is to encounter subjective everyday experiences and thus the life-world backgrounds of certain behaviors, decisions, and perceptions (Roßmeier and Weber 2021). However, in line with interpretative approaches, this is done by tracing the meanings of actions, not the causes (Davoudi and Strange 2009, p. 21; cf. Lüders 2004, p. 224).

In concrete, the interest and attention during the observations and walks is brought to practices in relation to leisure activities, such as sports or shopping, restaurant choices, housing decisions, the use of symbols and signs, and other (border-related) behaviors and routines that are selectively shown or omitted at specific places and neighborhoods. Thus, in combination with the qualitative interviews, it becomes possible to work out in concrete terms how the change processes in the urbanizing inner ring are experienced individually and how patterns of spatial appropriation and other practices adapt. For it is assumed that the progressing development in and around downtown San Diego, the recent social and aesthetic trends, as well as the politically or economically driven neighborhood branding efforts are facilitating new land uses, which then establish new interpretative links between spaces, uses, and different groups. Given the ongoing redevelopment, it is of interest how symbols and signs are used for the purpose of (socio-)spatial contrasting or, in other words, for b/ordering and othering process. Ultimately, from a border-theoretical stance, the neighborhood walks aim to observe physical borders as well as to uncover social and individual boundary drawings and thus approach the everyday border experiences, which underly the manifold distinctions in central San Diego.

During a research visit in the San Diego-Tijuana metropolitan region in the summer of 2019, the author has executed six observations and neighborhood walks (see Table 3.2). However, the pressing question arises what is meant by the terms participant field observations and phenomenological neighborhood walks and how have these methods been operationalized and performed in the present study. Generally, observations and neighborhood walks are each understood as "*flexible, methodologically plural and context-related strategies* that could incorporate widely different procedures" (Lüders 2004, p. 224 [original emphasis]). Furthermore, the observations have been carried out rather uncontrolled in order to be able to observe freely and without bias. Accordingly, following the methodologically 'vagueness' of phenomenological life-world analysis (Hitzler and Eberle 2004), the executed participant field observations of the present study included participating in planning and community meetings[3], strolling through art galleries, shopping in different stores, observing and enjoying food trucks, studying the clientele—as well as the menus—of restaurants and bars, and chatting with people wherever it seemed appropriate. Thus, the field research was not done in a classic ethnographic way of long-term observation but within a limited time frame of a couple of hours.

Furthermore, the field observations draw on a large 'walking component,' which the author terms phenomenological neighborhood walk. This part of the observations has been performed as attentive walking and partly cycling through specific neighborhoods, along large, prominent as well as small residential streets and through every park and green space central San Diego provides. Executing phenomenological neighborhood walks meant studying the infrastructure and the buildings, their use, sizes, and conditions as well as looking at the ways people live, where they do (which) sports, groceries, and other recreational activities (cf. Breitung 2011, p. 56). Thus, it became possible to not only pay attention to social

---

[3] Community planning groups (and their meetings) constitute a special way of citizen involvement in San Diego's planning processes. In the 1960 s and '70 s, the city adopted policies "that established and recognized community planning groups as formal mechanisms for community input in the land use decision-making processes. […] The recommendations of the planning groups are integral components of the planning process and are highly regarded by the City Council and by staff" (CofSD 2021, p. n.pag.). Each of the more than 50 (not overlapping and clearly demarcated) planning areas of the City of San Diego are represented by one group, which develops the given area's community plan. These plans constitute the land use element of the city's General Plan, the primary constitution for future development in San Diego and the policy frame for the City of Villages smart growth strategy. Thus, the present study is especially drawing on the 7 community plans covering San Diego's central neighborhoods as well as the General Plan itself in terms of neighborhood history, present and proposed future land uses, as well as planned projects and public improvements.

and individual bordering experiences but also to study how individuals interact with and react to borders and boundaries of the physical, political, and symbolic-aesthetic dimensions and what role they play in the organization of everyday life in central San Diego.

The chosen way of capturing the observed were field notes. These were taken whenever it seemed necessary and have been edited, contextualized, and extended several times after each observation. Inspired by Carter's (2014, pp. 21–22) field work in downtown Los Angeles, "most field notes were taken as raw data with the intent of building a narrative" and were supplemented with countless photographs, as outlined in the following. After the field trips, the notes have been compiled with the interviews into a triangulated corpus of own as well as gathered experiences, interpretations, and impressions (in line with the considerations on textual presentation by Van Maanen 1995; cf. Matt 2004), which have then been evaluated within a qualitative content analysis.

Table 3.2 List of Participant Field Observations and Phenomenological Neighborhood Walks

Neighborhoods	Date	Approximate Duration	Identifier
East Village, Gaslamp Quarter	2019/04/03	120 Minutes	PO-01
Barrio Logan, Logan Heights, Sherman Heights, Grant Hill, Stockton	2019/05/24	120 Minutes	PO-02
North Park, University Heights, Normal Heights, Kensington, City Heights	2019/05/27	180 Minutes	PO-03
Little Italy, Bankers Hill, Hillcrest	2019/05/28	150 Minutes	PO-04
Burlingame, South Park, Golden Hill, downtown	2019/07/09	180 Minutes	PO-05
Downtown Tijuana and beyond	2019/07/10	150 Minutes	PO-06

Source: Own representation and listing.

3.3.3 Cartographic Visualization and Photographic Documentation: Representing the Physical, Political, Social, and Aesthetic Dimensions of Hybrid Urban Borderlands

Integrating cartography and photography into qualitative social science means to solve the methodological task of incorporating rather positivist, representational methods into a social constructivist undertaking. Here, following a neopragmatic approach (cf. Kühne and Jenal 2020a, 2021; Kühne and Weber 2019), this task can succeed by understanding photographs and maps not as means of depicting inevitable and universal reality but as supplementation in the attempt to understand San Diego's infinite realities.

In this sense, photographs are integrated into the present methodical approach as "subjective statements rather than objective documents" (Harper 2004, p. 232). Because this is how the photographs came into being, as spontaneously occurring, subjective associations and observations, which have been captured visually in the field. For this reason, Harper (2004, p. 233) argues: "If we are aware of how the photograph is constructed, we can use it more successfully." Thus, it becomes obvious that the photographs used in the present work are merely the outcome of the author's subjective observation, which, however, can still enlarge the created window into the various social realities in and around downtown San Diego. For this purpose, the author has shot hundreds of photographs during the research visit in 2019 and several following recreational trips to San Diego between 2020 and 2022. On the one hand, the pictures are meant to showcase architecture and other physical structures or compositions, which indicate hybridization processes in the urbanizing inner-ring. On the other hand, the photographic documentation aims for the illustration of diverse ways of life, aesthetic-symbolic aspects, and generally—from the author's perspective—unique situations that help to tell the manifold (border and boundary) stories of central San Diego.

Cartographic representations, in turn, have been used mainly for two purposes: firstly, 'classic' two-dimensional cartographic illustrations have been designed to illustrate and localize physical-spatial structures including a variety of material borders but also political and other immaterial boundaries. Secondly, however, in a qualitative-oriented, experimental approach to cartography, aspects that were observed during the neighborhood walks or were highlighted in the interviews have been illustrated and compiled in the form of three-dimensional, digital sketches using icons, pictograms, arrows, and other pictural elements. Following a postmodern and relational understanding of scale (cf. Davoudi and Strange 2009, p. 60), these visualizations are meant to highlight functional networks, flows, and spatial (dis-)connections in and between different neighborhoods and communities in their fuzzy everyday life manifestations. Thus, it became possible to link the different used methods and their methodologies by translating qualitative findings into relational, three-dimensional maps. Generally, the various kinds of maps in this study have been created throughout different phases of the research process: Some constitute elaborated versions of ideas and sketches from the field notes, while others have been created during the evaluation process for the purpose of expanding the interpretation and presentation of the qualitative findings.

Another specialized, less common type of cartography integrated in the following is the figure-ground plan, a tool used in the field of urban planning to visualize solely the building structure of a certain area. By means of this specific illustration it becomes possible, for instance, to compare certain spaces of neighborhoods chronologically in order to underline overarching development trends and structural hybridization tendencies (also presented in Roßmeier 2019, p. 605; 2020, p. 468). In the present context, the given figure-ground plan allows for conclusions on redevelopment activity and tendencies of urban/rural hybridization in terms of the underlying housing structure of neighborhoods. More theoretically, from a relational perspective (Bosco 2015; Graham and Healey 1999; Nienaber and Wille 2020), figure-ground plans offer a way to 'bring the material back' into contemplation through contextualization with qualitative findings. Following this thought, the illustrations help to understand the analyzed borders and boundaries and their various meanings more comprehensively. Because through the complementation with cartographic illustrations, the present triangulation is able to direct the research focus to a variety of different bordering dimensions—to physical, political, social/individual, and symbolic processes.

3.3.4 Supplementation with U.S. Census Data: Considering Socio-Spatial Composition and Developments

The last part of the present methodical triangulation constitutes the implementation of survey data gained by the U.S. Census Bureau into the evaluation process. The primary way to access the data were online tools, such as the *2020 Census Demographic Map Viewer*[4], a GIS-based data tool provided by the United States

[4] https://www.census.gov/library/visualizations/2021/geo/demographicmapviewer.html.

Census Bureau, and *Data Surfer*[5], an online research tool by the San Diego Association of Governments (SANDAG). The tools provide census data and in part also estimate and forecast information for the entire San Diego County area. With Data Surfer, data can be retrieved for various geographical and political entities of San Diego County, such as council districts, subregional areas, community planning areas, zip code areas, and census tracts, thus allowing for versatile comparison possibilities, for instance, with regard to the relatively small-scale category of census tracts, which constitute smaller areas than the official boundaries of the neighborhoods in study and thus make detailed differentiation of and within various neighborhoods possible. For this purpose, recent data from the 2020 census (depending on availability) as well as from previous decades will be used, which allow insights into the development of population densities, ethnic composition, median age, educational attainments, housing values, and other socio-spatial details.

However, before the extracted data can be integrated into the present study, it is important to address methodological issues. In order to avoid the troubles of a methodological balancing act, the quantitative survey data will not be used for the purpose of *explanation* as it is common in positivist approaches. Rather, relevant data for the considered neighborhoods will be selectively supplemented in order to *underpin* and *understand* narratives expressed in the interviews as well as overarching development trends. Thus, despite the inclusion of perspectival disjunctive methods and data, the phenomenologically oriented social constructivist research and evaluation framework can be maintained by means of the neopragmatic triangulation logic.

3.4 Qualitative Content Analysis: Evaluation of the Qualitative Research Components

The empirical material of the present study was evaluated on the basis of a qualitative content analysis in the tradition of Mayring (2004, 2019). The advantage of this analytical method lies in its ability to evaluate large text corpora in a systematic but qualitative-interpretative manner (Schmidt-Lauber 2007). Its development was largely inspired by quantitative content-analytical procedures and their systematization but was adapted for the analysis of latent structures of meanings (Krakauer 1952, p. 634) and the individual context of the expressed. Qualitative content analysis and its reductive, summarizing procedure (see Figure 3.2) can

[5] https://datasurfer.sandag.org/

be regarded as a suitable evaluation approach for the present analysis because the research interest forms solely around the content level of the empirical material, not deeper psychological characteristics of the interviewees. Furthermore, regarding the deliberate openness and resulting variety of the different interview types in terms of thematic focus, scope, and degree of detail, the qualitative content-analytical approach has been selected for its ability "to make a cross-section of the material" (Mayring 2004, p. 269).

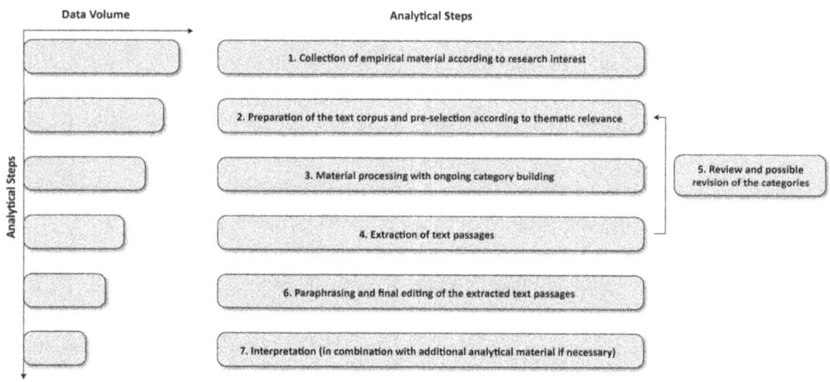

Figure 3.2 Qualitative Content-Analytical Procedure and Data Handling. (Source: Own representation in reference to (Mayring 2004, p. 268))

The evaluation process of the present study proceeded as follows: After the interview and observation phase, the recordings have been transcribed using the *f4transkript* software and combined with the field notes into a large text corpus. While building the corpus and reading through the material, text passages that appeared to be of little interest in terms of the research questions have already been marked as such using *f4analyse*, a computer-aided analysis program (Step 2 of the analysis). Thus, the data volume has already been reduced. At this stage, also the first loose categories have been formed, which built the preliminary frame for the coding system. Generally, codes or categories "refer to aspects within the text, which put the meaning of those aspects in a nutshell" (Mayring 2019, p. 3). They are developed step by step in interaction between the research questions and the text material.

In the present analysis, a mixture of deductive and inductive category building has been used, which draws on theoretical considerations of postmodern urban and social development trends and includes aspects of individual life-world perspectives and constructions from the interviews. While working further through the text corpus, the coding frame has been elaborated in more detail (Step 3) and thematically fitting parts of the material have been grouped together (Step 4). During this process, the thematic codes have been reviewed, revised, and regrouped multiple times whenever new relevant aspects or thematic connections became apparent during the analysis (Step 5). Subsequent to the categorization steps, using the memo function of f4analyse, the grouped material has been abbreviated and paraphrased and thus already filtered and pre-interpreted (Step 6). This step "seeks to reduce the material in such a way that the essential contents are preserved, but a manageable short text is produced" (Mayring 2004, p. 268; cf. Schreier 2014, p. 170) that sums up recurring argumentations and interpretations.

Lastly, in order to interpret the reduced text corpora further, unclear and ambiguous statements have been explained by means of additional material (Step 7). In line with the idea of neopragmatic triangulation, the individual methodical components—the observations and the walks, the cartographic and photographic illustrations, and the implemented census data—constitute the means to help construing vague or inexplicit interview passages. Thus, it became possible to relate physical-spatial developments to social aspects against the background of urban re-development, b/ordering processes, and changing individual and social urban landscape constructions.

References

Anderson, B. & Wylie, J. (2009). On Geography and Materiality. *Environment and Planning A: Economy and Space, 41* (2), pp. 318–335.

Balibar, É. (2004). *We, the people of Europe. Reflections on transnational citizenship.* Princeton: Princeton University Press.

Barnes, T. J. (2008). American pragmatism: Towards a geographical introduction. *Geoforum, 39* (4), pp. 1542–1554.

Bosco, F. J. (2015). Actor-network theory, networks, and relational geographies. In S. C. Aitken & G. Valentine (Eds.), *Approaches to human geography. Philosophies, theories, people and practices* (2 ed., pp. 150–162). Los Angeles: Sage.

Brambilla, C. (2010). Borders Still Exist! What are Borders? In B. Riccio & C. Brambilla (Eds.), *Transnational Migration, Cosmopolitanism and Dis-located Borders* (pp. 73–85). Rimini: Guaraldi.

Brambilla, C. (2015). Exploring the critical potential of the borderscapes concept. *Geopolitics, 20* (1), pp. 14–34.

Breitung, W. (2011). Borders and the City: Intra-Urban Boundaries in Guangzhou (China). *Quaestiones Geographicae, 30* (4), pp. 55–61.

Carter, N. D. (2014). *The Struggle to Create a Residential Community in Downtown Los Angeles.* (Doctor of Philosophy Dissertation), San Diego State University and University of California, Santa Barbara.
Chilla, T., Kühne, O., Weber, F. & Weber, F. (2015). ‚Neopragmatische' Argumente zur Vereinbarkeit von konzeptioneller Diskussion und Praxis der Regionalentwicklung. In O. Kühne & F. Weber (Eds.), *Bausteine der Regionalentwicklung* (pp. 13–24). Wiesbaden: Springer VS.
CofSD, P. D. (2021). Community Planning Groups. City of San Diego Planning Department. https://www.sandiego.gov/planning/community/cpg. Accessed: 14. June 2021.
Cooper, A., Perkins, C. & Rumford, C. (2016 [2014]). The Vernacularization of Borders. In R. Jones & C. Johnson (Eds.), *Placing the Border in Everyday Life* (pp. 15–32). London: Routledge.
Coulton, C. (2012). Defining Neighborhoods for Research and Policy. *Cityscape: A Journal of Policy Development and Research, 14* (2), pp. 231–236.
Coulton, C., Jennings, M. Z. & Chan, T. (2013). How Big is My Neighborhood? Individual and Contextual Effects on Perceptions of Neighborhood Scale. *American Journal of Community Psychology, 51* (1), pp. 140–150.
Coulton, C., Korbin, J., Chan, T. & Su, M. (2001). Mapping Residents' Perceptions of Neighborhood Boundaries: A Methodological Note. *American Journal of Community Psychology, 29* (2), pp. 371–383.
Cresswell, T. (2013). *Geographic Thought: A Critical Introduction.* Chichester: Wiley-Blackwell.
Curto, M. C. F. (2019). Neighborhood. In A. M. Orum (Ed.), *The Wiley Blackwell Encyclopedia of Urban and Regional Studies* (pp. 1–9). Chichester: Wiley Blackwell.
Davoudi, S. (2012). The Legacy of Positivism and the Emergence of Interpretive Tradition in Spatial Planning. *Regional Studies, 46* (4), pp. 429–441.
Davoudi, S. & Strange, I. (2009). Space and place in the twentieth century planning: an analytical framework and a historical review. In S. Davoudi & I. Strange (Eds.), *Conceptions of Space and Place in Strategic Spatial Planning* (pp. 7–42). London: Routledge.
Denzin, N. K. (2017). *The Research Act. A Theoretical Introduction to Sociological Methods.* New Brunswick: Transaction Publishers.
Flick, U. (2004). Triangulation in Qualitative Research. In U. Flick, E. von Kardoff & I. Steinke (Eds.), *A Companion to Qualitative Research* (pp. 178–183). London: SAGE.
Galster, G. (2001). On the Nature of Neighbourhood. *Urban Studies, 38* (11), pp. 2111–2124.
Gergen, K. (1999). *An Invitation to Social Construction.* Thousank Oaks: Sage.
Gerst, D. & Krämer, H. (2021). Methodologie der Grenzforschung. In D. Gerst, M. Klessmann & H. Krämer (Eds.), *Grenzforschung. Handbuch für Wissenschaft und Studium* (pp. 121–140). Baden-Baden: Nomos.
Girtler, R. (2001). *Methoden der Feldforschung.* Wien: UTB.
Graham, S. & Healey, P. (1999). Relational concepts of space and place: Issues for planning theory and practice. *European Planning Studies, 7* (5), pp. 623–646.
Greider, T. & Garkovich, L. (1994). Landscapes: The Social Construction of Nature and the En vironment. *Rural Sociology, 59* (1), pp. 1–24.
Harper, D. (2004). Photography as Social Science Data. In U. Flick, E. von Kardoff & I. Steinke (Eds.), *A Companion to Qualitative Research* (pp. 231–237). London: SAGE.

Hitzler, R. & Eberle, T. S. (2004). Phenomenological Life-World Analysis. In U. Flick, E. von Kardoff & I. Steinke (Eds.), *A Companion to Qualitative Research* (pp. 67–71). London: Sage.

Ingold, T. (1993). The Temporality of the Landscape. *World archaeology, 25* (2), pp. 152–174.

Ingold, T. & Vergunst, J. L. (2008). *Ways of Walking: Ethnography and Practice on Foot.* Aldershot: Ashgate.

Iossifova, D. (2015). Borderland urbanism: seeing between enclaves. *Urban Geography, 36* (1), pp. 90–108.

Krakauer, S. (1952). The challenge of qualitative content analysis. *Public opinion quaterly, 16*, pp. 631–642.

Kühne, O. (2018). Reboot „Regionale Geographie": Ansätze einer neopragmatischen Rekonfiguration „horizontaler Geographien". *Berichte zur deutschen Landeskunde, 92* (2), pp. 101–121.

Kühne, O. (2019). *Landscape Theories. A Brief Introduction.* Wiesbaden: Springer VS.

Kühne, O. & Jenal, C. (2020a). Baton Rouge (Louisiana): On the Importance of Thematic Cartography for 'Neopragmatic Horizontal Geography'. *KN – Journal of Cartography and Geographic Information*, pp. 1–9.

Kühne, O. & Jenal, C. (2020b). *Baton Rouge–the Multivillage Metropolis. A Neopragmatic Landscape Biographical Approach on Spatial Pastiches, Hybridization, and Differentiation.* Wiesbaden: Springer VS.

Kühne, O. & Jenal, C. (2021). Baton Rouge–A Neopragmatic Regional Geographical Approach. *urban science, 5* (17), pp. 1–23.

Kühne, O. & Weber, F. (2019). *Hybrid California. Annäherungen an den Golden State, seine Entwicklungen, Ästhetisierungen und Inszenierungen.* Wiesbaden: Springer VS.

Kühne, O. & Weber, F. (2020). Sozialwissenschaftliche Zugänge zu Landschaft. In R. Duttmann, O. Kühne & F. Weber (Eds.), *Landschaft als Prozess* (pp. 69–82). Wiesbaden: Springer VS.

Lee, B. A., Oropesa, R. S. & Kanan, J. W. (1994). Neighborhood Context and Residential Mobility. *Demography, 31* (2), pp. 249–270.

Lohmann, A. & McMurran, G. (2009). Resident-Defined Neighborhood Mapping: Using GIS to Analyze Phenomenological Neighborhoods. *Journal of Prevention & Intervention in the Community, 37*, pp. 66–81.

Lorimer, H. (2005). Cultural geography: the busyness of being 'more-than-representational'. *Progress in Human geography, 29* (1), pp. 83–94.

Lüders, C. (2004). Field Observation and Ethnography. In U. Flick, E. von Kardoff & I. Steinke (Eds.), *A Companion to Qualitative Research* (pp. 222–230). London: SAGE.

Matt, E. (2004). The Presentation of Qualitative Research. In U. Flick, E. von Kardoff & I. Steinke (Eds.), *A Companion to Qualitative Research.* London: Sage.

Mayring, P. (2004). Qualitative Content Analysis. In U. Flick, E. von Kardoff & I. Steinke (Eds.), *A Companion to Qualitative Research* (pp. 266–269). London: SAGE.

Mayring, P. (2019). Qualitative Content Analysis: Demarcation, Varieties, Developments. *Forum Qualitative Social Research, 20* (3, Art. 16), pp. 1–14.

Mein, G. & Pause, J. (2020). *Self and Society in the Corona Crisis.* Esch-sur-Alzette: Melusina Press.

Menand, L. (2001). *The Metaphysical Club: A Story of Ideas in America.* New York: Farrar, Straus and Giroux.

Mooney Melvin, P. (1985). Changing Contexts: Neighborhood Definition and Urban Organization. *American Quaterly, 37* (3), pp. 357–367.

Newman, D. (2006). The lines that continue to separate us: borders in our 'borderless' world. *Progress in Human geography, 30* (2), pp. 143–161.

Newman, D. (2011). Contemporary Research Agendas in Border Studies: An Overview. In D. Wastl-Walter (Ed.), *The Ashgate Research Companion to Border Studies* (pp. 33–47). Farnham: Ashgate.

Newman, D. & Paasi, A. (1998). Fences and neighbours in the postmodern world: boundary narratives in political geography. *Progress in Human geography, 22* (2), pp. 186–207.

Nicotera, N. (2007). Measuring Neighborhood: A Conundrum for Human Services Researchers and Practitioners. *American Journal of Community Psychology, 40*, pp. 26–51.

Nienaber, B. & Wille, C. (2020). Cross-border cooperation in Europe: a relational perspective. *European Planning Studies, 28* (1), pp. 1–7.

Putnam, H. (1995). *Pragmatism: An Open Question.* Hoboken: Blackwell.

Rorty, R. (1979). *Philosophy and the Mirror of Nature.* Princeton: Princeton University Press.

Roßmeier, A. (2019). *Quo vadis ‚Temporary Paradise'?* Urfsurbanisierung und räumliche Konflikte im StadtLandHybriden San Diego. In K. Berr & C. Jenal (Eds.), *Landschaftskonflikte* (pp. 591–616). Wiesbaden: Springer VS.

Roßmeier, A. (2020). Stadtlandhybridisierung als Prozess. Eine Typisierung aktueller (Re-) Urbanisierungsprozesse und Entwicklungszüge in Innenstadtrandlagen von Los Angeles. In R. Duttmann, D. Knitter, O. Kühne & F. Weber (Eds.), *Landschaft als Prozess* (pp. 455–474). Wiesbaden: Springer VS.

Roßmeier, A. & Weber, F. (2021). Hybrid Urban Borderlands – Redevelopment Efforts and Shifting Boundaries In and Around Downtown San Diego. *Journal of Borderlands Studies*, pp. 1–27.

Rumens, N. & Kelemen, M. (2016). American Pragmatism and Organization Studies: Concepts, Themes and Possibilities. In M. Kelemen & N. Rumens (Eds.), *American Pragmatism and Organization. Issues and Controversies* (pp. 3–24). London: Routledge.

Rumford, C. (2012). Towards a Multiperspectival Study of Borders. *Geopolitics, 17* (4), pp. 887–902.

Rumford, C. (2014). *Cosmopolitan Borders.* Hampshire: Palgrave Macmillan.

Schmidt-Lauber, B. (2007). Das qualitative Interview oder: Die Kunst des Reden-Lassens. In S. Göttsch & A. Lehmann (Eds.), *Methoden der Volkskunde. Positionen, Quellen, Arbeitesweisen der Europäischen Ethnologie* (pp. 219–248). Berlin: Reimer.

Schreier, M. (2014). Qualitative Content Analysis. In U. Flick (Ed.), *The SAGE Handbook of Qualitative Data Analysis* (Vol. London, pp. 170–183): SAGE.

Sidaway, J. D. (2009). Shadows on the path: negotiating geopolitics on an urban section of Britain's South West Coast Path. *Environment and Planning D: Society and Space, 27* (6), pp. 1091–1116.

Sperling, J. (2012). The Tyranny of Census Geography: Small-Area Data and Neighborhood Statistics. *Cityscape: A Journal of Policy Development and Research, 14* (2), pp. 219–223.

Suttles, G. D. (1972). *The social construction of communities.* Chicago: University of Chicago Press.

Taylor, R. B. (2012). Defining Neighborhoods in Space and Time. *Cityscape: A Journal of Policy Development and Research, 14* (2), pp. 225–230.

Thrift, N. (2008). *Non-representational theory: Space, politics, affect*. London: Routledge.

U.S. Census Bureau. (2021). Glossary. U.S. Department of Commerce. https://www.census.gov/glossary/#term_Censustract. Accessed: 10. April 2021.

van Maanen, J. (Ed.) (1995). *Representation in Ethnography*. Thousand Oaks: Sage.

Waterton, E. (2019). More-than-representational landscapes. In P. Howard, I. Thompson, E. Waterton & M. Atha (Eds.), *The Routledge Companion to Landscape Studies* (2nd ed., pp. 91–101). London: Routledge.

Wille, C. & Weber, F. (2020). Analyzing border geographies in times of COVID-19. In G. Mein & J. Pause (Eds.), *Self and Society in the Corona Crisis. Perspectives from the Humanities and Social Sciences* (Vol. 2, The Ends of Humanities, pp. 1–17). Esch-sur-Alzette: Melusina Press.

Wylie, J. (2005). A single day's walking: narrating self and landscape on the South West Coast Path. *Transactions of the Institute of British Geographers, 30* (2), pp. 234–247.

Wylie, J. (2007). *Landscape*. London, New York: Routledge.

4 From Native Land to America's Finest City: Historical Milestones and Leading Developments in San Diego

Subsequent to the outlined theoretical-perspectival foundation and the methodical approach of the present study, the following chapter is outlining key events and developments of San Diego's settlement history (see Table 4.1 at the end of this chapter). This illustration is providing a historical background for the analysis of the current urbanization and redevelopment processes in central San Diego.

Prior to its relatively recent development into the 8th most populous city of the United States as well as the arrival of European expeditions and settlers, San Diego has been populated by several Native American tribes and bands that migrated into the area at least 14,000 years ago. The tribal history of Southern California constitutes the longest era of human settlement in San Diego County, however, simultaneously also the part with the least remaining physical evidence. Despite their long settlement period, the local tribes have only developed few permanent physical structures and have not applied significant agricultural techniques. Instead, the stone age tribes were hunters or fishers and gatherers as well as nomads that merged with other tribes over the course of the time (Ford 2014, p. 176). Roughly between 12,000 BC and 5,000 BC, the San Diego area was populated by the San Diegito or Dieguiño people (as the Spanish settlers have later referred to them), which consisted of several smaller tribes. Historically, the Dieguiños have been split up by the San Diego river, which is running in east-west direction along present-day Interstate 8. The southern tribes—in Native language called the Kumeyaay people—have populated the areas from the river until south of Ensenada in today's Baja California, while the northern people—the Luiseño—have settled in the northern parts of San Diego County and in Orange and Riverside County (Kumeyaay 2022; San Diego State University 2022). Around 5,000 BC, the La Jollan tribe evolved from the Dieguiño people, which all together were later—between the years 0 and 1000 AD—intruded by Yuman-speaking as well as Shoshonean-speaking tribes. Several centuries later,

A. Rossmeier, *San Diego's Hybrid Urban Borderlands*,
Räume – Grenzen – Hybriditäten, https://doi.org/10.1007/978-3-658-42667-5_4

however, the solitude and the long-practiced ways of life of the various Native American people in California have been brutally put to an end by the arriving explorers, plunderers, and priests from Europe.

In the 1500 s, after the successful looting and destruction of the Aztecs and the Incas, the Spanish conquistadors have sailed farther north along the West Coast of North America in hope for more rich empires to conquer. The European history in California began in 1542 when the Portuguese explorer Juan Cabrillo sailed past Point Loma into San Diego Bay under the Spanish flag and named it San Miguel Bay. However, despite the first discovery of Cabrillo as well as a second arrival of a Spanish exploration led by Sebastian Vizcaíno in 1602, a permanent settlement has not been established in San Diego until the year of 1769. Even though Vizcaíno endorsed the establishment of a settlement along San Miguel Bay, which he renamed into San Diego de Alcalá at the time of his short stay, the Spanish continued to focus on the established trade with Southeast Asia, for the claimed land of Alta California has been viewed as "a dry, barren, and poor alternative to these riches" (Baker 2007, p. 11). However, throughout the 17th and 18th century, numerous treasure-filled Spanish galleons that have travelled across the Pacific from Manila to the port of Acapulco in Mexico have been seized by the British and the fleets of the seafarer Sir Francis Drake. Additionally, the English but also France and Russia have begun to show interest in Northern America's West Coast, which worried the weakened Spanish increasingly. Consequently, Spanish leaders commanded a small number of priests and soldiers to travel north from Mexico—both on land and on sea—in order to establish settlements along the Coastline, convert the Native Americans, and fortify the reign of the Spanish flag in Alta California. Thus, on July 16, 1769, the Franciscan priest Junípero Serra established the first of 21 missions along the Californian coast in San Diego. Subsequently, a presidio has been erected near the mission, which served the purpose of protecting the small and rather vulnerable settlement from the commencing violent conflicts with the local tribes. Despite the presidio's existence and the presence of soldiers, the settlement has experienced several attacks by Native Americans over the following 50 years.

Obeying the Spanish commands, the priests of the mission were dedicated to destroy the Native ways of life and convert the local tribe members into "loyal, tax-paying, hard-working Christian citizens" (Baker 2007, p. 13) through cultural and religious education. Additionally, some of the local soldiers have confronted the Natives with brutal violence, thus igniting ongoing tensions that have put the entire settlement under threat. After several attacks by the Natives tribes, the missionaries sought to physically distance the mission from the presidio, which was believed to be the target of the violent clashes. Consequently, in 1774, the mission had been relocated further inland to its present location with the help of a small number of mostly forcibly converted Kumeyaay members (Trafzer 2014, p. 51). However, in 1775, after the Spanish soldiers continued to mistreat the Natives, the local tribes united for their largest attack that struck the newly established mission instead of the soldiers and their presidio and ultimately led to the death of several missionaries and the burn down of the newly erected mission structures. Despite the devastating event, the priests have held on to the new location and began to rebuild the mission with the help of converted Natives. In 1784, mission San Diego de Alcalá had been completed and in the following years expanded by the church building that can still be found in Mission Valley today (see Figure 4.1). At the turn of the 19th century, the mission in San Diego had become the most populous of the missions in California and counted more than 1,500 converted Natives (Baker 2007, pp. 17–18). The captured Natives held important responsibilities and "contributed significantly to the success of the missions, serving as brick and tile makers, carpenters, cowboys, farmers and freighters" (Trafzer 2014, p. 51). However, despite the gained power and success, after roughly 50 years of Catholic conversion and settlement efforts, the Spanish mission era in San Diego and all of California came to an end.

Figure 4.1 The First Spanish Mission in California: Mission San Diego de Alcalá, completed in 1784. (Source: Picture Albert Roßmeier 2020)

By 1821, Mexico gained its long-sought independence from Spain and thus received large territories of the present-day southwestern United States, including California. Subsequent to the independence, in 1833, Pío Pico, who served twice as governor of California, enforced the Mexican decree of secularization and its intent to transform the 21 missions into parish churches and return the occupied land to the converted Natives. Thus, in 1835, the mission is San Diego has been secularized and the associated land has been parcellated. Despite the initial aim of the Mexican government to grant the land to the remaining converts, the mission land in San Diego was given to loyal and influential Mexican persons and families, such as Pío Pico himself. With these land grants, the mission lands have been turned into prestigious Mexican ranchos on the basis of a feudal-like system. The short rancho period "is fondly remembered as 'Old California,' the colorful era of fiestas, rodeos, opulent clothes, and bejeweled saddles" (Baker 2007, p. 21). Accordingly, also San Diego enjoyed great reputation during the Mexican period and was appointed capital of California in 1825 during José Echeandía's time in office as governor of California. Furthermore, the city prospered through cattle farming and the production of animal hides, which have been traded illegally with the increasing number of arriving smuggler ships from New England. Smuggling hides became the main economic sector of the region—hides were referred to as 'California banknotes' (Mills 1985)—because the large distance of both the Mexican and the American government impeded monitoring. However, due to the location of San Diego in the northern part of Alta California, away from the Mexican government and its troops, the vast mission lands have faced regular attacks, robberies, and killings by the remaining tribe members of the county, which outnumbered the rancheros and their families. Over the time, this hostile situation has led to a significant migration outflow of San Diego, which ultimately meant the loss of the city's capital status in 1838 and thus also of its important federal revenues.

Ten years after San Diego lost its status as the capital of California, the Treaty of Guadalupe Hidalgo was signed in order to end the Mexican-American war. 1848 marks the year in which California fell to the United States and the American era in San Diego began. Furthermore, the handover of the large territories to the Americans has created the Mexican-American border at its present-day location in California. Since then, the border has impacted the remaining Kumeyaay Natives strongly and divided the large group into two sections even though border patrol has only been introduced in 1924 and fences or walls have not been installed until the 1990 s (Johnson and Jones 2016 [2014], p. 2). In addition to the major political event, in 1848, gold has been found at Sutter's mill in northern California and get-rich-quick stories promoted an unprecedented population influx to the American West. Reacting to this boom, the rancheros of Southern California became eager to drive their cattle more north in order to sell it for beef to the thousands of hungry migrants that began to settle in the gold field territory and the northern Sierra Nevada. Additionally, many San Diegan rancheros took out investment loans in hope for the rush to last. However, with the reduction of mining activity in the 1950 s and the implementation and steady increase of federal and local taxes under the new flag, the local rancheros have neither been able to pay the taxes for their vast lands nor to pay back their loans. As a consequence, many rancheros began to sell their lands under value to solvent Americans that started to settle in the area. Thus, more and more of the San Diegan rancheros lost their once prestigious lands, and after less than 30 years, the Mexican rancho era came to an end. This situation has been exacerbated by severe droughts in the early 1860 s, which led the remaining Mexicans to drive their cattle into Baja California. Ultimately, "the once-great cattle industry of California [was] virtually destroyed" (San Diego History Center 2020b, n.pag.), and most of the rancheros had left Southern California.

Subsequent to the period of missions and Mexican rancheros, the era of bayfront settlement and urban development in today's downtown area slowly began to unfold. In the 1850 s, the excitement for economic prosperity, which was believed to come with the change of government, led to first attempts to move the small settlement of San Diego closer to the city's port and bay. However, the success of 'New Town' failed to appear until the arrival of Alonzo Horton in 1867, who acquired roughly 1,000 acres of land at the present downtown location. His efforts to develop the new settlement addition transformed New Town to the main commercial center of San Diego and led to the relocation of "the city's first bank, main newspaper, and several government buildings to this site" (CofSD 2016, p. 156; on the historic development of downtown San Diego see section 5.4.1). In the following years, a number of events triggered a first significant real estate and population boom in San Diego, which had not only led to the further development of New Town but also to the parcellation and development of new suburban neighborhoods in close proximity. First, the novel Ramona by Helen Hunt Jackson was published in 1884, "a tale of cross-cultural romance set near San Diego" (University of San Diego 2022, p. n.pag.), which acted as a world-wide advertisement of California and the romanticized way of life in the West. One year later, constituting even higher influence, the intercontinental railroad link between the East and the West was finally established, which enabled the immigration of thousands of new residents and began to establish San Diego as a sunny tourism destination on the Pacific Ocean. As a consequence, numerous hotels and other infrastructure has been built—the famous Hotel del Coronado opened in 1888—and by 1887 the city's population had grown to roughly 40,000. Additionally, eager state and local boosters helped to spur the development in San Diego by advertising Southern California in the eastern parts of the country through the notion of health and wealth in Mediterranean climate—they promoted the 'San Diegan lifestyle' through newspaper articles or the export of 'healthy' citrus fruit on the new rail connection (cf. Mills 1985; Sandul 2014). Due to their success, "large tracts of new development began to appear on the hills immediately adjacent to Downtown" (CofSD 2016, p. 156).

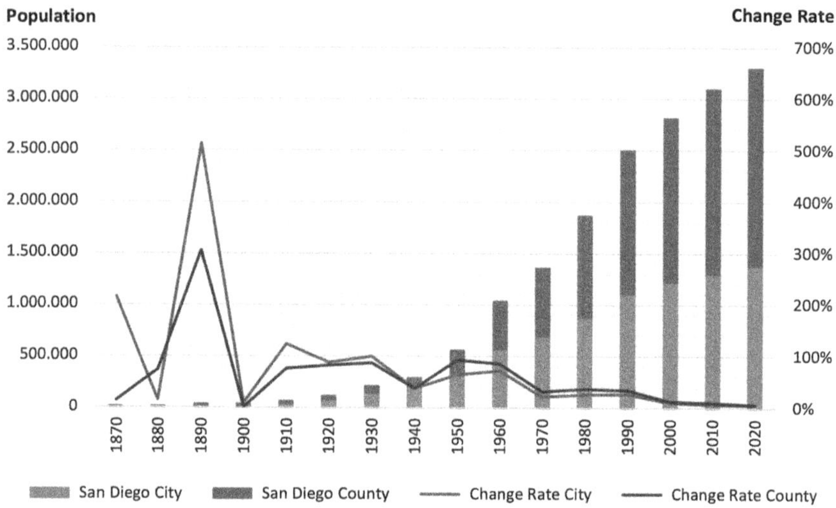

Figure 4.2 Comparison of the Population Development of San Diego City and County in Absolute Terms Plus Their Change Rates in Percent. (Source: Own representation and calculation based on data from Griffin and Weeks (2014) and U.S. Census Bureau)

However, in 1888, the inflated real estate bubble broke and the boom phase ended abruptly. Sudden credit tightening brought the local land speculators into financial woes, which changed the real estate market drastically. In order to pay back their loans on which the speculators had been operating, they were forced to quick land sales, which ultimately brought the "prices down and shattered land values" (Mills 1985, n.pag.). As a consequence, the population dropped again to roughly 16,000 in 1890 (see Figure 4.2). However, despite this sudden decrease,

San Diego reached a next boom phase in the ensuing years, which was strongly initiated by the eager investments of John D. Spreckels. In 1892, the San Franciscan entrepreneur acquired the local railroad system, electrified the horse and cable cars, and prompted the expansion of the various streetcar lines throughout the emerging suburbs. Additionally, the investor financed the construction of several buildings in downtown as well as of the 'impossible railroad,' which stretched eastward along the international border through the rugged terrain of the Laguna Mountain range. Thus, "Spreckels' widening investments locally were largely responsible for putting San Diego back into what has become its normal way of life—that of rapid growth" (Mills 1985, n.pag.). Next to Spreckels' influence, the arrival of water from the Cuyamaca Mountains in 1889 through the efforts of the San Diego Flume Company and the growing economic sector of citrus farming helped San Diego to overcome its first real estate crisis. However, with the close proximity of the populous Los Angeles, "San Diego was not able to develop into a very large city. Very disadvantageous for its development was the partial isolation in the southernmost corner of the state, being the 'last post' before entering the wilderness of Baja California and the Sonoran Desert. Growth of San Diego actually begins with the twentieth century" (Mirkowich 1941, p. 309) and the festivities of the Panama-California Exposition in 1915 in Balboa Park, which constitutes another major development driver in the city's history.

After rather insignificant growth in the last decade of the 19th century, the population of San Diego doubled each decade between 1900 and 1930. While the city's population was 16,169 in 1890 and only a few hundred more in 1900 (17,700), the numbers grew rapidly throughout the World War I period and until the Great Depression—to 39,578 in 1910, to 74,361 in 1920, and to 147,995 in 1930 (see Figure 4.2). On the one hand, the population growth is tied to the magnetic effect of the internationally acclaimed Panama-California Exposition and the continued success of local boosters and (suburban) land developers. On the other hand, economic and military developments—on land, water, and along the waterfront—have played a large role in the region's growth. During the first decades of the 20th century, the waterfront along the established port and the Barrio Logan area has faced major industrialization with the upcoming tuna boom. Numerous immigrants from Asia and Central and Southern American as well as European countries arrived in San Diego to work in the tuna canneries and packing houses—a development that began to shift the composition of the local population sustainably. In 1930, almost 15 percent of the local residents have immigrated from other countries (Mirkowich 1941, pp. 309–310). Furthermore, in 1919, the naval base has been established on almost 100 acres of bayfront area in Barrio Logan due to its unique and advantageous geographical location (Norris 1983; Pourade 1960–1967). Over the time, particularly in the course of World War II and the increased military operations in the Pacific, the facility grew drastically—today, the naval base occupies more than 1,600 land acres and 326 acres of water—and so did the economic impact of the local defense industry. In summary, the Army, Navy, and Marine activities in San Diego did not only act as a migration magnet but constituted also a significant contributor to the local economy—throughout World War II and until the present day (CofSD 2016, p. 157). "What had been a military presence became military domination in one of the nation's most important strategic centers" (Killory 1993, n.pag.).

The strong population growth during and after the war years in San Diego, however, posed serious challenges: "San Diego's growth as a military center far outpaced its ability to provide services and sufficient housing for the many thousands of war industry workers streaming in from every state in the union" (Killory 1993, n.pag.). Because, as Starr (2009, p. 60) describes vividly, "it was one thing to be a Navy town in peacetime, with an economy sustained by federal dollars. It was quite another matter entirely to be responsible for the creation of a civic infrastructure supporting a military mega-center." Shortly before the outbreak of the war, in 1930, the city was populated by 147,995 residents. By 1950, however, this number had increased by more than 100 percent to 333,865 residents. This spike, combined with rising incomes and mass-availability of automobiles, changed the growth pattern of the region drastically and with long-lasting effects. Within the urge for efficiency and quantity in terms of housing development, San Diego experienced a mass production of suburban housing in the postwar era and throughout the second half of the century, which resembled Fordist production approaches. This was not only fueled by private and military housing development efforts but also by "public policies that favored new home building over rehabilitation and highway construction over mass transit" (Herzog 2019, p. 2). Furthermore, the upcoming modern planning methodology of single-use zoning separated uses and created disjunct zones of single-family neighborhoods, office parks, and retail and commerce districts in the sprawling suburban landscape of San Diego County. Thus, the private automobile has been reinforced even more as the region's main and preferred mode of transportation in the postwar years. Accordingly, the dismantlement and the disappearance of the streetcars in 1949 "was greeted with as much pride as their debut" (Mills 1985, n.pag.) since it represented an increase in individual freedom (of mobility).

In terms of the settlement structure, the continued mass production of single-family housing at the urban fringes and the cherished dominance of the automobile changed the socio-spatial pattern and the morphology of the region drastically. Between the early 20th century and the 1970 s, the City of San Diego accounted for more than half of the population of the county. In the 1930 s, more than 70 percent of the county's population has lived within the boundary of the jurisdiction of San Diego. Equivalently, until 1940, the percentual growth of the city has been higher compared to the growth of the county—except for the time of San Diego's first bust phase in the late 1880 s, which led to a significant population exodus (see Figure 5.1). After the 1930 s, however, the city's percentual share of the county's population has decreased gradually, from 70,6 percent in 1930 to 42 percent in 2020 due to the strong growth of the county's suburban communities throughout the decades (Griffin and Weeks 2014, p. 63). The widespread white flight phenomenon has strongly shaped the urban development in San Diego throughout the 20th century until today. Particularly since the postwar years, the city's traditional shape of a single center-oriented settlement around an urbanized central business district gave way for a sprawling network of "scattered commercial nuclei, deliberately encouraged by the construction of vast new employment centers such as the linear Mission Valley complex, the 'Golden Triangle' in University City, and the newer Carmel Valley" (Ford 2014, p. 187). In this course, the relevance of downtown San Diego as the region's primary business and value location faded significantly. However, in the 1970 s and 80 s, city officials and private developers were eager to counteract this trend by means of the Horton Plaza Mall redevelopment, which aimed for the economic, planning, and social revaluation of the downtown area, as discussed in detail in section 5.4.1.

Table 4.1 Historical Milestones of European Settlement in San Diego

Date	Event
1542	Spanish explorer Juan Cabrillo discovers San Miguel Bay (renamed San Diego Bay in 1602 by Sebastian Vizcaíno)
1769	Establishment of Presidio of San Diego and Mission San Diego de Alcalá by Junípero Serra in today's Old Town (first permanent European settlement on the United States West Coast)
1777	Arrival of the first larger group of non-soldier settlers at the Presidio of San Diego
1784	Completion of the Mission San Diego de Alcalá (after attack in 1775) at new site further up the San Diego River

(continued)

Table 4.1 (continued)

Date	Event
1816	Completion of Mission (Padre) Dam, the first major irrigation project on the West Coast
1821	San Diego (home to around 600 residents) falls to the newly independent Mexico
1833	Mexican secularization act: secularization of the 21 European missions in California and privatization of the land through 33 land grants
1834	San Diego established as a pueblo
1838	Decrease of the population (to around 100–150) and revocation of the pueblo status
1848	Treaty of Guadalupe Hidalgo: End of the Mexican-American War (1846–1848), California falls to the United States; Advent of the California Gold Rush after James W. Marshall's discovery of gold at Sutter's Mill in Northern California
1850	William H. Davis buys 160 acres in today's downtown San Diego for $2,304; Incorporation of the City of San Diego; Compromise of 1850: California joins the Union as a free state, creation of San Diego County
1857	Establishment of the first overland mail route to San Diego
1867	Alonzo Horton arrives in San Diego and establishes New Town (buys 800 acres for $265)
1868	Ephraim Morse, Thomas Bush, and Alonzo Horton set aside 1,440 acres for today's Balboa Park, originally named City Park until 1910
1870	Horton set aside half a block of New Town for today's Horton Plaza (sold to the city in 1894 under the promise to keep it a public space forever)
1872	Fire in Old Town destroys several business buildings
1885	First transcontinental train reaches San Diego, connecting San Diego to the East
1886	Foundation of the San Diego Street Car Company and establishment of San Diego's first public transport service run by horsecars; Establishment of La Jolla
1887	Population peak and real estate boom: 40,000 inhabitants in the City of San Diego (first 'boom period' ends in 1889; population decreased to 16,000); First electric street railway system of the USA founded in San Diego by the Electric Rapid Transit Company; Villa Montezuma, the Victorian, Queen Anna style home of Jesse Shepard is built (see **Figure 5.27**); Incorporation of National City (second city in the county)
1888	Completion of Cuyamaca Dam; Hotel del Coronado opens; Incorporation of Escondido and Oceanside
1891	Splitting off and Incorporation of Coronado

(continued)

Table 4.1 (continued)

Date	Event
1892	The San Diego Cable Railway takes over the electric line and operates cable cars until 1892, when John D. Spreckels bought the city transit system
1897	Foundation of State Normal School for teacher training (integration in the California State College system in 1960, renamed San Diego State University in 1971)
1900	Formation of the Marine Biological Association of San Diego (later Scripps Institution of Oceanography)
1905	The dry Salton Basin (formerly containing Lake Cahuilla) fills with water from the Colorado River after a dike failure, creating Salton Lake
1906	Spreckel's San Diego & Arizona Eastern Railroad Company begins construction for a railway line to Yuma
1911	Incorporation of Chula Vista
1912	Incorporation of El Cajon and La Mesa
1915	Opening of the Panama-California Exposition in Balboa Park, which was designed in Spanish Colonial style; Opening of Balboa Stadium next to San Diego High School
1917	Banker Louis J. Wilde wins the prominent 'Smokestacks vs. Geraniums' mayoral campaign against 'Geranium-Candidate' George W. Marston; Military buildup in response to World War I
1919	Tijuana tourism booms in the wake of the prohibition; Decision to settle the Navy's Pacific Fleet in San Diego; Completion of Spreckel's San Diego & Arizona Railroad (abandoned in 1976 after severe storm damage)
1925	Establishment of regular airline service between Los Angeles and San Diego (first regular airline service in the US)
1927	El Cortez Hotel opens in Downtown San Diego
1936	The baseball team Padres (formerly known as Hollywood Stars) comes to San Diego (Lane Field) from Los Angeles
1940	Immense military buildup in response to World War II
1945	A recession sets in after the end of World War II
1947	Opening of the San Diego Aqueduct, transporting water from the Colorado River into the city
1952	Foundation of California Western University in Point Loma
1954	Foundation of University of San Diego in Linda Vista
1956	Incorporation of Imperial Beach
1958	Opening of Interstate Highway 8 through Mission Valley (Mission Valley shopping center opens in 1961)

(continued)

Table 4.1 (continued)

Date	Event
1959	Incorporation of Del Mar
1963	Incorporation of Vista and San Marcos
1964	Opening of the 1,000-acre UCSD La Jolla campus
1965	Mexican authorization of maquiladora factories operated by non-Mexican companies
1967	Opening of San Diego Stadium in Mission Valley (renamed Qualcomm in 1997)
1969	Opening of Coronado Bay Bridge and end of ferry service
1970	Population of the City of San Diego reaches 696,474, making San Diego the second largest city of California; Foundation of Chicano Park beneath Coronado Bay Bridge
1972	Mayor Pete Wilson (serving three terms, from 1971–1982) declares San Diego as 'America's Finest City' after cancelation of the plans for the Republican National Convention in San Diego
1975	Resolution for downtown redevelopment by Mayor Pete Wilson and creation of Centre City Development Corporation (CCDC)
1977	Incorporation of Lemon Grove
1980	Incorporation of Poway and Santee; Dedication of the San Diego Trolley, the city's new light-rail transit system
1981	Ground-breaking for the construction of Horton Plaza Mall in the course of downtown redevelopment (opening in 1985)
1985	Opening of Otay Mesa border crossing
1986	Incorporation of Encinitas and Solana Beach
1989	Opening of San Diego's Convention Center at the Waterfront (expanded in 2001)
1990	Opening of California State University, San Marcos
2003	Large wildfires (Cedar, Otay, and Paradise) destroy a cumulated area of 345,000 acres and 2,300 structures
2004	Opening of Petco Park ballpark in East Village
2016	Rejection of the plans to build a new football stadium in East Village by the San Diego voters

Source: Own Representation based on Amero (1990), Pryde (2014, pp. 10–11), and San Diego History Center (2020a, 2020b, 2020c, 2020d, 2020e, 2020f).

References

Amero, R. W. (1990). The Making of the Panama-California Exposition, 1909–1915. *The Journal of San Diego History, 36* (1), p. n.pag.

Baker, G. (2007). *San Diego. Another HarborTown History*. Santa Barbara: Harbortown Histories.

CofSD, P. D. (2016). North Park Community Plan. City of San Diego Planning Department. https://www.sandiego.gov/sites/default/files/amended_2016_north_park_community_plan_web.pdf. Accessed: January 11, 2022.

Ford, L. (2014). The Visions of the Builders. The Historical Evolution of the San Diego Cityscape. In P. R. Pryde (Ed.), *San Diego. An Introduction to the Region* (pp. 175–190). San Diego: Sunbelt Publications.

Griffin, E. & Weeks, J. R. (2014). Peopling the Region. San Diego's Population Patterns. In P. R. Pryde (Ed.), *San Diego. An Introduction to the Region* (pp. 63–76). San Diego: Sunbelt Publications.

Herzog, L. A. (2019). Suburbia/Postsuburbia. In A. M. Orum (Ed.), *The Wiley Blackwell Encyclopedia of Urban and Regional Studies* (pp. 1–7). Chichester: Wiley Blackwell.

Johnson, C. & Jones, R. (2016 [2014]). Where is the Border? In R. Jones & C. Johnson (Eds.), *Placing the Border in Everyday Life* (pp. 1–11). London: Routledge.

Killory, C. (1993). Temporary Suburbs: The Lost Opportunity of San Diego's National Defense Housing Projects. *The Journal of San Diego History, 39* (1 & 2), p. n.pag.

Kumeyaay. (2022). The Indians of San Diego County. Kumeeyay.com. https://www.kumeyaay.com/the-indians-of-san-diego-county.html#:~:text=The%20tribal%20groupings%20make%20up,Yuman%20language%20family%2C%20Hokan%20stock. Accessed: 7. July 2022.

Mills, J. R. (1985). *San Diego. Where California Began* (Vol. 5). San Diego: San Diego History Center.

Mirkowich, N. (1941). Urban Growth in the San Diego Region. *Economic Geography, 17* (3), pp. 308–310.

Norris, F. (1983). Logan Heights: Growth and Change in the Old East End. *The Journal of San Diego History, 29* (1), p. n.pag.

Pourade, R. F. (1960–1967). *The History of San Diego.* (6 vols.). San Diego: Union-Tribune Publishing Company.

Pryde, P. R. (2014). Introduction: The Uniqueness of the Region. In P. R. Pryde (Ed.), *San Diego. An Introduction of the Region* (pp. 1–14). San Diego: Sunbelt Publishing.

San Diego History Center. (2020a). Timeline of San Diego History: 20,000 BC to 1789. https://sandiegohistory.org/archives/biographysubject/timeline/bc-1798/. Accessed: 21. December 2020.

San Diego History Center. (2020b). Timeline of San Diego History: 1800–1879. https://sandiegohistory.org/archives/biographysubject/timeline/1800-1879/. Accessed: 21. December 2020.

San Diego History Center. (2020c). Timeline of San Diego History: 1880–1899. https://sandiegohistory.org/archives/biographysubject/timeline/1880-1899/. Accessed: 21. December 2020.

San Diego History Center. (2020d). Timeline of San Diego History: 1900–1929. https://sandiegohistory.org/archives/biographysubject/timeline/1900-1929/. Accessed: 21. December 2020.

San Diego History Center. (2020e). Timeline of San Diego History: 1930–1959. https://sandiegohistory.org/archives/biographysubject/timeline/1930-1959/. Accessed: 21. December 2020.

San Diego History Center. (2020f). Timeline of San Diego History: 1960–1999. https://sandiegohistory.org/archives/biographysubject/timeline/1960-1999/. Accessed: 21. December 2020.

San Diego State University. (2022). Land Acknowledgment. SDSU. https://sacd.sdsu.edu/diversity-resources/_pdfs/land-acknowledgement.pdf. Accessed: 7. July 2022.

Sandul, P. J. P. (2014). *Calfiornia Dreaming. Boosterism, Memory, and Rural Suburbs in the Golden State*. Morgantown: West Virginia University Press.

Starr, K. (2009). *Golden Dreams: California in an Age of Abundance, 1950–1963*. Oxford: Oxford University Press.

Trafzer, C. E. (2014). American Indians. The County's First Residents. In P. R. Pryde (Ed.), *San Diego. An Introdcution to the Region* (5th ed., pp. 47–61). San Diego: Sunbelt Publications.

University of San Diego. (2022). Chronology of the Indigenous Peoples of San Diego County. University of San Diego. https://www.sandiego.edu/native-american/chronology.php. Accessed: 13. July 2022.

5 When San Diegans are "Getting a Little Loose with their Naming Convention:" Illuminating Subjective Everyday Experiences and Interpretations of Neighborhood Change

After the outline of the historical milestones and guiding developments of the City of San Diego, the empirical results of the present study will be presented in the course of the following chapter. Within the aim of illuminating subjective everyday experiences and interpretations of neighborhood change, the chapter follows the spatial pattern of the elaborated 'redevelopment wave,' which is moving gradually through the various districts of central San Diego (see Figure 5.1). This trend has commenced in downtown and Little Italy and gained momentum in Uptown San Diego (section 5.1), spread into North Park and from there eastward into the Mid-City community (section 5.2) as well as southward into the Greater Golden Hill area (section 5.3), and penetrated from the former downtown warehouse district of East Village into adjacent neighborhoods, such as Barrio Logan and Southeastern San Diego (section 5.4), and now even further south and east into the next ring. Generally, the individual subsections follow the structure of planning policy boundaries around Balboa Park, which form different community planning areas and associated neighborhoods. Constituting planning policy associations of adjacent neighborhoods, the level of community planning areas is where official guidelines on the basis of public recommendations are prepared (in the form of community plans) in accordance with the city's General Plan and the City of Villages smart growth strategy to determine community goals and plan future land use. On the one hand, the breakdown of the chapter into community planning areas and neighborhoods allows to work out differences and

A. Rossmeier, *San Diego's Hybrid Urban Borderlands*,
Räume – Grenzen – Hybriditäten, https://doi.org/10.1007/978-3-658-42667-5_5

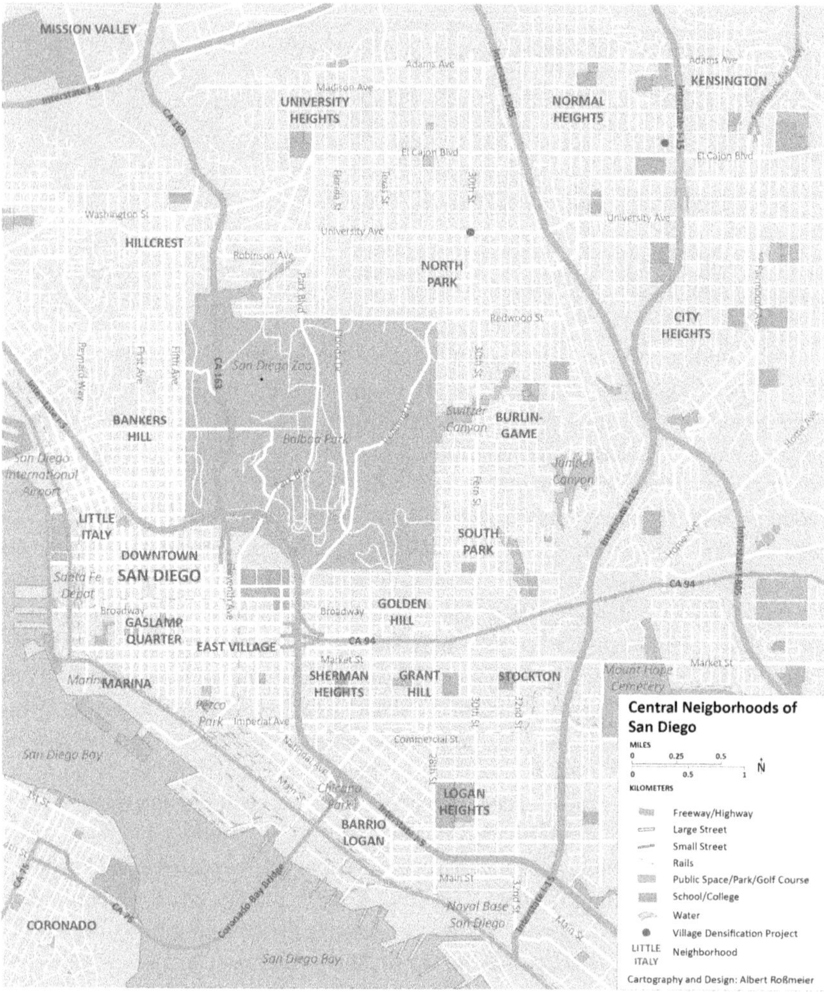

Figure 5.1 Study Area in the City of San Diego: The Central Neighborhoods Around Balboa Park. (Source: Own cartography and design)

specifics of the various planning areas. On the other hand, light will be shed on processes of social and physical hybridizations and multi-layered b/ordering processes in and between each neighborhood or community. However, only the neighborhoods that were addressed by the interviewees and lie within the area of central San Diego will be covered in the following. On top of that, areas and certain aspects or developments that were discussed more frequently or in more detail in the interviews are given special consideration. Such particularly addressed spaces are, for instance, the border areas of Greater North Park and the neighborhoods of Normal Heights/City Heights northeast of Balboa Park as well as of East Village and Barrio Logan on the southeastern downtown edge. In the last subsection (5.5), these border areas will serve as detailed examples for the synthesis of empirical insights and the underlying theoretical approach and thus the contextual application and derivation of the concept of hybrid urban borderlands (Figure 5.1).

5.1 Hybrid Influences in the Neighborhoods of Uptown San Diego: Between Historic Preservation and the Ongoing Development Trend

The following section covers significant neighborhoods and developments of Uptown San Diego (see Figure 5.2). The Uptown planning area is officially divided into the neighborhoods of Mission Hills, Bankers Hill (section 5.1.1), Hillcrest (section 5.1.2), the Medical Complex, and the Western parts of University Heights. However, in contrast to the communities of Bankers Hills and Hillcrest, the Medical Complex as well as Middletown have not been mentioned by the interviewees and are thus unsuitable for an in-depth analysis. The neighborhoods of Mission Hills and University Heights have been addressed briefly, however not in sufficient depth. After the detailed discussion of the Uptown neighborhoods, an urban landscape and border-theoretical summary and interpretation (section 5.1.3) will conclude this chapter.

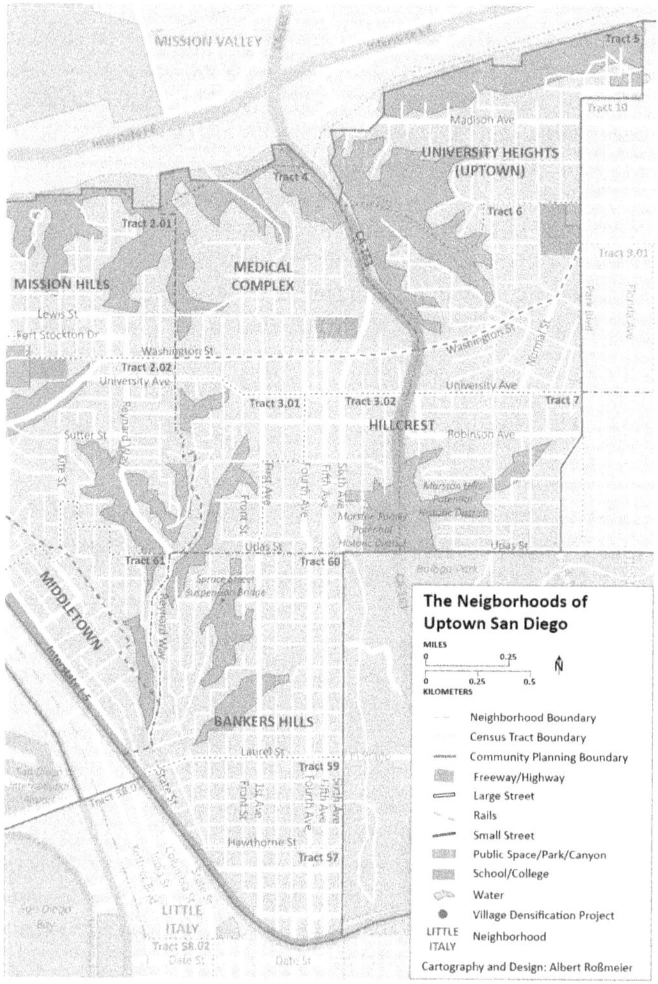

Figure 5.2 The Neighborhoods of Uptown San Diego. (The map shows the neighborhoods of the Uptown community planning area, their official boundaries, and the 2020 census tract subdivision. Source: Own cartography and design)

5.1.1 Delayed Development and Uneven Urbanization of the Historic Neighborhood of Bankers Hill

The neighborhood of Bankers Hill, which was developed in the late 19th century under the name of Florence Heights (after the Florence Hotel, built in 1884 by Alonzo Horton's brother-in-law) and later known as Park West, runs along the West side of Balboa Park, and lies between Interstate 5 and downtown San Diego in the south and the Uptown Neighborhoods of Middletown, Mission Hills, and Hillcrest in the north and northeast (see Figure 5.2). The historic neighborhood of Bankers Hill is the only district of the present study that has official boundaries that coincide with census tract boundaries. This makes the area particularly suitable for a detailed consideration of quantitative neighborhood change on the level of census tracts (see Table 5.1) as supplementation for the gained qualitative insights. Although all the different neighborhoods of Uptown San Diego have experienced an influx of new residents, new housing construction, and infrastructural diversification in the last decades, different dynamics and degrees can be identified, even within official neighborhood boundaries, as it will be elaborated in the following. By means of 2000 and 2010 census data, differences between the southern (census tract 57, see Figure 5.2), central (tract 59), and northern (tract 60) section of Bankers Hill will be worked out. First, the southern part will be discussed.

While the percentual increase of persons with college and higher degrees in the neighborhood of Bankers Hills (28 percent) is slightly higher compared to the entire Uptown planning area (24 percent) as well as to the City of San Diego (26 percent), the southern tract of the neighborhood records the highest increase (65 percent). However, compared to all the tracts of the Uptown planning area, the southern part of Bankers Hill just north of Little Italy has throughout the 1990s, 2000s, and 2010s been the portion of the neighborhood with the lowest rates of college and higher education (from 25 percent of the population in 2000 to 36 percent in 2010 (see Table 5.1), as revealed by the interactive map tool 'Urban Displacement Project' (Chapple and Thomas 2021). Correspondingly, the percentage of population with educational attainment of less than high school diploma has throughout the decades been higher in the southern tract compared to the other two census tracts of Bankers Hill (Chapple and Thomas 2021). In addition to this, the southern portion of Bankers Hill had a disproportionate growth in persons living below the poverty line: While the share of population below poverty status grew by 25 percent (to 402 persons) in tract 59 and by 41 percent (to 433 persons) in the tract adjacent to Hillcrest (tract 60), the numbers in tract

57 increased from 265 persons (18 percent) in 2000 to 853 persons (44 percent) in 2010 and thus by 222 percent.

During this period, also the total population changed disproportionately in the area south of Hawthorne Street. Between 2000 and 2010, the area recorded a population increase of 17 percent, which is higher than the development in the other two tracts (both 9 percent), the combined Uptown area (5 percent), and the City of San Diego (6 percent). The differences in the change of median age are striking as well: The entire neighborhood of Bankers Hill shows a higher median age (from 38.1 years in 2000 to 41.3 years in 2010) than the Uptown community (which developed from 38.6 in 2000 to 40 in 2010) and is particularly higher than the median age of the City of San Diego (which was 32.6 years in 2000 and changed to 33.8 in 2010). The southern portion of Bankers Hill shows a remarkable change here as well. The median age developed from 33.6 years in 2000 to 42.5 years in 2010, which describes a change of 8.9 years in that decade. Summing up, the data portrays the southern section of Bankers Hill as an aging but growing part of the neighborhood with a high share of population below poverty status and low levels of educational attainment, whereas the area north of Hawthorne Street up until the boundary to the trendy and "strongly developing" (EP-01) neighborhood of Hillcrest seems to comprise of a more educated and economically better situated population, which is, on the one hand, slightly growing but, on the other hand, also aging (median age in tract 60 in 2010 is 45.7 years). Thus, the neighborhood of Bankers Hill seems to comprise of several areas that underly different dynamics, as it will be further elaborated below.

In addition to the compiled census data, the Uptown community plan provides further information on the existing land-use in the different areas and neighborhoods (CofSD 2019, pp. LU24–LU27). The plan indicates that the southern tract of Bankers Hill comprises predominantly of office use and multi-family housing. However, 6th Avenue along Balboa Park provides denser housing (in the 45–73 du/ac range[1]) than the southwestern parts around Interstate I-5 (which lies in the 30–44 du/ac range). The central tract of Bankers Hill between Hawthorne Street and Laurel Street is structured accordingly but with higher densities of commercial uses. In general, the highest density in housing and commercial uses in Bankers Hill is found along 6th Avenue. Therefore, 4th, 5th, and 6th Avenue are labeled mixed-use corridors in Bankers Hill with the section around Laurel Street—which constitutes an important entrance to Balboa Park via Cabrillo

[1] For zoning and land-use purposes in California, density is usually indicated and calculated in dwelling units per acre (du/ac). One acre equals approximately 4,047 square meters.

Bridge—designated as neighborhood village (CofSD 2019, p. LU37). In this area, the community plan set the goal to incentivize development according to the 'City of Villages' growth strategy. Generally, "villages within Uptown have established themselves as community and neighborhood-oriented areas with local commercial, office, and multifamily residential uses, including some structures with office or residential space above commercial space. Within these neighborhood and community village areas, horizontal mixed-use developments along with stand-alone office and village supportive commercial uses contribute to the overall 'village characteristics' already exemplified in Uptown" (CofSD 2019, p. LU38).

This character is also described by an interviewee of the resident category who compares the developments of the historically residential neighborhood of Bankers Hill to the dynamics in Kensington and Golden Hill (see Figure 5.2) and points out the recent influx of younger generations into selected areas of central San Diego:

> "I've noticed it seems like there's an increase in, like, little pocket neighborhoods that are getting upgraded where young professionals are starting to not want to commute all the time, and they're starting to, you know, 25-, 30-year-olds start to have a career and have a little bit of money, they're choosing to go to these, I don't know what the cause of... I don't know whose idea it was or whatever, but places like Golden Hill, and Bankers Hill, and Kensington, they've had a resurgence of, like, young professionals that work in them, and now there's little restaurants popping up and stuff" (R-13).

In addition to this statement, an interviewee from the academic field compares the developments in Bankers Hill and Golden Hill as well and states that "they are closer to Balboa Park, which arouses interest" (AR-02).

Accordingly, in the central tract of Bankers Hill, where the community plan is intending to create a walkable village-style neighborhood center, the median property value has been growing the most compared to the rest of the neighborhood, the entire Uptown area, and the City of San Diego: While the median property value grew by 19 percent in the southern tract, by 58 percent in the northern section, by 85 percent in the entire Uptown community, and by 120 percent in the City of San Diego, the central tract of Bankers Hill records an increase of 138 percent. Along with this stark change went a disproportionate increase of median household income, median rent, as well as of the total number of housing units. Even though the real numbers are partly higher in the northern section of Bankers Hill, the change rates locate the most significant developments in the central tract of the neighborhood. Here, the median household income increased by 70 percent, the median rent by 63 percent (only the rate for the City of San Diego is higher with 68 percent), and the total number of housing units by 26 percent in the time from 2000 to 2010 (see Table 5.1).

Lastly, the northern part of Bankers Hill will be discussed below in order to underpin the following qualitative insights and to provide additional material for understanding the different realities of Bankers Hill. Unlike the southern and central portion of the neighborhood, the northeastern part—which borders the residential area of Mission Hills and is structured by unique canyon topography (Arroyo and Maple Canyon)—comprises mostly of detached single-family homes labeled as 'spaced rural residential' in the land-use map provided by SANDAG and SanGIS (2018)[2] and as 'low-density residential' in the 5–9 du/ac range in the recent Uptown community plan (CofSD 2019, pp. LU24–LU27). In 2010, this area recorded the highest median property value ($686,154) compared to the other tracts of Bankers Hill (tract 59 with $498,913 and tract 57 with $392,647), the combined Uptown community area ($659,085), as well as the City of San Diego ($506,107). In contrast to the suburban structured northwestern section, the eastern part of tract 60 along Balboa Park shows similarities to the central tract of Bankers Hill: The area constitutes of a mix of high-density office and commercial uses with multi-family housing in larger mixed-use building complexes and thus defines a transition zone between the 'urban village'-style areas of central Bankers Hill and Hillcrest. Eventually, what emerges from this is a closely spaced interplay of rural or suburban elements in the form of wooded canyons with backyard trails and single-family homes—in parts historic—with urban influences of higher density along the prominent Avenues, which connect Bankers Hill to downtown in the south and to Hillcrest in the north (see Figure 5.3).

In a next step, it becomes possible to link the quantitatively elaborated perspective on the grown connection of Hillcrest and the northern parts of Bankers Hill to the portrayals of certain interviewees that describe the development of Bankers Hill as being triggered by an early gentrification of the adjacent neighborhood of Hillcrest. A young interviewee from the interview category academia/research, who lives in North Park and grew up in Southeastern San Diego, emphasizes that "it was Hillcrest first. Hillcrest started earlier, like, back in the'90s, and then it would come down and it continued to, like, University Heights, Normal Heights, all that, Little Italy, downtown" (AR-07). Accordingly, also interviewee

[2] SanGIS (San Diego Geographic Information Source) " is a Joint Powers Authority (JPA) of the City of San Diego and the County of San Diego responsible for maintaining a regional geographic information system (GIS) landbase and data warehouse" (SanGIS 2012).

Table 5.1 Quantitative Neighborhood Change between 2000 and 2010 in Bankers Hill compared to Uptown San Diego and the City of San Diego

			Total Population	Median Age	Population 25 Years and above with College Degree	Population Below Poverty Status	Housing Units	Median Rent (in $)	Median Household Income (in $)	Median Property Value (in $)
Neighborhood of Bankers Hill	**Tract 60**	**2000**	3,251	44.4	1,397 (43%)	308 (10%)	1,950	$738	$43,027	$435,295
		2010	3,536	45.7	1,643 (46%)	433 (12%)	2,246	$1,095	$57,188	$686,154
		Change	**+9%**	**+ 1.2 years**	**+18%**	**+41%**	**+15%**	**+48%**	**+33%**	**+58%**
	Tract 59	**2000**	2,570	36.2	1,011 (39%)	322 (13%)	1,684	$591	$29,504	$209,650
		2010	2,821	35.8	1,268 (45%)	402 (14%)	2,123	$964	$50,243	$498,913
		Change	**+9%**	**−0.4 years**	**+25%**	**+25%**	**+26%**	**+63%**	**+70%**	**+138%**
	Tract 57	**2000**	1,668	33.6	423 (25%)	265 (18%)	1,095	$564	$22,252	$330,000
		2010	1,948	42.5	699 (36%)	853 (44%)	1,152	$870	$36,568	$392,647
		Change	**+17%**	**8.9 years**	**+65%**	**+222%**	**+5%**	**+54%**	**+64%**	**+19%**
	Combined	**2000**	7,489	38.1	2,831 (38%)	895 (12%)	4,729	$631	$31,594	$324,982

(continued)

Table 5.1 (continued)

		Total Population	Median Age	Population 25 Years and above with College Degree	Population Below Poverty Status	Housing Units	Median Rent (in $)	Median Household Income (in $)	Median Property Value (in $)
	2010	8,305	41.3	3,610 (43%)	1688 (20%)	5,521	$976	$48,000	$525,905
	Change	**+11%**	**3.2 years**	**+28%**	**+89%**	**+17%**	**+55%**	**+52%**	**+62%**
Uptown Community Planning Area	**2000**	35,772	38.6	14.642 (50%)	3,960 (11%)	21,661	$674	$39,328	$355,309
	2010	37,612	40.0	18.157 (57%)	4,550 (12%)	22,998	$1,043	$55,887	$659,085
	Change	**+5%**	**1.4 years**	**+24%**	**+15%**	**+6%**	**+55%**	**+42%**	**+85%**
City of San Diego	**2000**	1,223,400	32.6	305,953 (39%)	173,519 (15%)	469,689	$714	$45,825	$230,076
	2010	1,301,617	33.8	385,460 (45%)	177,838 (14%)	515,412	$1,202	$63,198	$506,107
	Change	**+6%**	**1.2 years**	**+26%**	**+2%**	**+10%**	**+68%**	**+51%**	**+120%**

Source: Own composition and calculation based on compiled census data by SANDAG (2003k, 2003l, 2003m, 2003o, 2003r, 2016e, 2016f, 2016p, 2016q, 2016r). The subdivision of the neighborhood of Bankers Hill into its three census tracts is shown in Figure 5.5.

R-13 shares: "I think Hillcrest and North Park both, I think both of those kinda led the way." Addressing the consequences of the development trends in Hillcrest, a developer from East Village explains that also "the Bankers Hill area has developed with nice restaurants, there has been quite a lot going on over the last five or six years" (EP-07). Similarly, a developer from Barrio Logan states: Bankers Hill "is being developed heavily" (EP-04). In the same vein, a business owner from Hillcrest illustrates the rapid development of Bankers Hill, downtown, and Hillcrest as follows: "Everything is growing. The area is changing fast and there's also a lot of development going on in Bankers Hill, like, rents are extreme" (EP-01). The pictured development direction from north to south, from Hillcrest to Bankers Hill, is also suggested by census data, which shows that the increase of median rent in Bankers Hill was most significant in the areas just south of Hillcrest in the 1990s and that the central tract of the neighborhood has only been affected in the following years, while the southern area still records the lowest total median rent (Chapple and Thomas 2021).

But not only the neighborhoods on the west side of Balboa Park are affected, as another interviewee states: "You see new development in Uptown, North Park, and Golden Hill but also new development near downtown" (AR-06). What is changing, however, is "where the money is coming from" (AR-06). The interviewee claims that the development in downtown is mostly financed by foreign developers, while the construction projects in Uptown are sourced rather locally. In this sense, interviewee EP-06, a resident from Bankers Hill, describes local efforts to design and erect a neighborhood sign on 5th Avenue where new development is on its way. This effort plays into the idea of creating visibility and neighborhood identity, which seems—in the case of EP-06—to be connected to economic interests. "I want Bankers Hill to be the Manhattan of the West Coast. Balboa Park is even bigger than Central Park" (EP-06), as the interviewee claims. Hillcrest, Bankers Hill, and Uptown in general "have never been disinvested like the [the areas] east" (AR-02) of Balboa Park, nevertheless the neighborhoods "are changing, more houses are being replaced by small condo units, and more high-rise towers" (AR-02) are going up. A resident from Ocean Beach (R-12) comments the changes in the neighborhood equivalently and deplores that Bankers Hill has lost most of its numerous Victorian-style homes, which were still prevalent in the 1960s and '70s (see Figure 5.3).

A different perspective is provided by R-13, who describes Bankers Hill as a formerly "run-down neighborhood with old houses that were all dusty for landscaping and, you know, it was just a kind of stale neighborhood 20 years ago." But, according to the interviewee, due to "a trend in California real estate to buy a house, and then upgrade it, and then sell it immediately, [...] Bankers Hill has

Figure 5.3 Diverse Building Structure in Bankers Hill—Hybrid Composition of Preserved Historic Architecture, Repurposed Single-Family Homes, and Large Condominium Buildings near Balboa Park. (Top: Looking south on the intersection of Nutmeg Street and Fourth Avenue. Bottom: The historic Victorian-style Britt-Scripps Villa next to a larger apartment building and a mixed-use building at the intersection of Maple Street and Fourth Avenue. Source: Pictures Albert Roßmeier 2019)

a lot of residences, a lot of single-family homes that have been targets for these people that flip houses." Similarly, interviewee R-11, a resident from East Village, who came to San Diego in the 1990s, describes Bankers Hill as a formerly disinvested area, which developed into a livable urban neighborhood. However, the biggest downside of the neighborhood, according to the interviewee (R-11), is the frequent airplane noise south of Laurel Street due to its proximity and location in the flight path of San Diego's international airport 'Lindbergh Field' (cf. CofSD 2019, p. LU23). The airplane noise and the imposed height limit in the flight path, as argued by interviewee EP-12, seem to be additional reasons for the delayed development and urbanization of the southern part of Bankers Hill (see Figure 5.4). However, according to interviewee R-12, the noise is increasingly tolerated; the positive reputation and the neighborhoods proximity to downtown begin to outweigh it. "I'm sure some people don't like it, but everybody that I've known that is there, kind of, even doesn't notice or, kind of, likes it. And I... Even the people I would rent my condo to... There was actually a film that would collect on the porch from the place, like, specs of exhaust or something, it's crazy. But I mean look at Bankers Hill, it's packed, and it's nice. I mean, obviously somebody likes it there."

Figure 5.4 Airplanes Approaching San Diego's International Airport in the Flight Path over the Neighborhood of Bankers Hill. (Source: Pictures Albert Roßmeier 2019)

In addition to the presented census data and various statements of interviewees from different perspective groups, subjective impressions and experiences from the conducted participant observation (PO-04) in the neighborhoods of Little Italy, Bankers Hill, and Hillcrest will be presented in the following (see

Figure 5.1). Against the shared insights of interviewee R-11, Bankers Hill did not appear to be a 'packed neighborhood' with bustling street life and a large number of shopping and gastronomic opportunities. Pedestrians appeared rarely in the streets, the neighborhood seemed not to attract tourists of any kind, and homelessness was less prevalent compared to Little Italy and especially Hillcrest. However, certain differences and development tendencies became apparent in the different sections of the neighborhood, which allow for conclusions on the everyday life relevance of (neighborhood) borders and boundaries in and around Bankers Hill.

Starting the neighborhood walk in the touristic district of Little Italy meant crossing Interstate 5 into the southern part of Bankers Hill, which immediately revealed a stark contrast between the two neighborhoods. While Little Italy accommodates many touristic and gastronomic offers as well as retail with storefronts to walk by, the southern and southeastern part of Bankers Hill comprises mostly of large office and commercial buildings and smaller apartment complexes. Accordingly, interviewee R-06, who lives in the official border area of Hillcrest and North Park, describes Bankers Hill as a rather uninteresting area with little development and construction. In the same vein, interviewee R-10 perceives the neighborhood as predominantly residential and as quiet home of "downtown families." Furthermore, in terms of opportunities for exploration and discovery, the walkability decreases abruptly when crossing Interstate 5 from south to north, which means that the freeway not only describes a hard physical-spatial border but also a distinct functional boundary. In line with Haselsberger (2014), it can be concluded that Interstate 5 describes a set of various, overlapping boundaries (functional, political, social, and symbolic) in the border area of Little Italy and Bankers Hill, which consolidate into a thick and unambiguous physical border that severs the communities of Uptown San Diego from downtown.

In contrast to this, north of Laurel Street and Maple Canyon, the neighborhood seems to underlie different influences. Here, remaining historic Victorian-style building stock is located next to newer apartment, office, and mixed-use building complexes with high densities and interspersed repurposed single-family homes, which creates a hybrid mix of urban and rural/suburban elements (see Figure 5.3). As indicated by the land-use maps in the Uptown community plan (CofSD 2019, pp. LU24–LU27), the western parts around the canyon topography seem to comprise mostly of single-family housing, while the eastern parts closer to Balboa Park accommodate larger structures—"all shiny new buildings," as described by interviewee EP-12. On top of that, it became apparent during the neighborhood walk that not only new housing opportunities are constructed in Bankers Hill but sporadically also trendy, new restaurants, breweries, and cafés are being established in the northern and northwestern parts of the neighborhood, which draw on cosmopolitan, social media-influenced design concepts and aesthetics addressed to younger generations.

Thus, the developing Avenues along Balboa Park connect Bankers Hill to the village-style neighborhood of Hillcrest, on the one hand, functionally and symbolically-aesthetically through the commercial and gastronomic offers, which are penetrating further south and, on the other hand, also through planned and existing bicycle infrastructure in both neighborhoods. According to interviewee AR-02, the expansion of bicycle paths is "part of the changing atmosphere" in the inner-ring neighborhoods, which goes along visibly with an increasing number of cyclists in the streets around Balboa Park (equivalently arguing EP-06). Following these observations, the boundary between Bankers Hills and Hillcrest seems to be rather fluid and open compared to the hard border separating southern Bankers Hill from Little Italy (see Figure 5.9). This is not only due to the physical presence of the highway but especially because of the uneven development and urbanization tendencies in the different sections of Bankers Hill, which create hybrid situations of simultaneity and ambiguity (cf. Kühne and Schönwald 2015, p. 165).

5.1.2 Gentrification Waves in the LGBTQ-Neighborhood of Hillcrest

Established in the late 19th century, the neighborhood of Hillcrest is one of the early first-tier neighborhoods constructed outside of downtown to accommodate the significantly growing population of San Diego's booster years and the following Panama-California Exposition boom of the early 20th century (see Figure 4.2, cf. Ford 2014, pp. 180–181). However, in the post war era, after the neighborhood had been left behind for thriving new housing and shopping developments along the expanding freeways, such as Mission Valley and others, "Hillcrest struggled to find its identity among the communities of San Diego. The community was unable to generate any spark of its own and was widely recognized as a community of elderly and low-income residents living in run-down housing" (Dillinger 2000, n.pag.). This image began to change in the 1970s and '80s, when the local LGBTQ community discovered and 'repurposed' the then outdated neighborhood (Hennessey 2000). Slowly replacing the elderly residents of Hillcrest and thus significantly lowering the median age of the area, the influx of younger queer citizens marked "the emergence of a new community in Hillcrest" (Dillinger

2000, n.pag.). Today, the neighborhood is—according to the San Diego Tourism Authority (2021, n.pag.)—known as "the heart of San Diego's LGBT community, Hillcrest exudes urban charm and a lively local atmosphere. Many of San Diego's best restaurants, boutiques and nightlife can be found here." Thus, the neighborhood describes the evolution from a sleepy suburban community of detached single-family homes to a neighborhood with an 'urban village'-style center. However, today, it seems that the neighborhood of Hillcrest is again structurally and socially transitioning and undergoing new dynamics in the course of progressing efforts of inner-ring urbanization in San Diego. In the following, it will be elaborated how the recent upheavals in the community are individually perceived and experienced and to what extent the restructuring of the neighborhood is promoting hybridization processes on various levels. First, however, descriptive and quantitative insights will be given in order to supplement the following qualitative interpretations.

Today, the neighborhood of Hillcrest lies in the center of the Uptown community planning area, shaped by canyon and mesa topography, two large streets and highways, and Balboa Park on its edges (see Figure 5.2). In the east, Hillcrest borders the community area and neighborhood of North Park along Park Boulevard. In the south, the official neighborhood boundary runs along Upas Street and Balboa Park. In the west, the official neighborhood boundary follows a unique canyon structure as well as Curlew and Dove Street. In the north, the boundary runs along the important four to six lane major arterial road of Washington Street, which constitutes a distinct obstacle for pedestrian traffic, especially east of its intersection with Highway 163 (Kayzar 2003). In terms of land use, the southwestern (westward of Front Street) as well as the southeastern (eastward of 6th Avenue) neighborhood sections around the finger-like canyons of Hillcrest are represented in the current Uptown community plan (CofSD 2019, pp. LU24–LU27) as low-density residential areas (5–15 Du/Ac). In the plan, the southeastern neighborhood section is designated as being two separate potential historic districts, Marston Family on the west side of Highway 163 and Marston Hills on the east (see Figure 5.2), named after the preserved early 20th century home of the influential Marston family (see Figure 5.5). However, the section in between the two suburban-style areas in the southeastern and the southwestern neighborhood corner, which ranges from Front Avenue to 6th Avenue, constitutes a higher density commercial (0–109 Du/Ac) and residential (45–73 Du/Ac) corridor with large condominium and mixed-use buildings. Within this corridor lie 4th and 5th Avenue, which are labeled as areas with 'village qualities' of significance for the entire Uptown community due to the high commercial density, which confers Hillcrest a certain degree of centrality within the community area and beyond (CofSD 2019, p. LU37). It is particularly this commercial corridor and its centrality that connects the neighborhoods of Bankers Hill and Hillcrest and thus reduces the possibility of—at least functional—everyday life boundary experiences and encounters (cf. Wille and Nienaber 2020) at this rather fluid and therefore vague neighborhood boundary (PO-04; see also Figure 5.12). In addition to 4th and 5th Avenue in north-south direction also University Avenue as well as the western section of Washington Street, which constitute both important connections to the Greater North Park area, are labeled mixed-use 'community villages' (cf. CofSD 2019, p. UD68).

Figure 5.5 The Historic Arts and Craft Home of the Influential Marston Family in Hillcrest Built in 1905. (Source: Picture Albert Roßmeier 2021)

Unlike in Bankers Hill, the neighborhood boundaries of Hillcrest are not aligning with census tract boundaries, which impedes the quantitative comparison of different neighborhood sections. Here, the census tract boundaries miss large portions of the neighborhood and combine these sections with areas of the surrounding districts, such as the Medical Complex, University Heights, and in part also North Park. This is the case, for instance, for the large 'Uptown District' development project at the northern neighborhood edge, which accommodates community commercial (0–73 Du/Ac) and residential uses (45–73 Du/Ac) in high densities. Interestingly, however, the census tract 3, which covers most of the western parts of Hillcrest, has been split into two separate tracts (3.01 and 3.02) since the 2020 census, which indicates substantial population growth in the previous decade in the area between the western neighborhood boundary and Highway 163 (cf. U.S. Census Bureau 2021b). Additionally, 2020 census data shows that the housing units in the entire Uptown community are between 85 to 90 percent occupied—except for the tracts 3.02 and 60 (the northern tract of Bankers Hill), which are areas undergoing extensive development and densification. In these two tracts, the occupancy rates have decreased compared to the previous decades and are currently only 70 to 85 percent. In 2000, tract 3 (now 3.01 and 3.02) recorded an occupancy rate of 95 percent, which declined to 91 percent in 2010 (SANDAG 2003b, 2016g).

According to the graphic presentation of the median gross rent in the Hillcrest area from the decades of 1990 to 2015 by Chapple and Thomas (2021), tract 3 seems to be the community section that recorded rent increases first, followed by the adjacent tract 60 in Bankers Hill in the 2000s and later by tract 7 (which comprises of large portions of eastern Hillcrest) in the time frame from 2010 to 2015. Concerning the median household income between 2010 to 2015, the western part of Hillcrest (tract 3.01 and 3.02) was characterized by higher incomes compared to the rather suburban-style eastern section (tract 7). Looking at the housing composition of tract 3 in 2010 reveals that the majority of this neighborhood section comprised of buildings with 10 or more units (53 percent in 2010, SANDAG 2016g). In terms of the demographic development between 2000 and 2010, the share of underage residents of tract 3 in Hillcrest made up only 5 percent in 2010. In comparison, the share of residents under the age of 18 in the jurisdiction of San Diego was 22 percent in 2010. Lastly, tract 3 also shows the highest share of college-educated residents compared to all the tracts of the Uptown and North Park community areas in the given time frame (Chapple and Thomas 2021).

Summing up, the presented census data allows to start painting the picture of the neighborhood section between First and 6th Avenue as a popular and quickly evolving part of Hillcrest and Uptown in general, which consists mostly of professional and childless non-family households (73 percent in 2010, SANDAG 2016g) living in a relatively urbanized area of large apartment buildings. In the following, this picture will differentiated by means of the collected subjective interpretations and experiences or, in other words, the various realities of the ongoing urbanization processes in and around downtown San Diego.

When told to speak freely about the neighborhood of Hillcrest, it's peculiarities, and the individual perception of recent local developments, several

Figure 5.6 Making Use of the Socially Constructed Identity and Reputation of the Neighborhood of Hillcrest: A Replica of the Historic Neighborhood Sign on University Street Hanging in a Grocery Store in Hillcrest. (Source: Picture Albert Roßmeier 2021)

interviewees put their descriptions in context with historical and social developments of the second half of the 20th century, which transformed the neighborhood into "the center of gay life in San Diego" (Dillinger 2000, n.pag.). Starting at the earliest point in time, interviewee AR-07 from the perspective category academia/research traces the establishment of the Sears department store in the 1950s in Hillcrest, which brought economic activity to the neighborhood and attracted purchasing power from adjacent residential communities, such as University Heights. However, as stated by AR-07, following the opening of Fashion Valley in the 1960s, the neighborhood of Hillcrest and the offers provided along University Avenue fell from favor. "By the early 1980s, large, self-contained shopping malls with a variety of stores, free parking, and easy freeway access had drawn customers away from the retailers located along older automobile strips" (Kayzar 2003, p. 6), such as University Avenue in Hillcrest. As a result, "University went declining and if you go past the 805 freeway, you can still see, like, on University, there is still lots of shops but past here they have gotten a little more run-down" (AR-07). Thus, interviewee AR-07 is linking the unfavorable past of University Avenue in Hillcrest with its current condition and indicates the everyday significance of Interstate 805 as a functional and symbolic boundary in today's neighborhood structure of central San Diego (see Figure 5.1).

Addressing the decay of Hillcrest and University Avenue as well, interviewee R-12 shares biographical insights and experiences from their school days in the 1970s. Back then, the neighborhood was perceived as unsafe and dilapidated. However, "that has of course changed completely until today. Today, Hillcrest is the home of the gay community" and enjoys positive reputation (see Figure 5.6). In addition to this temporal comparison, the interviewee portrays the recent establishment of several sidewalk cafés and other businesses serving the demands and expectations of "millennials." According to R-12, some older and less trendy businesses were forced out of the neighborhood and made way for new uses and shops, especially along today's 'community village' areas of 4th, 5th, and University Avenue (see Figure 5.5). Regarding this development, interviewee AR-10 expresses the term of "gay gentrification." In accordance with AR-07 and R-13, interviewee AR-10 reports that "Hillcrest was one of the first neighborhoods [around Balboa Park] to gentrify" and that the area underwent substantial change in the early 1990s due to the arrival and emergence of the local LGBTQ community. Interviewee R-05 goes further back in time and marks the beginning of the upheavals in the 1980s. Yet, one of the reasons for the success of the neighborhood, according to interviewee AR-10, is the social and economic way of life of many LGBTQ community members as childless and career-oriented professionals achieving higher incomes.

On top of that, it is striking how interviewee AR-10 is linking the developments in Hillcrest to the changing housing market and infrastructure in the neighboring communities of North Park and City Heights. In addition to the statements of other interviewees (AR-07 and R-13) who depict the advance of the development dynamics of Hillcrest southward into Bankers Hill, AR-10 is describing a sequential push and displacement of residents with a lack of economic capital from west to east. The described push began with the onset of gentrification in Hillcrest in the 1990s and was followed in the 2000s by a 'rediscovery' of the then predominantly low-income community of North Park. Accordingly, interviewee EP-06 claims: "When downtown, Little Italy, and Hillcrest got expensive, people started moving to North Park." There, aroused interest in the neighborhood and emerging development pressure led to a significant increase of real estate prices. In a next step, according to interviewee AR-10, this has led to the progression of gentrification processes from Hillcrest over North Park into the adjacent communities of Mid-City (Normal Heights and City Heights) east of the Interstates 805 and 15. There, the recent initiation of redevelopment and reinterpretation processes manifests itself in the form of changing real estate marketing strategies, the inflow of displaced residents from Hillcrest and North Park, and new construction projects (discussed in detail in section 5.2). Summing up, AR-10 is picturing a certain linkage that has established over time beyond the official neighborhood boundaries of the communities north of Balboa Park in terms of an alignment of the local real estate market and business infrastructure. At the same time, the described push can also be interpreted as a shift of the boundaries of affordable or accessible space from west to east—even beyond the physical borders of the Interstates 805 and 15. Following the argumentation of Liao et al. (2018), this development constitutes a form of rebordering on the neighborhood level, which is promoting exclusion and (socio-)spatial reordering.

In this context, interviewee EP-02, who is affiliated with one of the 17 Business Improving Districts (BID)[3] of San Diego, generalizes: “If it gets too expensive, people go a little further. This is how certain developments in the city are spreading.” According to the interviewee R-13, this shift is promoting the creation of “little pop-up neighborhoods” with their own neighborhood signs. The interviewee strongly emphasizes the connection between the erection of neighborhood signs and the re-development of neighborhoods: “That seems to be the milestone when one of these neighborhoods has successfully arrived and re-birthed itself, they get a sign. Like, Hillcrest has one, North Park has one, Kensington has one. It becomes a landmark. And it’s kind of a way the neighborhood puts a stamp on their accomplishment of re-birthing themselves.” In border-theoretical terms, the erection of neighborhood signs in redeveloping communities can be interpreted as a strong visual othering process in the symbolic dimension within which the ‘own’ and the ‘alien,’ the ‘here’ and the ‘there’ are defined and at the same time separated from each other (see also Figure 5.6).

Concerning the neighborhood of Hillcrest, the portrayed push from west to east has not leveled the (re)development dynamics. Even though interviewee AR-10 states that Hillcrest might have lost some popularity since the success of the development of North Park, “it [Hillcrest] is still getting more expensive” (AR-10). Equivalently, Interviewee EP-01, a business owner from Hillcrest, states that the neighborhood is strongly developing and that there is a lot of change happening in the area after all. According to R-10, rent in Hillcrest has increased by 20 percent in the last 10 years, thus also adding more pressure on local businesses. In the same vein, interviewee R-04 deplores that the overall trend of increasing

[3] Administered by the Economic Development Department of the City of San Diego, Business Improving Districts are designated commercial areas and associations in which local business owners are actively recovering or contributing to the attractiveness of a determined territory in terms of public improvements and maintenance, event organization, and other measures.

real estate costs in Hillcrest is promoting the replacement of less profitable businesses by more cost-intensive competitors. Concerning new development, both interviewee R-10 from Hillcrest and interviewee R-04, an active member of several Uptown community planning and governing committees, discuss updated height limits for new constructions in Hillcrest and interpret the change as a significant catalyst for the ongoing urbanization process. As argued by the two interviewees, due to these legal adjustments, more demolition works and construction sites are seen in the neighborhood. "A lot of development is happening here. They break down a lot of apartments and build big condos" (R-04). Both interviewees view the local business improvement district, the 'Hillcrest Business Association,' as a big driver of the legal adjustment. According to interviewee R-04, the association comprises solely of businesspeople that are not residents of Hillcrest but are interested in its development and have thus pushed for the repeal of the height limit. In the same vein, interviewee R-10 discusses the increasing interest of investors in Hillcrest due to positive reputation the neighborhood has gained over the years.

Furthermore, interviewee AR-07 discusses a perceived significant transformation of the neighborhood, its purpose, and social composition. As argued by the interviewee, the importance of Hillcrest as a center for the local LGBTQ community is decreasing in recent years, thus creating a need to reinvent itself in this period of extensive residential development (on future perspectives for gay neighborhoods see Bitterman and Hess 2021):

> "There is this narrative that the gay neighborhoods are declining and the markers for that tend to be a lot of historic gay bars and clubs are closing down, they're losing business. A lot of people blame the social media apps, there is no need for a physical space. But then also because there is more legal equality and more social acceptance of the LGBT community. There's not a need to physically congregate in one place because that's, kind of, why they popped up; as a community to support itself. But now, there's no need for that, [...] it kind of served its purpose" (AR-07).

Additionally, interviewee AR-07 is linking the argumentation for the loss of importance of Hillcrest as a LGBTQ neighborhood with the assumption that the role of these communities is generally changing for different audiences. For, as experienced by AR-07, traditional nightlife businesses addressed to homosexual audiences are increasingly visited by heterosexual female groups and enjoyed as 'safe space.' Along with this goes that "the city advertises it [Hillcrest] now as being like a tourist destination, a place to go eat brunch, to go to watch the Super Bowl" (AR-07). Thus, the neighborhood of Hillcrest seems to enjoy even greater attention, among other aspects, because of increased social acceptance of sexual diversity. Further, a social and functional transformation of Hillcrest becomes apparent with the shift to extensive advertising and marketing of the neighborhood by different stakeholders that make use of the neighborhood's positive reputation and strive to change its external perception.

Figure 5.7 Physical Manifestation of Ongoing Urbanization and Development Pressures in Hillcrest—the Conversion of Small Residential Buildings for Commercial Uses. (Source: Picture Albert Roßmeier 2021)

In the following, subjective experiences and 'in the landscape' observations (Kühne 2019, p. 33; cf. Wylie 2007) from the neighborhood walk (PO-04) through Hillcrest are complementing the social perceptions discussed above. The composition of buildings and businesses in the designated 'urban village' areas of Hillcrest was particularly striking during the walk. Here, increasing rents and a high development pressure in general seem to promote the renovation and conversion of smaller residential buildings for commercial purposes (see Figure 5.7). For instance, along 5th Avenue owner-operated businesses in converted single-family homes sit in between large, new apartment complex buildings and create a hybrid mix of commonly suburban and urban structural elements. Along with the preservation efforts at the neighborhood's canyon-structured edges (such as the proposed historic neighborhood quarters of Marston Family and Marston Hills) comes the described development hype in the designated community commercial districts along its prominent Avenues (CofSD 2019, pp. LU38–LU39). In urban landscape-theoretical terms, this confers the neighborhood its transitional, hybrid appearance of an urbanizing former inner-ring suburb in which old and small residential structures are gradually replaced by larger multi-use buildings (cf. Kühne et al. 2016; Kühne and Weber 2019; Kühne et al. 2020; Roßmeier 2019).

Accordingly, during the neighborhood walks (PO-03, PO-04) stark differences in the concentration of businesses on University Avenue in Hillcrest and North Park became apparent. The majority of the bars, clubs, gyms, restaurants, and cafés in Hillcrest are located in the immediate neighborhood area around the intersection of University and 5th Avenue. East of Hillcrest's physical border of Highway 163, University Avenue continues to function as a commercial corridor, however, with lower density. Further eastward, toward the official boundary between Hillcrest and North Park, the number of smaller, owner-operated gastronomic businesses, such as hip cafés or ice cream shops, has decreased even more; this was particularly striking east of the intersection with Normal Street. In North Park, storefront-seamed sidewalks, cafés, and beer tasting rooms or breweries on University Avenue have only reappeared east of the intersection with Utah Street and in closer vicinity to the trendy neighborhood center. In between the intersections with Normal and Upas Street, University Avenue features merely a fragmented selection of fast-food restaurants, liquor, cell phone, and convenience stores as well as gas stations and car repair shops. Thus, against the example of Hillcrest and Bankers Hill, which are functionally connected through a diverse composition of businesses along 5th and 6th Avenue, the walkability between the neighborhoods of Hillcrest and North Park in terms of exploration potential and storefront density is only partly established (see Figure 5.9).

Summing up, during the participant observations (PO-03, PO-04) the neighborhood centers of Hillcrest and North Park appeared with busy sidewalks filled with window shoppers, dog walkers, and other pedestrians enjoying the 'village centers.' However, the section in between the two neighborhood centers, the boundary area of Hillcrest and North Park has not been highly frequented by pedestrians and cyclists. Additionally, parts of the sidewalks in this section were in poor condition and the street was partly missing bicycle lanes. Thus, it can be concluded that even though University Avenue constitutes an important gateway into the neighboring community area of North Park, the functional and infrastructural connection between the two communities seems only partly established due to this larger car-centered section with reduces walkability. Despite the absence of a physical border line, a functional boundary area prevails between Hillcrest and North Park, which appeared to be of significant everyday life relevance for the residents during the observations 'from the inside' of this boundary (cf. Rumford 2014).

In addition to the differences in the business structure along University Avenue, the target group orientation of various businesses in Hillcrest was striking. During the neighborhood walk (PO-04), it has become obvious that the variety of consumption offers in Hillcrest targets mostly younger, postmodern generations with aesthetic-oriented and 'urban' lifestyles and higher stocks of economic and symbolic capital. For instance, the only branch of the trendy grocery store Whole Foods Market in central San Diego lies in Hillcrest on University Avenue. Additionally, four branches of Starbucks Coffee Company are located in the neighborhood next to several owner-operated cafés, bakeries, and a weekly farmers market. With their large selection of fair trade and organic produce, these businesses serve the emerging 'sustainable lifestyle' of "rich millennials" (R-02) in the 'rediscovered' neighborhoods of central San Diego, as perceived by interviewee R-02 and EP-01. During the neighborhood walk (PO-04), the discussed social trend toward sustainable consumption and non-motorized transportation became apparent physically-materially in the form of installed bike sharing stations and construction sites for new bicycle lanes. Accordingly, interviewee EP-01 shares that they no long own a private vehicle since moving to Hillcrest, which reflects the discussed lifestyle change in the redeveloping neighborhoods of San Diego. Interviewee R-04, in contrast, an active member of the Uptown parking committee, deplores that more and more parking spaces in Hillcrest are being converted into bicycle stands. Next to these infrastructure updates, "a lot of new development is being built without parking because [in addition to the adjustment of the height limit] they are allowed to build without parking restrictions." Thus, the interviewee's official efforts and complaints to obtain parking spaces in the neighborhood demonstrate how the spatial practices and demands of newer, younger residents are getting into conflict with the preferences of long-term residents. In conclusion, this example bespeaks that the redevelopment of the neighborhoods in and around downtown San Diego is individually perceived and interpreted in markedly different ways.

As a final point of the collected experiences and interpretations from the fieldwork in Hillcrest changes of the use of symbols and signs in the neighborhood will be addressed in the following. During the neighborhood walk through Hillcrest, a multitude of rainbow flags became obvious that are hanging inside and outside of restaurants, bars, and shops in order to symbolize tolerance against social diversity. Generally, the use of ornaments is very pronounced in Hillcrest—roadside plantings, trees, walls, and other infrastructural elements, such as electric transformers, feature a wide array of decorations and painted murals. Furthermore, private residential buildings are frequently equipped with small signs and messages, which reflect the postmodern urge of self-expression (for instance, the political slogan 'no human is illegal' has been recognized multiple times, which can be interpreted as a dissenting response to the U.S. immigration policy, especially in the San Diego-Tijuana border region). Thus, with regard to the portrayal of the interviewee R-12, the demographic shift that Hillcrest underwent in the recent decades becomes legible symbolically: "aggressive graffiti" (R-12) from the late 20th century, when the neighborhood was perceived as unsafe and run-down by several interviewees, give way to different symbols of a diverse, postmodern society of the 21st century. As described by EP-06, Hillcrest "became very artistic" and evolved because "it is ok to be gay now," which is reflected in the rainbow flags and small messages in the front yards. Today, the neighborhood is perceived as "very safe. Hillcrest is my nice small 'gayborhood,'" as interviewee R-04 describes. In this sense, the display of colorful symbols and signs is promoting the (re-)creation of neighborhood identity in Hillcrest and can be interpreted as an active symbolic b/ordering and othering practice, which establishes the neighborhood as a unique community within the neighborhoods of San Diego. Summing up, border and boundary experiences and encounters in Hillcrest take especially place in the social and symbolic, only partly in the functional, and even less pronounced in the physical dimension (see Figure 2.5).

5.1.3 Urban Landscape and Border-Theoretical Conclusions for Uptown San Diego—Border Lines, Transition Zones, and Hybridization Tendencies in the Redeveloping Neighborhoods

Before moving on to the next subsection on the community area of Greater North Park, in the following a brief summary and conclusion will complete the discussion of the neighborhoods of Uptown San Diego. This purpose is served not only by the following remarks but also by the three-dimensional sketch of the neighborhoods of Bankers Hill and Hillcrest (Figure 5.9), which constitutes a graphic summary and illustration of the gained qualitative insights. Generally, following a postmodern and post-structuralist position rather than positivist logic, the provided sketch is highlighting relations and networks, not exact locations of borders or boundaries (Davoudi and Strange 2009, p. 56).

The neighborhood of Bankers Hill sits on the northern edge of downtown between Interstate 5, Balboa Park, and Hillcrest. As discussed above and illustrated in the sketch, the neighborhood comprises of different sections which have established different kinds of relationships with their surrounding neighborhoods and communities. The southern part of Bankers Hill, census tract 57, constitutes an area with a relatively high share of population below poverty status and a low share of college-educated residents. Thus, it stands in contrast with the neighborhood of Little Italy in downtown as well as the northern parts of Bankers Hill and Hillcrest (see Table 5.1). The central and northern parts of Bankers Hill, census tracts 59 and 60, are facing extensive development in terms of new housing construction and infrastructural diversification as well as disproportionate increases in rent and median housing value—especially along 4th, 5th, and 6th Avenue. As described by several interviewees (AR-7, AR-10, R-12, R-13), the ongoing redevelopment of the Uptown area has started with the emergence of the LGBTQ community in Hillcrest, which then led to the progression of the development trend into adjacent neighborhoods—in the south toward Bankers Hill and in the east toward North Park and Mid-City (cf. Joassart-Marcelli 2021; Joassart-Marcelli and Bosco 2020).

In conclusion, it became apparent during the participant observation and the conducted interviews that the neighborhood of Bankers Hill is characterized by a gradient of development from north to south with the north being a trendy and vibrant area populated by higher-educated and economically better situated residents and the south being a part into which the developments have only just begun to penetrate. On top of that, the highest population densities can be found in the eastern sections of the neighborhood next to Balboa Park, where street life is more and more establishing. In contrast, the western neighborhood sections are mostly comprising of 'sleepy' low-density residential areas without pedestrian or bicycle traffic in the widened, car-centered streets (PO-04).

Additionally, the uneven redevelopment of Bankers Hill can be tied to the different border and boundary situations at the neighborhood edges and the established networks to the adjacent communities. As illustrated in the sketch, the southern part of Bankers Hill is bordered by Interstate 5, which constitutes a hard physical border preventing or reducing the connection between the neighborhood and downtown San Diego. During the participant observation (PO-04), when crossing Interstate 5 from Little Italy into Bankers Hill, a stark decrease of storefronts and gastronomic or tourism offers became apparent. Generally, Little Italy can be ascribed a high walkability due to its density of retail, restaurants and cafés, and the semi-public squares of simulated ethnicity (Comer-Schultz 2011, p. 310). In contrast, the southern section of Bankers Hill comprises mostly of larger office and commercial buildings as well as smaller apartment complexes (in the east) or of detached single family homes (in the west) and is thus not providing sufficient walkability. In terms of neighborhood identity and symbolic bordering, Bankers Hill does not feature a neighborhood sign or does not have printed neighborhood logos in the streets, on benches, and waste bins, such as the neighborhood of Little Italy. Apparently, Bankers Hill it is not (yet) branded like other redeveloped inner-ring neighborhoods. According to interviewee EP-06, this is due to the fact that Bankers Hill is one of the few neighborhoods in central San Diego that does not have a maintenance assessment district (MAD[4]) to bear the expenses of a sign.

Summing up, the southern area of Bankers Hill is not only physically bordered by Interstate 5 but also immaterially in the functional, symbolic, and social dimension through stark contrasts with its surroundings. This bespeaks the everyday relevance of the southern official neighborhood boundary; for in contrast to the development trend, which progresses from Hillcrest into northern Bankers Hill, the dynamics of Little Italy do not seem to spill over into the southern parts of the neighborhood. On top of that, even though business associations are not (yet) actively pursuing processes of identity creation, Bankers Hill is currently severed by the 'erected' symbolic boundaries at the neighborhood's southern edge in terms of the efforts of symbolic contrasting in the adjacent community of Little Italy.

Concerning the functional dimension (as illustrated in the sketch), the neighborhoods of Bankers Hill and Hillcrest have established a strong infrastructural connection along 4th, 5th, and 6th Avenue. This is where the so-called 'village centers' of the neighborhoods are located, which feature a diverse mix of retail, restaurants, bars, and other uses. Additionally, University Avenue constitutes Hillcrest's walkable commercial center and an important gateway into the adjacent community of North Park. While the infrastructural composition along the Avenues next to Balboa Park has created a strong functional link between Hillcrest and Bankers Hill, the transition to the redeveloped neighborhood of North Park appears less consistent. The walkability in between the two neighborhood centers of Hillcrest and North Park decreases significantly due to deficient pedestrian and bicycle infrastructure and change of business composition from

[4] Managed by the Park and Recreation Department of the City of San Diego, maintenance assessment districts are legal associations of local property owners dedicated to paying and acquiring benefits used for the construction and maintenance of neighborhood signage, landscaping, and other decorations and improvements. As a sort of public–private partnership, San Diego's MADs as well as BIDs reflect the local tendency to privatize public services, which makes development projects dependent on the availability of local resources.

pedestrian-oriented to car-centered and from aesthetic-oriented to function-driven. Thus, despite the portrayed progression of gentrification dynamics from Hillcrest into North Park and their shared image of 'vibrancy,' the neighborhoods are severed from each other functionally and spatially by an area with poor walkability and low pedestrian-oriented commercial density, which can be described as an in-between zone in the form of a borderland (cf. Anzaldúa 2012 [1987]; Banerjee and Chen 2012; Iossifova 2013; Roßmeier 2020; Roßmeier and Weber 2021).

Figure 5.8 Urban/Rural Hybridity in the Neighborhood of Hillcrest—Current Mixture of Old and New, Small and Large Building Structures in Direct Vicinity to Each Other along 6th Avenue and Ivy Lane. (A sub-divided, historic single-family home can be seen on the left, a medium-sized apartment building sits in the center of the picture, and a large, multi-unit apartment complex building is shown in the right corner of the image. Source: Picture Albert Roßmeier 2021)

As a last remark on the recent changes in the Uptown communities, the aspect of hybridity needs to be addressed. The onset of redevelopment processes in the neighborhoods of Hillcrest and Bankers Hill has promoted the emergence of spatial pastiches as well as manifold hybridization tendencies on different levels and in varying extents. Since its establishment, the neighborhood of Hillcrest has been structured into the commercial area of University Avenue and the residential, low-density areas at its edges, which entailed the still prevalent adjacency of larger commercial buildings and smaller single-family homes. However, with the progression of the redevelopment processes and the following adaptions to local height limits (R-04, R-10) as well as the substitution of older, smaller houses with larger condominium buildings, the resulting mixture of various structures and uses is becoming more complex (see Figure 5.8). The different spatial compartments fragment and fray, creating a hybrid neighborhood structure of typically urban and suburban or rural elements in direct vicinity to each other, which presents the community as neither fully urban nor typical suburban (cf. Kühne 2012).

In the case context of Bankers Hill, the redevelopment processes are provoking an at least two-fold hybridization. On the one hand, the juxtaposition of (historic) single-family homes on spacious properties along woodened canyons and mixed-use buildings in typical urban density is describing the neighborhood's transitional state in the overall process of urban/rural hybridization. On the other hand, hybridity is emerging in Bankers Hill through the demographic disparities produced by the portrayed uneven redevelopment of the neighborhood. This creates an ambiguous image of an 'interesting,' young, and evolving neighborhood on the downtown edge (EP-01, EP-04, EP-06, EP-07) and simultaneously also of a quiet and aging community of families (R-06, R-10). Epistemologically, this individually produced ambiguity exemplifies the contingency of social reality and its construction character. From an urban landscape perspective, it can be concluded that the portrayed ambiguity and the uneven redevelopment of Bankers Hill provokes processes of socio-spatial compartmentalization as well as the emergence of spatial pastiches through symbolic reframing and functional reinterpretation of the neighborhood (see generally Aitken and Zonn 1994; Hoesterey 2001; Kühne and Weber 2019; Vester 1993). The formation of spatial pastiches is, from a border-theoretical stance, the result of simultaneous dissolution and emergence of demarcations, the regrouping of differences, and, in other words, the outcome of the renegotiation of boundaries between the neighborhoods of Bankers Hill and Hillcrest (Figure 5.9).

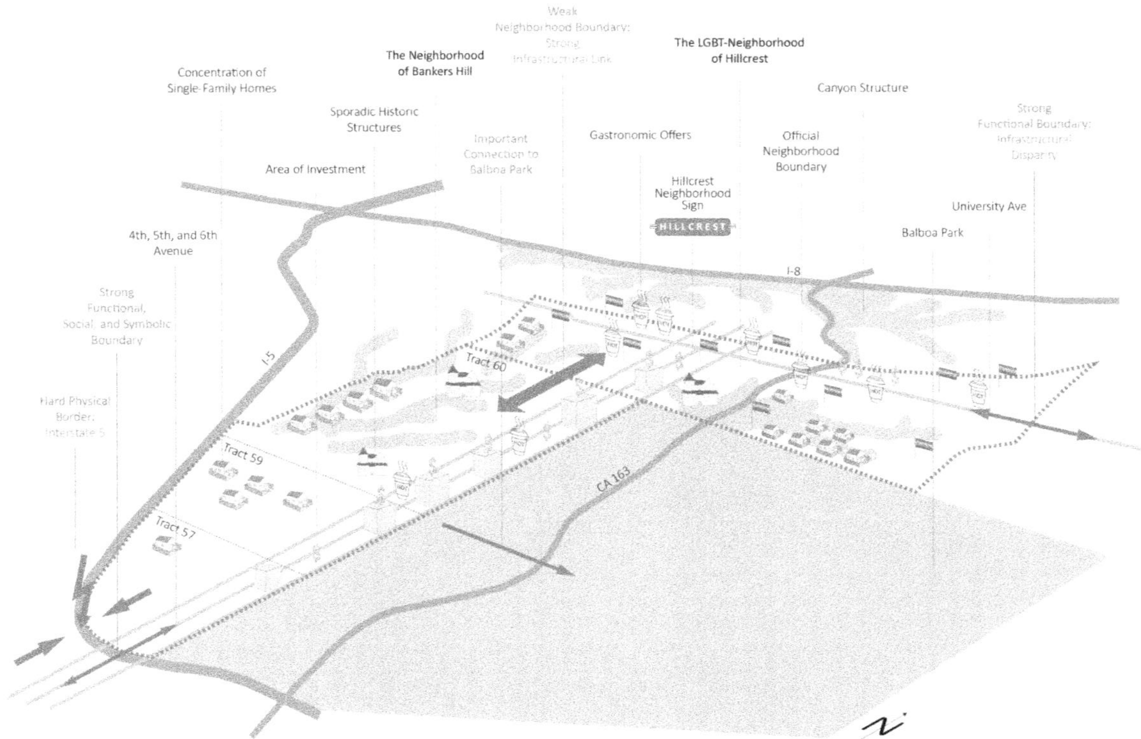

Figure 5.9 Digital 3D Sketch—Networks, Flows, and Demarcations within and around Uptown San Diego. (General information is provided in black, border-related conclusions are displayed in red. Thick arrows indicate strong neighborhood connections beyond official boundaries, thin arrows indicate weak links. A flash symbol in combination with two arrows pointing toward each other is used to display physical demarcations with high everyday relevance. Source: Own cartography and design)

5.2 Redevelopment Trend from West to East: The 'Revitalization' of the Greater North Park Area and the Penetration of the 'North Park' Concept into the Community of Mid-City

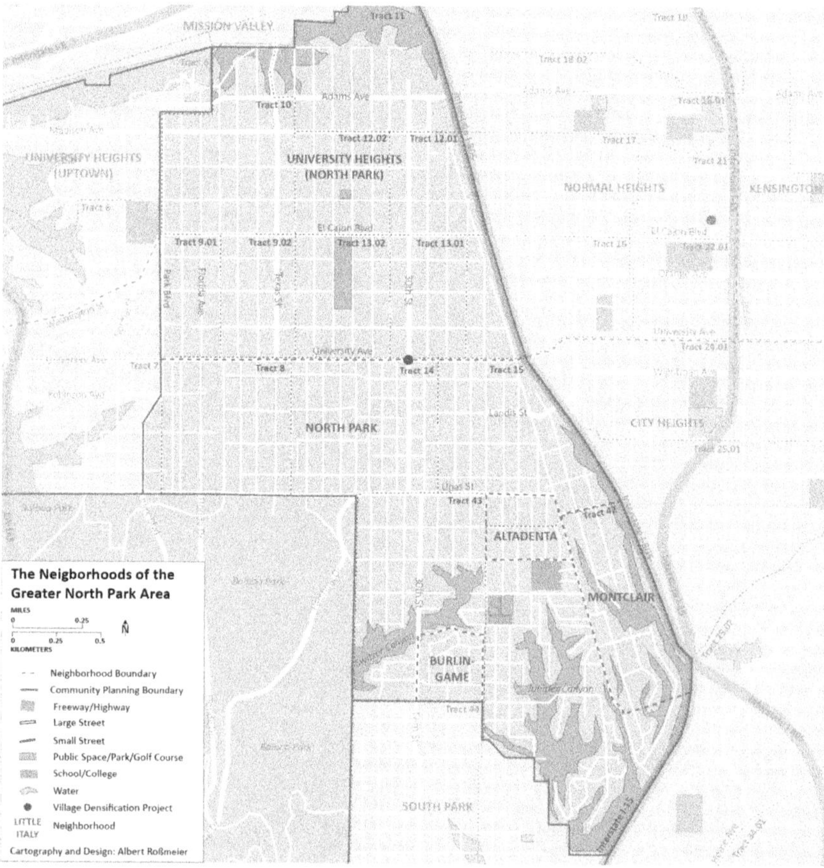

Figure 5.10 The Neighborhoods of the Greater North Park Area and Their Vaguely Defined Official Boundaries. (Source: Own cartography and design)

In the following chapter, the focus will be shifted onto the Greater North Park community (see Figure 5.10). Due to strong thematic emphasis of various interviewees on the individual neighborhood of North Park compared to the other official sections of the community, the following remarks will particularly deal with the area around the intersection of University Avenue and 30th Street, which constitutes the commercial community center of Greater North Park. The goal of this chapter is to illuminate the recent redevelopment processes in the community and their social interpretations in order to gain a fuller understanding of San Diego's ongoing inner ring 'resurgence.' The first section (5.2.1) will present historic insights into the development of Greater North Park and address the community's recent planning trajectory in light of the 'City of Villages' smart growth strategy. Additionally, a quantitative perspective on the recent transformation of the area will be given by means of census data. Subsequently, building on the conducted interviews and participant observations, qualitative insights will be presented (section 5.2.2). The discussion of the qualitative results will demonstrate the change of individual neighborhood interpretations over the years and discuss the commencing progression of the 'North Park' concept and its gentrification dynamics into the adjacent neighborhoods of Normal Heights and City Heights in the Mid-City community. Lastly, the chapter ends with an urban landscape and border-theoretical conclusion (section 5.2.3), which serves as a theoretical interpretation of the elaborated insights.

5.2.1 Historic and Quantitative Perspectives on the Recent Transformation of a Neglected Inner-Ring Community into a Vibrant 'Urban Village'

In the second half of the 19th century, the broad mesa top in the immediate area north and northeast of Balboa Park—today's Greater North Park community planning area—has been developed into an agricultural community by a small number of settlers, who cultivated citrus fruits in large orchards. In 1887, following the population and building boom triggered by the city's (delayed) connection to the intercontinental railroad network and the end of the northern Californian gold rush (Griffin and Weeks 2014), the first suburban settlement in North Park has been subdivided and established under the name of University Heights. One year later, in "1888, streets were being graded, water and sewer service was established, and lots were offered for sale" (Mengers 2017, p. 24). Within the following decade, the area south and east of University Heights has been developed gradually—initially spurred by the streetcar connection of the

Electric Rapid Transit Company and later by the Class 1 streetcar of John D. Spreckel's San Diego Electric Railway Company, which went along University Avenue and Adams Avenue in east-west direction. Residential construction in the most southern section of today's North Park community area, in the neighborhood of Burlingame, has started years later in 1912 after the erection of the streetcar bridge over Switzer canyon, which established the north-south connection along 30th Street and thus a continuous railway network around Balboa Park and between the flourishing new suburbs (CofSD 2016, pp. 4, 156; cf. Covington 1993; Crawford 1993).

Radiating out from downtown, the streetcars prompted commercial districts and neighborhood centers along large avenues and important crossings in the emerging suburbs (cf. CofSD 2016, p. 6). Consequently, the popular intersection of University Avenue and 30th Street in North Park flourished into a significant commercial center in the course of the 1920s and '30s. Additionally, the ongoing efforts of eager investors as well as the prestige of the international exposition in 1915 attracted new wealthy residents to the developing areas around Balboa Park and the city in general (see Figure 4.2, cf. Joassart-Marcelli 2021, pp. 46–52). During this period, "North Park was considered the fastest growing neighborhood in San Diego" (CofSD 2016, p. 156). Relatedly, also the purchasing power in the community grew dramatically during the first half of the 20th century and despite the dismantlement of San Diego's streetcar system in the late 1940s, the community of North Park continued to thrive until after World War II. Eventually, in the early post-World War II era, the North Park community developed two significant suburban shopping districts with a great number of department and other specialty stores along the old streetcar route on University Avenue and various car-oriented and drive-in businesses on the wider-built El Cajon Boulevard. Compared to the streetcar and pedestrian shopper dimensioned University Avenue, El Cajon Boulevard "grew up with the automobile […] and features an entirely different ambiance, density, and architecture" (Ford 2014, p. 185) up until today.

However, with increasing car ownership and road construction in San Diego, unlike before, new commercial developments started to follow larger streets instead of prioritizing locations along the former streetcar routes. Thus, in the 1960s, in response to the erection of multi-lane freeways throughout the county, large automobile dependent shopping centers have been established in the emerging next rings of the growing city. Consequently, newly developed shopping complexes, such as the large mall in Mission Valley, had gradually turned the attention of shoppers away from the older suburban commercial areas of Uptown and North Park. Additionally, the construction of Interstate 8, which constitutes the main connection to the Mission Valley shopping center and runs along the valley just north of North Park, had reduced vehicle traffic in the streets of the community considerably (CofSD 2016, p. 158). At the same time, the mass production of family homes in new neighborhoods of San Diego's thriving suburbia drew the city's affluent white population out of the denser downtown area and the older suburbs. As a result, in the 1960s and '70s, Greater North Park evolved more and more "into a bedroom community with a population primarily dependent upon other areas of the City for jobs, services and goods. It was during this time that North Park entered a period of decline with failing business districts and deteriorating neighborhoods," as illustrated in the area's current community plan (CofSD 2016, p. 4). Throughout the 1980s, the building stock and the infrastructure in North Park gradually decayed, while the large employment centers in the northern section of the county began to emerge (Ford 2014; Griffin and Weeks 2014). Over time, criminal activity in the neighborhood increased and El Cajon Boulevard became stigmatized as a 'cruising strip' for drugs and prostitution. Eventually, by the early 1990s, North Park had developed into a low-income neighborhood, neglected by economic and planning forces and thus exemplified the widespread demise of U.S.-American inner-ring suburbs of the late 20th and early twenty-first century (cf. Hanlon 2008, 2010; S. Lee and Leigh 2005; Vicino 2008).

It was not until the late 1990s that the community of North Park had been reached by the redevelopment dynamics, which ignited in Hillcrest decades before in the late 1970s and early '80s. Eventually, in North Park, the period of decade-long distress and neglect has been followed by a phase of renewed economic, planning, and social interest into the community, which has promoted the 'resurgence' it is experiencing today. The community's proximity to prominent areas, such as "downtown and Balboa Park, its old-fashioned main streets lined with broad sidewalks and storefronts, and the quaint of architecture and relative affordability of its craftsmen- and Spanish-style cottages attracted college-educated people in search for urban lifestyles" (Joassart-Marcelli and Bosco 2020, pp. 39–40; cf. Pryde and Stutz 2014, p. 198). However, the recent redevelopment phase of the community is not only being promoted by the current social trend toward 'urban living' or by real-estate or other economic interests but also by local planning policies dedicated to the creation of walkable mixed-use 'urban village' areas. In the early 2000s, a small number of pilot projects and areas throughout the more urban areas of San Diego (see Figure 1.1) have been proposed for further densification and urbanization measures as part of the adopted 'City of Villages' smart growth strategy of the city's recent General Plan (CofSD 2018; cf. Erie et al. 2011; Ford 2014; Griffin and Weeks 2014). By means of this strategy, the City as well as the County of San Diego set the goal to incentivize developments that counteract the decadelong sprawl of low-density neighborhoods into the backcountry of the county and the outward thrust of the so-called 'urban-wildlife interface' (Erie et al. 2010; Radeloff et al. 2005). Thus, among four other selected projects, North Park and the area around its community center (the intersection of University Avenue and 30th Street) have been proposed for 'revitalization' (CofSD 2004, p. 1) through extensive development that is intended to strengthen walkability and transit access, ease individual car dependency, and create a vibrant mix of civic uses, housing, jobs, retail, and schools. The following remarks are tracing the quantitative development the neighborhood of North Park experienced since the late 1990s and its designation as village pilot project in 2002.

First, however, in order to be able to look deeper into the North Park area, it is important to address the community's particularity in terms of its neighborhood boundaries. Although the Greater North Park community consists of several individual neighborhoods, the boundaries of these distinct sections are not officially defined (yet they are displayed in a map by CofSD 2016, p. 5, which serves as a reference for Figure 5.10). Instead, the boundaries are specified to be approximate—they are solely "based upon historical documents, county assessor's parcel maps, property deeds, subdivision maps, police beat maps, the existence of active neighborhood organizations, and residents' perceptions about where they live within North Park" (CofSD 2016, p. 4). Furthermore, remarkably embracing the processuality of neighborhoods, the community plan emphasizes the future possibility that new "neighborhoods can emerge as local residents in other areas of the North Park community organize and promote their own neighborhood identity" (CofSD 2016, p. 4). As a consequence of the absence of official neighborhood boundaries, a comparison of the entire community area as well as of selected census tracts in the immediate area around the commercial community center (census tracts 12.01, 12.02, 13.01, 13.02, 14, and 15; see Figure 5.10) by means of census data of the last two decades will be used to create a quantitative perspective on the demographic changes throughout North Park's redevelopment phase (see Table 5.2). However, two conditions limit the extent of the following quantitative comparison. On the one hand, 2020 census data[5] can currently only be retrieved for a very limited number of topics. On the other hand, the results of the 2020 census are only available for the tract level, not for the level of the

[5] 2020 census data is retrieved from the official '2020 Census Demographic Data Map Viewer' tool (https://www.census.gov/library/visualizations/2021/geo/demographicmapviewer.html).

community planning area. Thus, a detailed comparison of census tracts for the cohorts of 2000, 2010, and 2020 will be presented in the following, supplemented by data on the community area for the years of 2000 and 2010.

In the time frame from 2000 to 2010, the Greater North Park community area recorded a slight decrease of its total population by two percent. Between 2000 and 2020, the selected census tracts around the community center have faced minor decreases as well, with a maximum of five percent of population decline in tract 15 (from 4,010 residents in 2000 to 3,827 in 2020). This development stands in contrast, for instance, to the increasing population of the neighborhoods of Hillcrest and Bankers Hill in the adjacent Uptown community. In the time frame from 2000 to 2020, Hillcrest grew by 11 percent, while Bankers Hill recorded an even higher increase of 17 percent (see Table 5.1). Additionally, also the underage population in the Greater North Park area is decreasing, from 7,680 residents under the age of 18 in the year 2000 to 6,057 in 2010, which equals a decrease of 21 percent. This development is above average compared to the City of San Diego, which has recorded a decrease of underage residents of only five percent in the same time frame. In the following decade, from 2010 to 2020, this trend has continued—especially in the selected tracts north of University Avenue (tracts 12.01, 12.02, 13.01, and 13.02), which feature denser housing structures compared to the two tracts south of the community center (tract 14 and 15; see Figure 5.11). In 2020, the formerly connected tracts 12.01 and 12.02 (which have been split for the 2020 census) have recorded a cumulated strong decrease of underage residents by 45 percent compared to the year of 2000, while tract 13.01 and 13.02 have experienced an even higher cumulated decrease of 57 percent—from 1,098 residents under the age of 18 in 2000 to 475 in 2020. In comparison, tract 14 and 15, which are characterized by low-density single-family housing, have recorded less drastic decreases (24 percent decline in tract 14 and 36 percent in tract 15). Consequently, the share of adults is highest in close proximity to the denser 'urban village' area of North Park around University Avenue and El Cajon Boulevard and lowest at the community edge in the north, which is structured by the mesa edges toward the canyon of Mission Valley, as well as in the southern edge, which is bordering the Greater Golden Hill community and its canyon-structured lower density residential areas.

In equivalence with the development of the total population of the Greater North Park area, also the community's population density has decreased slightly between 2000 and 2010. In the following decade, from 2010 to 2020, this trend has continued and the population density has further declined in the selected tracts. In equivalence to the development in the Greater Golden Hill area, the decrease in density might indicate the shift toward new residents with higher

Table 5.2 Demographic Development in Selected Census Tracts of the Greater North Park Area: Comparison of the Cohorts of 2000, 2010, and 2020

			Total Population	Total Population Under 18	Population Density (Persons per sq. mi.)	Total White Population	Share of White Residents (In Percent)
Tract 12	**2000**		5,641	1,082	21,303	2,397	42%
	2010		5,660	840	21,375	2,399	42%
	Change		**0%**	**−22%**	**+72**	**0%**	–
	2020	Tract 12.01	2,235	246	18,175	1,107	50%
		Tract 12.02	3,216	345	22,676	1,678	52%
		Cumulated	5,451	591	20,585	2,785	52%
	Change (In Relation to 2000)		**−3%**	**−45%**	**−718**	**+16%**	–
Tract 13	**2000**		6,068	1,098	19,845	2,555	42%
	2010		6,197	811	20,267	2,708	44%
	Change		**+2%**	**−26%**	**+422**	**+6%**	–
	2020	Tract 13.01	2,696	219	19,827	1,509	56%
		Tract 13.02	3,331	256	19,618	1,810	54%
		Cumulated	6,027	475	19,711	3,319	55%
	Change (In Relation to 2000)		**−1%**	**−57%**	**−556**	**+30%**	–

(continued)

Table 5.2 (continued)

		Total Population	Total Population Under 18	Population Density (Persons per sq. mi.)	Total White Population	Share of White Residents (In Percent)
Tract 14	**2000**	3,283	502	11,393	1,924	59%
	2010	3,084	401	10,702	1,950	63%
	Change	**−6%**	**−20%**	**−691**	**+1%**	–
	2020	3,205	384	11,122	2,203	69%
	Change (In Relation to 2000)	**−2%**	**−24%**	**−271**	**+14%**	–
Tract 15	**2000**	4,010	743	13,091	2,031	51%
	2010	3,934	621	12,843	2,090	53%
	Change	**−2%**	**−16%**	**−248**	**+3%**	–
	2020	3,827	476	12,494	2,402	63%
	Change (In Relation to 2000)	**−5%**	**−36%**	**−598**	**+18%**	–
Greater North Park Area	**2000**	46,631	7,680	13,216	24,808	53%
	2010	45,728	6,057	12,961	24,783	54%
	Change	**−2%**	**−21%**	**−256**	0%	–

Source: Own composition and calculation based on compiled census data (SANDAG 2003c, d, e, f, 2016h, i, j, k; U.S. Census Bureau 2021a). 2020 census data has not been available for the level of the community planning areas. The subdivision of the community into the various census tracts is shown in Figure 5.10. The tracts 12 and 13 have been split for the 2020 census, which typically indicates population growth in the respective area.

levels of economic capital and hybrid lifestyle preferences of suburban housing (reconverted subdivided single-family homes) in relatively urban environments. Generally, the four analyzed tracts differ strongly in terms of population density—the tracts 14 and 15 are significantly less densely populated than the tracts 12 (12.01 and 12.02) and 13 (13.01 and 13.02). The figure-ground plan below (see Figure 5.11) visualizes these differences by means of the outline of the built environment of a selected portion of the community, which allows for certain conclusions on the density and use of the various sections. The housing structures in closer vicinity to Balboa Park (tract 14) and other open green spaces in the community are remarkably fine-grained and thus testifying to residential single-family areas with wide streets and lower density. Accordingly, as census data shows, the share of single-family housing units (detached and attached combined) is relatively high in the two tracts. In tract 14, the share lies at 52 percent in 2000 and 63 percent in 2010 and for tract 15 at 60 percent in 2000 and 65 percent in 2010. In contrast to the significant increase, the total number of housing units has slightly decreased in the two tracts between the years 2000 and 2010 (by one percent in tract 14 and 15). This is indicating a trend of de-densification in the areas closer to Balboa Park.

In comparison, the commercial neighborhood center around the intersection of University Avenue and 30th Street features mostly coarse-grained structures, which are indicating higher density and mixed use. This difference is mirrored also quantitatively: The 2020 census records a population density of 11,122 residents per square mile for tract 14, while tract 13.01 is almost twice as densely populated with 19,827 persons per square mile. In terms of the total number of housing units, compared to the tracts 14 and 15, the values in the tracts 12 and 13 are higher, while the shares of single-family units are significantly lower. In tract 13, the number of single-family housing units has decreased by more than 11 percent between 2000 and 2010, while the total number of housing units has increased by more than 8 percent, as shown above. The comparison of the two cohorts is indicating a densification trend in tract 13, which is connected to the construction of multi-unit buildings.

One of the few currently available topics of the 2020 census is race. In terms of the racial composition of the neighborhood of North Park, the decades from 2000 to 2020 show a significant development. In the first decade, despite the overall population decrease, North Park experienced a commencing slight increase of the share of white residents, from 53 percent in 2000 to 54 percent in 2010. In relation to these numbers, at the turn of the millennium, the population of all the four analyzed tracts was predominantly white. Thus, from 2000 to 2010, the total number and the share of white residents has started to increase slightly in all

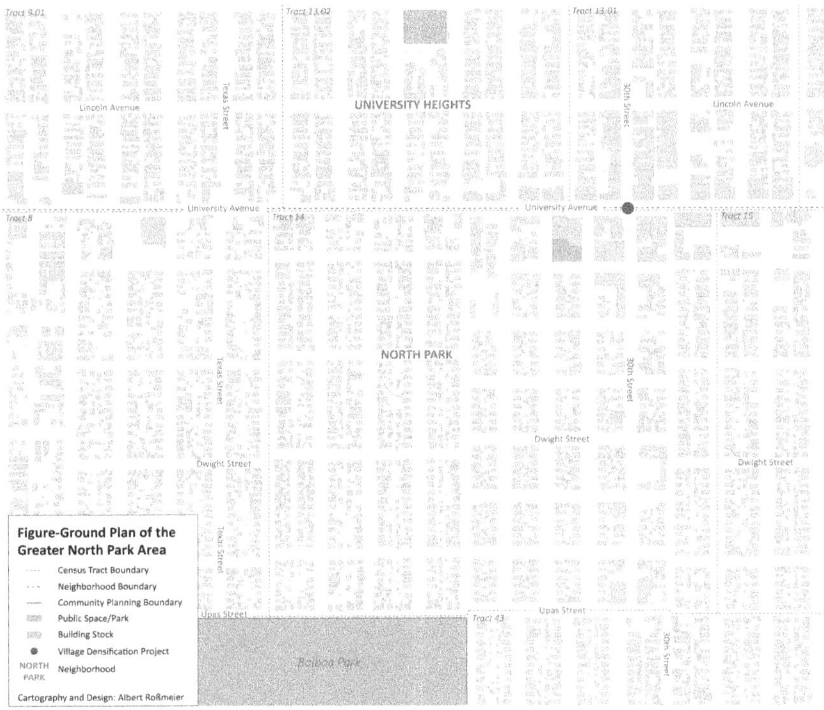

Figure 5.11 Figure-Ground Plan of the Greater North Park Area. (The plan shows the grain size of the built environment, the public green areas, and selected official boundaries of large parts of the community area of Greater North Park. The housing structure in closer vicinity to Balboa Park is fine-grained, consisting mostly of single-family homes in lower density. In contrast, the official neighborhood center around the intersection of University Avenue and 30th Street features mostly coarse-grained commercial and mixed-use structures in higher density. Source: Own cartography and design)

four tracts. While the population of tract 12 was roughly 42 percent white both in 2000 and in 2010, the more 'suburban'-style tract 14 at the edge of Balboa Park recorded a percentual increase of the share of white residents by 4 percent in this time frame (from 59 to 63 percent). In the following decade, however, the influx of white population into the Greater North Park area became more pronounced, resulting in an increase of the share of white residents by at least 10 percent in each tract (except tract 14) compared to 2010. From 2010 to 2020, especially the two denser populated tracts north of University Avenue recorded a strong cumulated increase of the share of white population, from 44 percent in 2010 to 55 percent in 2020 in the combined tracts 13.01 and 13.02 and from 42 to 52 percent in the combined tracts 12.01 and 12.02. Thus, in the last decade, the share of the white population is changing more drastically in the rather coarse-grained (combined) tracts 12 and 13 compared to the less densely populated areas of the community (tracts 14 and 15). In summary, this quantitative development is mirroring the commencement of the redevelopment processes in the 2000s by means of a beginning social shift and the following success of the community's 're-establishment' in the 2010s indicated by a more significant influx or 'return' of white residents. Despite the drastic shift in the (combined) tracts 12 and 13, the areas south of University Avenue feature the highest shares of white residents in 2020 with 63 percent in tract 15 and 69 percent in tract 14.

Table 5.3 Quantitative Neighborhood Change between 2000 and 2010 in Selected Tracts of the Greater North Park Area

		Population 25 Years and above with College Degree	**Median Household Income (in $)**	**Median Housing Value (in $)**
Tract 12	**2000**	1,145 (30%)	$27,297	$138,935
	2010	1,483 (35%)	$35,534	$343,841
	Change	**+5%**	**+30%**	**+147%**
Tract 13	**2000**	1,152 (27%)	$28,330	$158,558
	2010	1,181 (38%)	$39,483	$261,667
	Change	**+11%**	**+39%**	**+65%**
Tract 14	**2000**	1,073 (43%)	$37,872	$225,216
	2010	1,569 (64%)	$67,168	$595,032
	Change	**+21%**	**+77%**	**+164%**
Tract 15	**2000**	1,032 (35%)	$31,505	$180,649
	2010	1,356 (45%)	$54,777	$489,674
	Change	**+10%**	**+74%**	**+171%**

Source: Own composition and calculation based on compiled census data (SANDAG 2003c, d, e, f, 2016h, i, j, k). The tracts 12 and 13 have been split for the 2020 census, which typically indicates population growth in the respective area.

In addition to the presented topics, the development of the educational attainment, the median household income, and the median housing value in the selected tracts will be included in the present data comparison (see Table 5.3). Despite census data on these developments being (currently) only available for the cohorts of 2000 and 2010, it seems beneficial for the quantitatively painted picture of North Park's redevelopment phase to include the three topics, as they illuminate the differences of the various neighborhood sections further. In terms of the educational attainment of residents in the selected tracts, census results from the year 2000 show lower shares of population with college degrees in the denser populated tracts north of University Avenue (tracts 12 and 13) than for the two low-density neighborhood sections (tracts 14 and 15). In 2000, the highest share of residents of age 25 and older with a college degree is found in tract 14 with 43 percent. Taking the 2010 census results into consideration, the relatively 'suburban'-style tracts 14 and 15 show the largest growth rates (21 percent growth in tract 14 and 10 percent in tract 15). In 2010, the tract adjacent to Balboa Park, tract 14, again recorded the highest shares of population with college degrees (64 percent in 2010) even though the total number is the lowest compared to the other three tracts (2,455 residents with college degrees in tract 14). The lowest growth rate from the year 2000 to 2010 (5 percent) and the lowest share of college-educated residents is found in tract 12 (35 percent in 2010). In conclusion, tracts 12 and 13 constitute denser populated areas with a relatively mixed population in terms of educational attainments, while the tracts 14 and 15 are characterized by a largely white population with high educational attainments living in low-density single-family homes and households with higher shares of underage residents.

The development of the median household income from 2000 to 2010 in the selected tracts shows a similar tendency. While the incomes in the four tracts have been roughly comparable in the year 2000 (ranging from $27,297 in tract 12 to $37,872 in tract 14), these values have drifted apart strongly during the following decade. The median of the household incomes has risen by 30 percent in tract 12 and by 39 percent in tract 13, however in the lower density neighborhood sections the values have increased by 74 percent (tract 15) and by 77 percent (tract 14). Accordingly, in 2010, the lowest median income has been recorded for tract 12, which lies between University Avenue and El Cajon Boulevard with $35,534 and the highest value was found in tract 14 with $67,168. Consequently, the denser populated northern tracts around the commercial corridors record lower shares of residents with college degrees as well as lower median incomes and stand in contrast to the tracts 14 and 15, which are characterized by higher incomes and higher shares of college-educated residents. Lastly, the development of median housing values in the various tracts from 2000 to 2010 paints a similar picture. The values in the tracts south of University Avenue have risen by more than 150 percent between 2000 and 2010—by 164 percent in tract 14, from $222,216 to

$595,032, and by 171 percent in the adjacent tract 15, from $180,649 to $489,674. Even though tract 12 has experienced a relatively high growth rate as well with an increase of 147 percent, the total values of both tracts 12 and 13 are not as high as in the low-density areas. In 2010, tract 12 has recorded a median housing value of $343,841, while tract 13 featured a median of only $261,667.

In conclusion, the community planning area of Greater North Park and all the selected tracts have recorded a slight decrease of the total population number as well as of population density in the last two decades. In contrast to the overall decrease, the neighborhood has experienced an influx of white residents, especially since the 2010s, which shifted the racial composition of the area drastically. Additionally, between 2010 and 2020, the share of underage residents in North Park has dropped significantly (see Table 5.2). Thus, similarly to the development of the neighborhood of Hillcrest in the late 20th century, the entire Greater North Park planning area underwent a demographic shift toward an almost exclusively white and adult inner-ring community. However, certain differences can be worked out for the various tracts: The data comparison has shown that the areas south of University Avenue differ strongly from the neighborhood sections north of the community center and that these differences have intensified in the course of the considered decades. Tracts 14 and 15 feature low population densities and a higher share of single-family housing units, the highest shares of white residents, and less drastic decreases of underage residents throughout the last two decades. Additionally, in the timeframe from 2000 to 2010, the shares of college educated residents, of median housing value, as well as of median household income have risen fastest in tract 14 compared to the other selected tracts, followed by the adjacent low-density area, tract 15. In contrast to this, the formerly connected tracts 12.01 and 12.02 as well as 13.01 and 13.02 are characterized by higher population densities, a higher share of multi-family units, and a more drastic influx of white residents in combination with a stronger decrease of the number of underage residents. However, in the timeframe from 2000 to 2010, the growth rates of the share of college educated population, the median housing value, and the median household income have been relatively low in these tracts. In summary, the recent developments in the Greater North Park area are affecting the various neighborhood sections differently and are thus perpetuating their existing structural differences. This is not only due to increasing social and economic interest in the community but also due to varying planning objectives and land use goals for the different sections of Greater North Park.

On the one hand, the recent community plan for Greater North Park aims to "encourage mixed-use development along its major commercial corridors with higher residential densities where commercial goods and services and public transit would be directly accessible" (CofSD 2016, p. 12). This strategy is building on the commercial past of the former streetcar corridors of University Avenue and El Cajon Boulevard, which are already housing coarse-grained physical structures (as seen in Figure 5.11). On the other hand, however, the future land use strategy for the community area also proposes to maintain the area's large "traditional single family residential neighborhoods [...] at low densities" (CofSD 2016, p. 12) despite the General Plan's overall objective to increase density in central locations. These two planning principles bear a certain contradiction, which testifies to increasing difficulties concerning the realization of planning goals in San Diego—especially in terms of infill development—because, as experienced by the urban planner EP-11, "people just like the communities the way they are, and they are very afraid of change in the form of new development happening in those areas and we hear that a lot" (EP-11). Throughout the county's recent planning history, density increases and other development projects in residential neighborhoods (such as in the urbanizing neighborhood of Linda Vista in the next or second suburban ring) have in part been subject to bitter opposition of San Diegans, forcing projects to scale down or to fail completely (discussed by interviewee AR-09 and AR-11, both scholars of urban planning and development and illustrated in detail by Erie et al. 2011, pp. 149–153; for a variety of conflicts involving planning and different spatial uses in the context of landscape see Berr and Jenal 2019). These two different planning approaches enhance the contradictions of the various neighborhood sections and thus the aspects of physical-spatial, lifeworld, and social hybridity in the community, which will be discussed thoroughly in the following by means of selected qualitative research results.

However, before turning to the qualitative components of the research, it is necessary to address an observed looming trend at the community's eastern boundary that becomes visible quantitatively and has been discussed repeatedly in various interviews. In recent years, the described redevelopment wave, which has progressed from Hillcrest into North Park, seems to spread further eastward over the official community boundaries and the physical borders of Interstate 805 and 15 into the western parts of the large Mid-City area. Taking into consideration 2020 census data, significant demographic gradients in west-east direction become obvious—for instance, in terms of the share of adult population and the ethnic composition of the two adjoining communities. Interestingly, the relation between the areas in the west and the east seems to differ strongly along the north-south axis of the official-physical demarcation, thus creating three different sectors.

Figure 5.12 Private Efforts of Branding and Identity Creation and the Community's Aspiration of Spatial Participation: The Little Saigon Neighborhood Mural at the Intersection of Menlo Avenue and El Cajon Boulevard in City Heights. (Source: Picture Albert Roßmeier 2019)

The most northern sections of the two communities—the northern parts of North Park (University Heights), Normal Heights, and the neighborhood of Kensington—are characterized by comparable demographics, they constitute areas with low shares of Hispanic and Asian population and are predominantly white and adult. In this sector, the community boundary and the physical borders of Interstates 805 and 15 do not seem to constitute significant demographic demarcations. In the central sector, which runs along the two commercial corridors, sustained gradients become visible from North Park over Normal Heights into City Heights and the semi-official Little Saigon district (see Figure 5.12). Here, 2020 census data shows a strong decrease of white residents from west to east, from the combined tracts 12 and 13 eastwards. Additionally, strong increases in the shares of Hispanic and Asian population as well as of residents under the age of 18 are recorded along this axis from west to east. Quantitatively, the community boundary and the physical borders—the two Interstates—seem to demarcate the various neighborhoods in this sector in a more pronounced way than at the northern ends of the communities. Lastly, the northern and the central sectors and the ways they are connected to the Mid-City community differ strongly from the southern section of the Greater North Park community next to the intersection of the Interstates 805 and 15 (east of the low-density tracts 15 and 42, see Figure 5.17). In this area, the Interstates 805 and 15 create pronounced demographic boundaries between two increasingly distinct communities in terms of their ethnic composition. Instead of gradients, the freeways constitute strong cut-off lines between a predominantly white area on the west (the low-density tracts 15 and 42) and a Hispanic neighborhood of City Heights on the east. Additionally, in this sector along the intersection of Interstates 805 and 15 and south of Landis Street, no pedestrian or street crossings are possible, which enhances the harsh urban border between southern North Park and City Heights (the next possible crossing is Broadway in Golden Hill). In the following, this initial, quantitatively elaborated perspective on the border and boundary situations between Greater North Park and the Mid-City community will be triangulated with qualitative insights from the conducted interviews and participant observations.

5.2.2 Qualitative Insights into the Redevelopment of Greater North Park and the Community's Multi-Dimensional Expansion into the Mid-City Community

Below, the first paragraphs deal with individual interpretations of the neighborhood change of the Greater North Park area with a particular spatial focus on the community center around the intersection of University Avenue and 30th Street. Subsequently, the multi-dimensional progression of the 'redevelopment wave' from North Park into the western parts of the Mid-City community will be discussed. The sub-section closes with subjective observations and conclusions from the neighborhood walks (Figure 5.13).

Questioned about North Park, its history, and current dynamics, interviewee R-12, who grew up and attended an elementary school in the neighborhood in the 1970s and '80s, claims that the population of North Park as well as of the areas along the entire eastern side of Balboa Park (including South Park, Golden Hill, and Southeastern San Diego) used to be predominantly black and Hispanic during these decades. The interviewee remembers North Park as an 'unsafe' neighborhood, "it was rough down there" (R-12) and the local residents were "aggressive" (R-12), especially toward the city's white population. In the 1930s, as interviewee R-12 explains, when the commercial corridor of University Avenue was thriving, a great number of newcomers to San Diego bought houses and properties in North Park. As these residents aged and passed away in the second half of the 20th century, the building stock had become outdated and run down, according

Figure 5.13 Efforts of Branding and Identity Creation: The Re-Erected North Park Neighborhood Sign at the Prominent Intersection of University Avenue and 30th Street as a Practice of Symbolic Neighborhood B/Ordering and Othering. (Source: Picture Albert Roßmeier 2019)

to the interviewee. However, in the late 1990s, "the young hipsters came and began to buy these homes" (R-12). The Ocean Beach resident describes the onset of a 'neighborhood resurgence,' which they view as "unbelievable" (R-12; similarly voiced by EP-05). They state further that North Park and the intersection of University Avenue and 30th Street have faced the most radical change of all the areas of central San Diego (interviewee R-06 argues equivalently). Referencing the impressions of their wife, this intersection "is the heart of the change, you know, it used to be really scary and dangerous and now you see women in high heels walking alone at two in the morning" (R-12).

Interviewee EP-11, an urban planner in San Diego shares memories about their childhood in North Park, too. Equivalently to interviewee R-12, EP-11 is as well referring to the previously disinvested state of the neighborhood and its commercial corridors in the 1980s. Interestingly, their interpretation seems to take place within a combination of a resident-specific lifeworld standpoint of subjective remembrance and a professional planning approach. This includes the interpretation of food availability and access (for a geographical perspective on and conceptualization of foodscapes see, Sedelmeier et al. 2021) in and around the neighborhood as well as of the community's building stock and its condition: "I remember living when I was, I think I was about 11 or 12, my family moved from Northern San Diego County down to North Park and I remember that kind of commercial corridor in North Park being a lot less invested, you know, more kind of low cost shopping but it wasn't easy access to supermarkets. The supermarkets around were run down, the homes themselves were less well maintained" (EP-11).

Similar to the interviewees R-12 and EP-11, the young college graduate AR-07 describes the recent neighborhood history of North Park by means of biographical insights as well. In their illustration the redevelopment processes in North Park are tied to the reactivation of empty retail space and the replacement of specific businesses along University Avenue and El Cajon Boulevard, which have made way for new offers and demands of the newly arriving residents:

> "I grew up here in North Park. My parents came here from Mexico in the 80's. First, they lived in Little Italy back when whole downtown was still pretty sketchy, it wasn't what it is now. They lived in Little Italy and then eventually they came to North Park, it was cheap, and it wasn't what it is now. It was way cheaper, it was still, El Cajon Boulevard and also University Ave were really sketchy because there is a whole history behind why these streets kind of went under. And then there is also 30th street, all these [AR: the interviewee pointed toward a building on 30th Street during the face-to-face interview], I think one of these was a bike shop or a hardware repair store, and they are definitely not now. These used to be really run down, and a lot of them were closed when I was growing up here. And this didn't really start, so I lived here in North Park from when I was born, and then around 2002 we moved to City Heights, and then we moved to Skyline, that's a little more ghetto area, and then we came back to North Park in 2010" (AR-07).

In contrast to the illustrations of the interviewees R-12 and EP-11, due to their difference in age, the childhood and teenage observations and experiences of interviewee AR-07 in North Park took place at a later point—in the 1990s and 2000s. Accordingly, Ocean Beach resident R-12 records the onset of the neighborhood 'resurgence' of North Park as early as the late 1990s, while interviewee AR-07 states that the redevelopment dynamics have begun in the late 2000s: "From 2008 or 2009, all this started really to change. I don't really know what the catalyst was. Oh yeah, it was Hillcrest first. Hillcrest started earlier, like '90s, and then it would come down, and it continues on that way to, like, University Heights, Normal Heights, all that, Little Italy, downtown" (AR-07).

Theoretically, when comparing the three interpretations of the interviewees R-12, EP-11, and AR-07—all of which agree on the profound redevelopment of the community—neighborhood perception and construction do not only seem to be tied to different 'modes' of interpretation (such as resident, expert, or common-sense perspectives) but are also highly dependent on subjective memories and—as in this case—the comparison of the memorized past with interpretations of the presence and potentially also ideas about the future. This underlines the temporal dimension of neighborhood perceptions in which the evolution of individual and common social patterns of interpretation takes place—similar to the social process of landscape construction and valuation. An interesting example for the temporal evolution of social landscape interpretation is the postmodern appreciation of and fascination for physical structures and sites of the industrial age (such as the German coal mine industrial complex *Zollverein*, a designated UNESCO World Heritage Site), which have not experienced broad social aestheticization into 'fascinating industrial landscapes' until a few decades ago.

Concerning the pace of the redevelopment trend and the described reactivation of retail space in North Park, interviewee R-10 shares: "ten years ago, it was just less, you know, it was less developed, lesser restaurants and interesting people. There was more vacancy, and the neighborhood was just not as popular as it is now. The brewery culture hadn't arrived yet." According to the interviewee, the successful establishment of breweries and tasting rooms and the growth of the craft beer scene in North Park are connected to the increased use of social media apps, such as Instagram, in which younger people use location tags in combination with photo or video documentation in order to "be seen in the bars" (R-10). In the same vein, interviewee R-12 is linking the successful comeback of the intersection of University Avenue and 30th Street to the arrival of "all the breweries, which attract the hipsters." As illustrated by the San Diego Tourism Authority (2022, n.pag.), over the years the Greater North Park community has developed into a destination "where hipsters, craft beer, families and music converge [...]. The epicenter of North Park is the bar-hopping intersection of University Avenue and 30th Street," which had in the early 2000s still been stigmatized by crime and perceived as a sketchy and run-down area by various interviewees.

Summing up the discussed changes concisely, interviewee EP-06 comments: "North Park is changing." Additionally, drawing on biographic memories, the regional planner EP-12 shares: "I remember, like, North Park is very different now. This neighborhood's changing, I think, more housing gets put in there and I think it has gentrified" (similarly interpreted by EP-01, a business owner in Hillcrest). In terms of new development, the developer EP-04 from Barrio Logan is describing the replacement of smaller single-family homes with larger multi-unit and even high-rise buildings—"the houses just get absorbed there [...]. They take one, two, three family homes and turn them into multi-family options with no parking." The described replacement of single-family housing units with larger structures coincides with the picture painted above by means of census data for the years of 2000 and 2010, which shows a slight increase of the total number of housing units combined with a significant increase of the share of multi-family housing units in the residential area between University Avenue and El Cajon

Boulevard (census tract 13). Additionally, in the course of this conversion trend, single-family housing units in North Park do not only seem to be transformed for residential purposes but also for commercial uses due to increasing pressure on the neighborhood and its real estate and corresponding zoning changes (see Figure 5.14). Typically, this constitutes a phenomenon of early and medium-stage gentrification and urban/rural hybridization in inner-ring neighborhoods (Kühne et al. 2016).

Figure 5.14 Urban Redevelopment Progress in Greater North Park—the Conversion of a Small Older Residential Building on University Avenue for Commercial Uses. (Source: Pictures Albert Roßmeier 2017 (top) and 2019 (bottom))

Today, however, interviewee AR- 08 sees "North Park, kind of, on another level of gentrification where it's already past, it's past Barrio Logan, it's past City Heights." According to interviewee R-10, North Park is becoming more expensive and experiences gentrification. As a consequence of increasing rents and other costs of living in North Park, "the people who used to be able to afford the inner ring now have to move more outwards. The development is definitely spilling over, we will see where it goes next. [...] I see a lot of people buying places in the next ring" (EP-04). In this statement, the developer EP-04 is referring to two particular destinations, which they view as part of the 'next' or second ring: The large Mid-City community east of Greater North Park and National City, an incorporated city, which lies southeast of the developing community of Southeastern San Diego. Similarly, in terms of the direction of the redevelopment trend, interviewee R-08 emphasizes that "North Park has changed already, so it's going south because you cannot go north, there is only family homes" (R-08). Arguing equivalently, Interviewee R-04, a long-term resident from Hillcrest interprets the current displacement tendencies in North Park and South Park as consequences of the preceding redevelopment of Hillcrest. According to their interpretation, over the last decades lower income residents have been pushed from Hillcrest into the North Park area and now further eastward into the Mid-City community. In terms of the displacement trend, the urban scholar AR-02 is emphasizing the connection between the socio-spatial organization of central San Diego and the historic freeway constructions: "The freeways have always been isolating neighborhoods, in parts intentional, causing blight and separation of communities. [...] [But now,] people have been moved over the freeways."

In addition to these statements, several interviewees describe not only the push of residents but also the gradual progression of the redevelopment pressures over the two Interstates 805—the official and physical border of the Greater North Park area—and 15 into the Mid-City community, which are causing "a similar process of beautification in City Heights" (AR-08). Sporadically, breweries, coffee shops, and other trendy businesses are opening up in Normal Heights and City Heights along the commercial districts of El Cajon Boulevard and University Avenue, as observed by interviewee AR-10 as well as by the author during one of the neighborhood walks (PO-03; see Figure 5.16). Interviewee R-06, a senior resident from Hillcrest perceives a beginning transition of the neighborhood of City Heights—"you see new apartment buildings going up in City Heights as well." Currently, the area is still "more laid back than North Park, it's not as hip, you know, also rents are lower compared to North Park" (R-10). According to several interviewees, the described lower rents in the Mid-City community arouse the interest of younger people that work in San Diego's central areas. Interviewee EP-05, who is involved in a place-making project in downtown San Diego, is living in "Normal Heights, because it's more affordable." For the same reason, interviewee R-03, a barista in a trendy café in North Park, moved to City Heights a few years ago. The interviewee perceives the neighborhood as an urban, low-income area, "but within the last 10 years people with higher incomes started to move to City Heights because it's still pretty central" (R-03).

In addition to affordability, important aspects of housing and location preferences are centrality and proximity to business-lined streets and trendy neighborhoods, as discussed by several interviewees. The real estate developer EP-04 is convinced that "people wanna live close enough to things [...]. We wanna be part of the scene, but we can't afford houses." Another reason for the commencing success of the inner and partly also the second ring neighborhoods, such as North Park, Normal Heights, and City Heights, is that "a lot of people don't want to be downtown, they don't like homelessness, but they wanna be close to downtown and close to Hillcrest, it is an attractive neighborhood," as interviewee EP-07 explains. This bespeaks the hybridity of the lifestyle and housing preferences that are currently promoting the 'resurgence' of the central inner ring neighborhoods. Aspects such as urbanity, density, and social and cultural diversity are valued but only to a certain extent and in a very selective way in which the new residents want to "feel like there is a little buffer between them and downtown. [...] It's like you need a little separation" (R-13). This interpretation can be linked to the provided quantitative illustration of the redevelopment of the more suburban areas in the North Park community (tract 14 and 15), which have recorded an increase of the share of white residents and single-family housing units in combination with a decrease of the total number of housing units, the total population, as well as of the population density throughout the last two decades.

This is interesting from two aspects. First, not only housing and location choices in San Diego are following this principle of spatial separation, but also the partly strong social opposition to planning efforts of densification and diversification in the forms of urban infill, retrofitting, or affordable housing projects (as mentioned above) is emerging from this very conflict about proximity. In planning and spatial sciences, this social attitude is often referred to as NIMBYism, which is especially found in residential low-density areas and applied when "they [the residents] fear that it's going to negatively impact the character of the suburbs" (AR-11; on the phenomenon of NIMBYism in the context of wind energy production see Roßmeier and Weber 2021 [2018]; F. Weber, Jenal, et al. 2017). Second, looking at the regional scale, the 'necessary spatial separation' between different cultures and groups in San Diego—illustrated by the interviewees EP-04, EP-07, and R-13—has historically run along Interstate 8. This freeway on the northern end of North Park "just seems to be a real significant border between what's considered residential San Diego and urban San Diego. And also, the Hispanic influence stops at the [Interstate] 8, or at least it slows down significantly," as interviewee R-13 from Carmel Valley illustrates (equivalently described also by the local border scholar AR-05). However, as shown by census data, the gradual redevelopment of San Diego's central neighborhoods is causing a significant shift of the ethnic composition in the trendy communities south of Interstate 8 (such as Uptown, Greater North Park, and Greater Golden Hill) toward higher shares of white residents. Border-theoretically, this is less a form of a beginning debordering of the cultural and ethnic boundary that has run along Interstate 8 for decades, but more of a push and relocation of this everyday life demarcation toward the south and southeast as further discussed below.

Concerning the population of the Mid-City community, interviewee AR-08 describes City Heights as "a cool neighborhood, really diverse, there's a lot of refugees from all over the place," which they regard as "vulnerable population in terms of economics" (AR-08). In the same vein, Interviewee R-11, who is an activist opposing gentrification in East Village, shares that Normal Heights is home to a large immigrant population, which is currently facing displacement. As illustrated by interviewee AR-11, the "immigrants and refugees" have settled in City Heights over the decades and "between, like, 1960 and 1995, the neighborhood had gone from maybe 90 percent Anglo kind of middle class to 90 percent non-white working class" (AR-11). Today, as shown above by means of census data, the socio-economic and ethnic composition of the neighborhoods to the west and east of the Interstate 805 is in certain parts very different. While the northern sections of the Greater North Park and the Mid-City communities—the neighborhoods of University Heights, northern Normal Heights, and Kensington—consist of a relatively homogenous population of white college-educated residents with higher incomes, the population in the section around the commercial corridors of El Cajon Boulevard and University Avenue and especially the parts south of this area differ strongly in terms of their ethnicity and income levels. In these two community sections, the white middle-class population of North Park in the west contrasts with the Latino and Vietnamese lower income population of City Heights to the east of the Interstates 805 and 15. Thus, over the time, the two freeways have not only constituted a strong physical border but have grown into a pronounced socio-economic and socio-cultural boundary of central San Diego, which are of significant everyday life relevance, as further elaborated below.

Additionally, the discussed spatial divergence of the two communities is also reflected in terms of their infrastructure, as Interviewee AR-07 explains: "If you go past the 805 freeway, you can still see, like, on University [Avenue], there is still lots of shops, but past here they have gotten a little more run down." Interviewee R-03 is describing the differences between the two communities accordingly: "Once you pass the [Interstate] 805, everything is less developed, and you can totally see the difference." This perception goes along with own subjective impressions from the participant observation conducted in the respective communities. The current state of the overall infrastructure and the building stock in the Mid-City community resembles the discussed previous state of the neighborhood of North Park and indicates s large-scale development efforts have not (yet) taken place. Unlike in Hillcrest and North Park, the majority of the private homes in the western Mid-City area have not been renovated and the local businesses cater predominantly to the local immigrant and lower income population (PO-03; see Figure 5.18). Interviewee AR-11 is shedding more light on the recent development trajectory of the western Mid-City community and the disparity with North Park. Their statement bespeaks both the individual's expert knowledge and their experiences as a long-term resident of City Heights:

> "In the 2000's when San Diego's real estate market got so overpriced, so people are moving an hour away from the city to buy houses and then even those houses were 300,000 and 400,000$, and so people were kind of like, 'I can't figure out what to do.' So, they started looking at these inner neighborhoods differently and being more willing to accept them, which, of course, realtors start to play a role in, in presenting the neighborhoods differently. And a lot of those apartment complexes that were seen as really sub-standard housing suddenly became candidates for condo conversion. Then you have rental units coming off the market and homeowners moving into City Heights. So, there was a brief period, probably like 2002 to 2007, where people were really concerned about gentrification in City Heights, [...] but the apartment complexes are really low quality. Virtually, the character of the housing that was added, I think, now holds gentrification at bay. Because there is a few of these really 1920's, 1930's houses, but they are surrounded by low quality apartment complexes that are jammed onto single family home lots. So unlike, like, North Park that has just exploded with value and new, kind of, trendy residents, City Heights really hasn't experienced that yet" (AR-11).

The interviewee shares further that the Mid-City community has been attributed a bad reputation for two reason: on the one hand, because of its high population density—the neighborhood "has become incredibly overcrowded" (AR-11) over the decades—and, on the other hand, "basically, in every case in San Diego, any kind of ethnic identity or cultural association with a place devalues it" (AR-11). Little Italy is specifically excluded in this statement, as the interviewee explains further, because compared to the Latino culture, the "Italian identity is very marketable."

However, despite the historically negative reputation of City Heights and El Cajon Boulevard in general (as deplored by EP-02), interviewee AR-07 emphasizes that now "even City Heights is getting more expensive" and slowly starting to see new development and an image change. Efforts of the local business improving association, place-making projects, and newly opening businesses are promoting a "growing sense of community" along El Cajon Boulevard, as argued by the local 'place maker' EP-02. The commercial corridor that is connecting North Park, Normal Heights, Kensington, and City Heights is "attracting more people from different backgrounds now" (R-03) because "people are celebrating this new era of locally oriented economic development" (EP-02). In the same vein, also the urban planner EP-11 is connecting the changing neighborhood image to the perceived commencing influx of new residents:

> "So, now that City Heights, you know, over time has been getting perceived to be safer, that's all why people moved in for example, so that started to increase now to the east of the [Interstate] 15 as well, you know, I've heard of a lot of people and professionals buying homes in the eastern City Heights area where there are still a lot of those kind of 1940's, 1950's suburbs. And it started to change all throughout City Heights, you know, kind of, at least, I think" (EP-11).

Narrating more from an expert perspective, interviewee EP-09 connects the 'resurgence' of San Diego's inner-ring neighborhoods to the realization in urban studies and planning that "people have this preference to move back to these urban neighborhoods where there's a lot of local restaurants, a lot of local bars, breweries, where it's definitely walkable but it's not as dense as downtown. So, the North Parks, and the Barrio Logans, there is so many, Kensington and Normal Heights, all those urban neighborhoods are more desirable and that's also, the market then is driving the developers, right? They develop more and more in these areas" (EP-09). Thus, the interviewees are describing how the Mid-City community, among others, is beginning to change socio-spatially, physically, and in terms of its reputation due to changing housing and location preferences of younger generations. The change processes that have taken place in Hillcrest and North Park—in parts decades ago—are thus slowly becoming traceable in the Mid-City community as well. For, as discussed above, City Heights is beginning to attract new residents and businesses and is slowly gaining the interest of real estate developers. Ultimately, not only former residents of Hillcrest and North Park are crossing the freeways in their search for more affordable housing, also new businesses and housing developments are becoming visible beyond the official eastern boundaries of Greater North Park. Thus, the strong everyday life relevance of the Interstates 805 and 15 as symbolic and life-world demarcations between two distinct communities in terms of social, economic, and infrastructural aspects is partly eroding—at least along the two commercial corridors of El Cajon Boulevard and University Avenue (see Figure 5.19)—in favor of the continuously developing and reconnecting first and second rings around San Diego's downtown, which have historically been split up into fragmented communities in the course of the freeway constructions of the 20th century.

In addition to the discussed social and infrastructural tendencies, several interviewees have illustrated a specific economically driven practice of breaking down the multi-dimensional everyday life demarcations between Greater North Park and Mid-City. According to EP-11 and AR-11, for sale properties and condominiums in Normal Heights and City Heights east of the official community boundary of Greater North Park are regularly promoted in real estate advertisements as being located in North Park or in a fictitious section of the neighborhood referred to as 'East North Park.' Interviewee EP-11 is interpreting these marketing practices as a consequence of the successful redevelopment of North Park and the value it added to the local real estate. Furthermore, the interviewee is using this example to discuss how everyday life boundaries between neighborhoods are subject to change and that these changes might not be recognized on the level of official city planning despite their everyday relevance for local residents:

> "Our generation wants to be in more livable, walkable, kind of, urban type areas with the activity and resources and so that's, even for the last 10 or 15 years, it has been affecting the surrounding areas as well. So, I remember looking for homes or places to rent and the North Park area, even like 10 years ago, seeing places that were advertised as being in North Park, but when you looked at the map they were East of the 805 in City Heights, so people were trying to already project the value of North Park and its attractiveness into these less invested areas to the East. [...] I think we see that a lot in San Diego currently, these flexible perceptions of boundaries, like, the idea of just a lot of city planners that we have certain barriers and defined boundaries for our planning documents, you know that 'this is this community, this is another community,' but we see people, you know, either creating communities between or stuff that's not kind of visible on the level of our city planning documents, or people trying to kind of extend the imaginary boundaries of very desirable communities into adjacent communities for the purposes of capitalizing on the attractiveness, the kind of soft value of those names, and the kind of image they project. [...] So, it seems to be there, like, where people are trying project, kind of, the quality of desirable and that is when they kind of start getting a little loose with their naming convention" (EP-11).

The interviewee AR-11, a long-term resident of City Heights, states to be "very irritated by the incursion of North Park on our western border" and discusses the eastward progression of the redevelopment trend against the background of a critical historic perspective:

> "The less valuable the land was, the less investment there was in bounding it and naming it. So, City Heights became this kind of amorphous area, large area in the center of the city. Some people in the city used the term Mid-City kind of synonymously with City Heights. So, well I'm irritated by this marketing strategy of rebranding at the margins, the truth is that City Heights is kind of vulnerable to it because it's such

> a large area that's really not as easily understood as one neighborhood. So, the part that's being called North Park is between Interstate 805 and Interstate 15 and that's a really distinct wedge because of that freeway completion, so they've been kind of 'City Heights is east of the 15, not east of the 805, and you could just claim this big chunk of territory'" (AR-11).

As a consequence of the described marketing and rebranding, interviewee AR-11 deplores that "public understanding of North Park will shift and then they [the residents of the western Mid-City community] will be confused when they go to a planning meeting and they're not included in the North Park plan because they're not North Park, you know. But in the meantime, they all gonna think they're North Park" (AR-11). This statement allows for several theoretical conclusions and enables distinct perspectives on social and spatial identity and the concepts of neighborhood and landscape. Arguing from a post-structuralist perspective, the interviewee's interpretation is highlighting the processuality of social and individual identity, which is affected by symbolic neighborhood changes and changes of spatial identity in the form of official, social, or private renaming and branding. In line with social constructivist landscape theory, the (attempted) symbolic changes of the neighborhoods of Normal Heights and City Heights gain particular relevance in everyday life because "cultural groups come to acknowledge them through a redefinition of themselves" (Greider and Garkovich 1994, p. 21), as predicted by interviewee AR-11. On the one hand, this is highlighting the interdependency of social and spatial identity. On the other hand, this example is revealing the changeability and processuality of neighborhoods and landscapes and thus also the everyday life connection between the two concepts because, as argued by social constructivist landscape theory, in any case "the open field is the same physical thing, but it carries multiple symbolic meanings" (Greider and Garkovich 1994, p. 21). In line with this understanding, it can be concluded that the meaning of both landscape *and* neighborhood is "about continuous change of the interplay between individuals, society and nature" (Jauhiainen 2003, p. 397) or space in general.

In addition to interviewee AR-11, other interview partners from various perspective categories, including resident-specific everyday life perceptions as well as interpretations through expert knowledge, are discussing the connection between social and spatial identity as well. In the context of downtown redevelopment and the installation of neighborhood signs and banners in the various neighborhoods of the city center, the place-maker EP-10 emphasizes the social demand for spatial uniqueness and distinct identity of communities. "There is East Village, which is a lot different than Gaslamp, a lot different than City

Center and people who live in those areas are, you know, they want to be associated with their neighborhood and want some kind of ties to it." Generally, this interpretation of location preferences and identification demands of residents in downtown San Diego is reflecting the postmodern search for and creation process of individual identity, which is increasingly fragmented and patchworked, fragile, and ultimately a 'project of everyday life' (Holstein and Gubrium 2000). Similar to interviewee EP-10, R-10 describes that "every neighborhood of San Diego somehow has their own lifestyle, you know. Many millennials move here and choose the neighborhood that fits their lifestyle best." Thus, the interviewee is describing the spatial, social, and symbolic fragmentation and compartmentalization of the City of San Diego into various unique neighborhoods and lifestyles, which contrast each other. According to interviewee AR-11, this spatial and social dissipation is promoted by a variety of aspects:

> "People are very attached to this idea of a neighborhood within the city. There are several reasons for this that I can think of. One is the topography and the canyons, canyon and mesa, which tend to create these kind of well-connected small neighborhoods that are divided from other neighborhoods by canyons and that, so there's a little bit of that. There's a little bit of 'I hate the city, so I wanna live in a small town. Tell me what small town I live in.' You know, everybody that lives in the beach towns say they live in Ocean Beach, Pacific Beach, Mission Beach, they don't say they live in San Diego, they live in North Park, they live in, you know. So, San Diegans area really attached to that" (AR-11).

In this statement, interviewee AR-11 is connecting the success and social valuation of distinct neighborhoods in San Diego to topographic and physical features, such as canyons, mesas, and highways, which enhance the symbolic and social separation. As stated in the current community plan for the Greater North Park area, "these characteristics also provide a sense of seclusion from adjacent communities not uncommon for San Diego's neighborhoods. Adjacent freeways reinforce this relationship as they have usually followed canyons and other low points in San Diego" (CofSD 2016, p. 6). In addition to these physical borders, interviewee AR-11 is describing social ordering and othering process, which are establishing immaterial boundaries between the various neighborhoods through the attribution of distinct lifestyles.

However, in contrast to this multi-dimensional compartmentalization of San Diego into a 'city of neighborhoods and villages,' various communities are experiencing a certain (re-)connection through processes of neighborhood assimilation and uniform redevelopment approaches. In this process, spatial pastiches and hybrid in-between spaces emerge at the neighborhood edges, which simultaneously divide and connect the various communities. In the present case context, the 'unique lifestyle' that is perceived to characterize Greater North Park is gradually attributed to the neighborhoods of Normal Heights and City Heights. Thus, a certain everyday life connection beyond the official neighborhood boundary and the physical borders of the Interstates 805 and 15 is slowly being established by means of this pastiche-like social reconceptualization of the Mid-City community. However, the commencing reinterpretation is not only promoting socio-economic, infrastructural, physical (in terms of renovations and new constructions), and image changes, as discussed above, but is also going along with initial updates of the symbols and signs in the community's public space. Building on insights from the participant observation (PO-03), in the following, subjective impressions and thoughts about symbolic aspects and changes in Greater North Park and Mid-City will be presented.

During the participant observation in the neighborhoods of North Park, University Heights, Normal Heights, Kensington and the large City Heights area, symbolic or aesthetic differences between the various neighborhoods became apparent. Summarizing the subjective impressions from the phenomenological neighborhood walk, the differences between the research areas in the west and the east of the Interstate 805 are (still) very pronounced. The Greater North Park community is home to several breweries and tasting rooms, trendy cafés and vegetarian restaurants but features also other unique offers, such as cooking and dancing classes (PO-03). Along with these activities and businesses, new aesthetics have been introduced in the neighborhood, which are intended to amplify the 'trendy atmosphere' visually. The new businesses are drawing on colorfully designed murals and logos as well as new landscaping and other contemporary, cosmopolitan design elements—"you know, Instagram-able stuff," as interviewee R-10 phrases it. In comparison, the business structure in City Heights and the represented aesthetics and symbolics differ strongly. Along the commercial corridors, mostly car dealers and repair or tire shops, liquor and other inexpensive stores—often barred –, pawn shops, and a large number of Hispanic and Asian markets and fast-food restaurants with foreign signs can be found next to a significant number of vacant or abandoned buildings and storefronts. The way the businesses along El Cajon Boulevard and University Avenue in Normal Heights and City Heights are advertised is remarkably different from the aesthetic-focus symbols and signs used by the restaurants of North Park and Kensington (see Figure 5.15). Furthermore, the difference is also mirrored in the provided furniture of the local businesses. While some restaurants and cafés in North Park feature wooden and unique, handmade interior, businesses in City Heights are

Figure 5.15 Varying Symbolics and Signs Representing Different States of Redevelopment Progress—Murals and Business Advertisments in the Neighborhoods of Kensington and City Heights. (Differences in restaurant advertisement: emotional, aesthetic-focus murals and logos in Kensington (top) and more rationally designed, information-focused signs in City Heights (bottom). Source: Pictures Albert Roßmeier (2019))

often equipped with low-cost options, such as plastic chairs and tables. Thus, in a phenomenological sense, the disparity of the two communities has become apparent in terms of the experienced smells, tastes, and feels of the various restaurants, shops, and other features in the streets. Generally, it can be concluded that the environment in North Park is subject to postmodern values in terms of the focus on aesthetics and emotionality, while the symbolic and physical environment of City Heights and Normal Heights is (still) following a largely modern, functionality-oriented imperative.

Furthermore, also the condition of the housing stock in the various neighborhoods varies strongly. While North Park, northern Normal Heights, and Kensington feature mostly new but also renovated historic, larger single-family buildings with well-kept gardens, the southern section of Normal Heights and the large neighborhood of City Heights consist mostly of unrenovated, simple, and smaller residential buildings, which suffer from a lack of maintenance. Additionally, a large number of front yards in the City Heights area contains broken furniture or other junk items, while the fences or walls of private homes and businesses are often times vandalized with graffiti. The front yards in the low-density area of North Park, in turn, are decorated with ornaments and signs with political, inclusive slogans. Additionally, the described difference between the neighborhoods is continued in terms of the condition of the public infrastructure, such as the landscaping, playgrounds, and sidewalks—the areas east of Interstate 805 and south of El Cajon Boulevard appear to be less well maintained compared to the neighborhoods of Hillcrest and North Park. Accordingly, the streets and the playgrounds of the neighborhoods of North Park, northern Normal Heights, and Kensington are more frequented than the public space along and in the areas south of El Cajon Boulevard. The well-maintained commercial street of Adams Avenue in Normal Heights and Kensington, for instance, is frequented by pedestrians and bicyclists and—relatively uniform—young and white users with a higher stock of economic and symbolic capital compared to the remarkably low number of pedestrians in the streets of City Heights. In contrast to the redeveloped and trendy areas, homelessness is more prevalent on as well as south of El Cajon Boulevard in the neighborhoods east of Interstate 805. There, the local population appears to be ethnically more mixed and is rather characterized by lower stocks of economic and symbolic capital.

In conclusion, during the participant observation (PO-03) the elaborated commencing progression of the redevelopment trend from Hillcrest over North Park into the neighborhoods of Normal Heights and City Heights appeared to take place especially in the areas north of El Cajon Boulevard, in the neighborhoods of northern Normal Heights and Kensington. Even though new businesses and the construction of new, large apartment complex buildings as well as a trend of increasing housing costs are now—according to several interviewees—also increasingly found eastward of Interstates 805 and 15, these developments are quantitatively still small in number. As a result, the neighborhoods in study do not appear uniformly in terms of their displayed symbolics, signs, and aesthetics. Thus, it can be concluded that despite the descriptions of several interviewees about the commencing progression of the redevelopment trend over the two freeways into the Mid-City community, this development is currently taking place only along the axes of the commercial corridors and—more importantly—only in the form of isolated, individual projects or businesses, which enhances the hybrid, pastiche-like transition state of the community.

5.2.3 Urban Landscape and Border-Theoretical Conclusions for the Communities of Greater North Park and Mid-City—Simultaneous Tendencies of Spatial and Social Separation, Assimilation, and Hybridization

The following paragraphs will summarize the presented analysis of neighborhood change processes in and between the Greater North Park and the Mid-City community. The results will be discussed against the present theoretical background of urban landscape and border theory, which allows to address the communities' simultaneous tendencies of separation, assimilation, and hybridization. In addition to the written synthesis of the empirical insights and the underlying theories, a three-dimensional sketch is provided, which aims to summarize and illustrate the results in an alternative but concise and theory-oriented cartographic way (see Figure 5.16).

The Greater North Park community is situated at the northeastern edge of Balboa Park, officially framed by Interstate 8 and Mission Valley in the north, the Interstates 805 and 15 as well as the Mid-City community in the east, the neighborhood of South Park in the south, and the urban recreational area of Balboa Park as well as the trendy LGBTQ-neighborhood of Hillcrest in the west. Established in the late 19th century as a citrus farming community, North Park has developed into a thriving streetcar neighborhood in the course of the beginning 20th century. As a result of its proximity to Balboa Park and the success of the international exposition held in San Diego in 1915, today's Greater North Park community has continuously attracted residents in their pursuit of the emerging suburban dream. In the 1920s and '30s, the streetcar corridor of University Avenue has developed into a significant shopping and commercial street of San Diego, which underlined the importance of the community. However, as a result of the transition of the means of transportation from the streetcar to private automobiles in the mid-century period, the relatively narrow dimensioned streetcar corridors and their commercial structures had been left behind for new shopping malls along the emerging multi-lane freeways of San Diego County. Even though commercial activity on University Avenue and the more car-centered corridor of El Cajon Boulevard had been thriving until the early post-World War II era, the neighborhood of North Park finally began to decline in the 1960s and '70s. Large parts of the affluent white population of the once prosperous inner-ring suburb had begun to leave the central area of San Diego for the spacious and economically as well as ethnically homogenous suburbs at the urban fringes and along the coast line of Southern California. As a results of restrictive covenants in the emerging neighborhoods of the next rings, economically less well-situated households as well as ethnic groups had become contained in the more urban communities of San Diego, such as Greater North Park and Mid-City. Additionally, the outward drift of purchase power had prompted the decline of the commercial corridors of University Avenue and El Cajon Boulevard.

However, in the late 1990s, the redevelopment dynamics of the neighboring community of Hillcrest had begun to progress eastward into the North Park area. Renewed economic and broadened social interest in the neighborhood have promoted the renovation of the dilapidated, partly historic building structure in the sections bordering Balboa Park and at the southern and northern community edges as well as the construction of new, large mixed-use structures along the redeveloping commercial corridors of University Avenue, El Cajon Boulevard, and 30th Street. Thus, the historically developed physical-structural difference of the various neighborhood sections of North Park has become even more pronounced since the turn of the millennium. According to census data and own calculations (see Table 5.2), the more 'suburban'-style areas of North Park along Balboa Park and the community edges have gradually de-densified. Additionally, even though the total number of housing units has decreased in these neighborhood sections, the share of single-family homes has increased. In comparison, the commercial neighborhood center and the immediate surrounding have faced a decrease of the number of single-family housing units and a significant increase of the total number of housing units in the last two decades. This trend has intensified the divergence of the "multi-character neighborhood" (CofSD 2016, p. 70) section around University Avenue and 30th Street and the "traditional character neighborhood" (CofSD 2016, p. 70) parts at the northern and southern community edges, as described in the Greater North Park community plan.

It can be concluded that the different neighborhood sections—the dense, walkable, and 'urban village'-style neighborhood center (tracts 12 and 13) on the one hand and the low-density, more suburban single-family areas (tracts 14 and 15) on the other hand—react in different ways to the approaching redevelopment wave. To varying degrees, all of the four tracts have recorded a population shift toward a higher share of white and college educated residents combined with a decrease of the share of underage as well as black and Hispanic residents. However, it seems that the residents of the tracts south of University Avenue strive to maintain their 'suburban flair' and lifestyle of relatively low density during the redevelopment processes, while the residents and businesses along the commercial corridors appear to embrace the emerging density and mix of uses more. This contrast is not only reflected socially (through different household structures) and physically (concerning the architecture and urban fabric) but is also promoted by different community planning goals for the distinct sections, which are ultimately enhancing the established physical-spatial, social, and lifeworld contradictions of the adjacent neighborhood areas or—in other words—the urban/rural hybridity found in North Park. Furthermore, the planning approach of preserving the low-density single-family character of selected areas of North Park itself stands in contrast with the overarching 'City of Villages' planning strategy, which is—according to the planner EP-11—striving to create "areas of more density and mixed-use throughout the city to, kind of, meet climate and transportation goals and to, you know, meet housing goals as well."

In addition to the quantitative considerations, especially the qualitative components of the underlying analysis have provided unique insights into the redevelopment of the neighborhood and its social interpretation. Various interviewees have discussed physical as well as social changes of the neighborhood, which have been interpreted and experienced in varying ways and modes—reaching from resident-specific everyday life experiences to interpretations through specialist knowledge on the basis of individual academic, planning, or development backgrounds. On a theoretical level, concerning the social construction of neighborhoods and neighborhood change processes, the qualitative insights allow for a twofold conclusion. On the one hand, neighborhood change has been discussed by means of physical aspects, e.g., the trend to renovate and reconvert historic family homes as well as the commencing replacement of older housing structure with large multi-unit buildings. In these individual experiences and argumentations, neighborhood change is taking place on the basis of the professional as well as the individual everyday life perception, differentiation, and interpretation of material aspects. Theoretically, this interpretation process resembles the subjective construction of landscape and home, which can be summarized as contingent social selections, combinations, and interpretations of physical elements into landscape (Greider and Garkovich 1994, p. 1). Consequently, this reveals the strong theoretical as well as everyday life connection of the concepts of landscape, home, and neighborhood, which are each—from a social constructivist perspective—merely symbolic meanings attributed to physical elements.

On the other hand, neighborhoods and neighborhood change are strongly tied to social aspects and their individual interpretations. Various interviewees have connected the perceived changes of the Greater North Park community to the shifting social and ethnic composition of the neighborhood. For several interviewees, the individual perception of neighborhood change has been strongly connected to (everyday experiences and professional considerations of) the arrival of white and young urban professionals in the neighborhood as well as the displacement of ethnic groups and lower-income residents to communities in the east and southeast, such as Mid-City. Additionally, various interviewees have compared the community of North Park to other neighborhoods and assessed its status quo in the described redevelopment wave on the basis of perceived social changes. Accordingly, it can be concluded that the individual and social construction of North Park—or of urban landscape in general—is not only taking place on the basis of physical elements in space but also by means of social aspects and the symbolic meanings attributed to them.

In addition to the analysis of neighborhood change in North Park, the relation of the Greater North Park community with the adjacent Mid-City area and the progression of redevelopment dynamics from west to east have been discussed in this chapter. The analysis is illuminating a multi-dimensionally ambiguous border and boundary situation between the two communities, which is different at various sections of the official-physical demarcation line and currently characterized by a variety of connections, gradients, and cut-offs due to the advancing redevelopment trend from Hillcrest over North Park into the adjacent neighborhoods of Normal Heights and City Heights. Even though the neighborhoods of North Park and the large Mid-City area are (still) showing strong differences—in terms of the social composition, business structure, as well as concerning physical-structural, functional/infrastructural, and symbolic-atmospheric aspects—certain overlaps and assimilations of the disinvested area in the east to the redeveloped neighborhoods in the west (Hillcrest and North Park) are becoming conceivable.

The official boundary between the two communities is running in north-south direction along a strong physical border of the Interstates 805 and 15—from their intersection with Interstate 8 in the north almost to their intersection with the State Highway 94 in the south. Concerning the connections between the two adjacent communities, the neighborhoods in the northern section of the demarcation—University Heights, Normal Heights, and Kensington—appear rather uniformly, in quantitative as well as qualitative terms. In this section of the official-physical demarcation, the smaller dimensioned commercial street Adams Avenue is providing the only freeway crossing and physical link between the two communities. Nevertheless, this part of the Greater North Park and the Mid-City community is characterized by relatively comparable demographics with low shares of Hispanic and Asian population and predominantly white, adult residents living in renovated and well-kept single-family homes. Additionally, also the infrastructure and the business composition appear uniformly in the northern border section of the two communities, which reduces the border experiences and demarcating effects of the official boundary and physical border strongly. In other words, the everyday life relevance of the demarcations in this section seems to be of low significance.

The southern portion of the official-physical demarcation between Greater North Park and Mid-City, however, is standing in stark contrast to the discussed northern section. In this area, the Interstates 805 and 15 are intersecting and creating a strong cut-off line between the redeveloped west and the rather disinvested east. Physically, the absence of bridges and underpasses in this area is creating a nonpermeable everyday life demarcation between the southern parts of North Park on the western side and City Heights on the eastern side. Symbolically, this is promoting a strong sense of seclusion. Additionally, in this border section, the areas to the west and the east differ strongly in terms of demographics: the southern area of North Park is predominantly white, college-educated, and features only a small share of low-income households. The southern area of City Heights, in turn, is home to a predominantly Hispanic population with significantly lower shares of college educated residents and higher shares of low-income households living in outdated and unrenovated family homes. Thus, the two Interstates form not only a strong physical border between two distinct communities but constitute the location of a significant socio-spatial, economic, and infrastructural boundary with strong relevance in everyday life. Following the argumentation of Haselsberger (2014), the set of various overlapping boundaries that is running along the respective section of the two Interstates can be described as a 'thick border.'

Lastly, a third sector along the border of the two communities has been identified, which includes the commercial corridors of University Avenue and El Cajon Boulevard. The area along the corridors and their immediate surrounding is equipped by several freeway crossings, allowing frequent access over the physical border lines of Interstate 805 and 15. Nevertheless, unlike in the northern section, the central neighborhood parts are not characterized by a strong similarity in terms of their business structure, demographics, infrastructure, as well as their image. The multiple boundaries are not dissolving or losing its everyday relevance as emphasized by several interviewees that have highlighted the differences of both sides of the border. However, the central border section is also not showing the strong demarcating function of the 'thick' and hard southern border area that is enhanced by the multi-dimensional contrariness of the southern parts of North Park and City Heights. Instead, this section is defined by various sustained gradients on a west-east axis and thus by 'softer' and less defined demarcations in the sense of a *'stretched' urban border* or an *urban borderland* (see Figure 5.16). Demographically, from west to east, the communities are characterized by a decrease of white residents as well as an increase of the share of Hispanic and Asian population (especially toward the unofficial Little Saigon district in City Heights) and of underage residents. Additionally, economically (concerning the business structure and the real estate marketing) and in terms of the reputation of the neighborhood areas, the central border section is also structured in the form of gradients due to the discussed eastward expansion of the 'North Park concept' and the redevelopment dynamics in general. As a result, the immediate areas east of the Interstates 805 and 15 along the commercial corridors of University Avenue and El Cajon Boulevard are strongly characterized by aspects of hybridity and ambiguity, by the simultaneity of newly arriving young residents, trendy businesses, and rising rents, on the one hand, and the local low-income population, ethnic businesses, and outdated infrastructure, on the other hand. Currently, this area is undergoing pastiche-like transformations and adaptions that blur the demarcations between west and east and are ultimately creating a *hybrid urban borderland*. This section of the border between North Park and Mid-City resembles a fluid transition zone of simultaneous connection and differentiation and a place of ongoing negotiation about participation and identity with a social future yet to be determined (cf. Roßmeier 2020; Roßmeier and Weber 2021). The provided digital 3D sketch (see Figure 5.16) is illustrating and summarizing this temporarily ambiguous border and boundary situation in the form of a subjective graphic interpretation in order to supplement the presented qualitative results and further the understanding of the current change processes.

5.3 The Greater Golden Hill Area: Redevelopment and Social Reinterpretation of a Victorian-Era Streetcar Community

The following section focuses on the neighborhoods of Golden Hill and South Park in the Greater Golden Hill community area (see Figure 5.17). Due to limited availability of census data on the neighborhood level, the discussion of Golden Hill and South Park is combined into one section. First, descriptive remarks, historic illustrations, and quantitative data on the demographic, socio-economic, and physical structure of the entire community area will serve as detailed introduction of the chapter (section 5.3.1). In a next step, individual interpretations

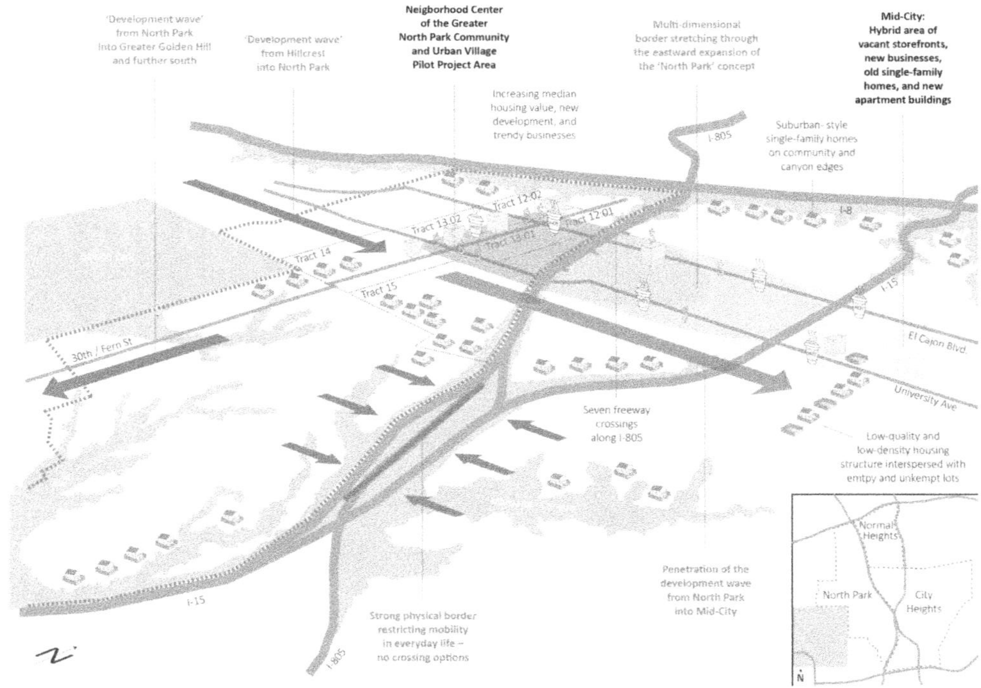

Figure 5.16 Digital 3D Sketch—Networks, Flows, and Demarcations between Greater North Park and Western Mid-City. (General information is provided in black, border-related conclusions are given in red. The red arrows are used to display the direction of the elaborated development wave, the transparent arrow indicates the direction of its commencing current progression. The multi-dimensional boundary stretching and dissolution, in other words, the hybrid urban borderland between Greater North Park and Mid-City is displayed through the red gradient along the commercial corridors of University Avenue and El Cajon Boulevard. Along the intersection of the Interstates 805 and 15, arrows pointing toward each other are used to display the hard or 'thick' demarcation with high everyday relevance. Source: Own cartography and design)

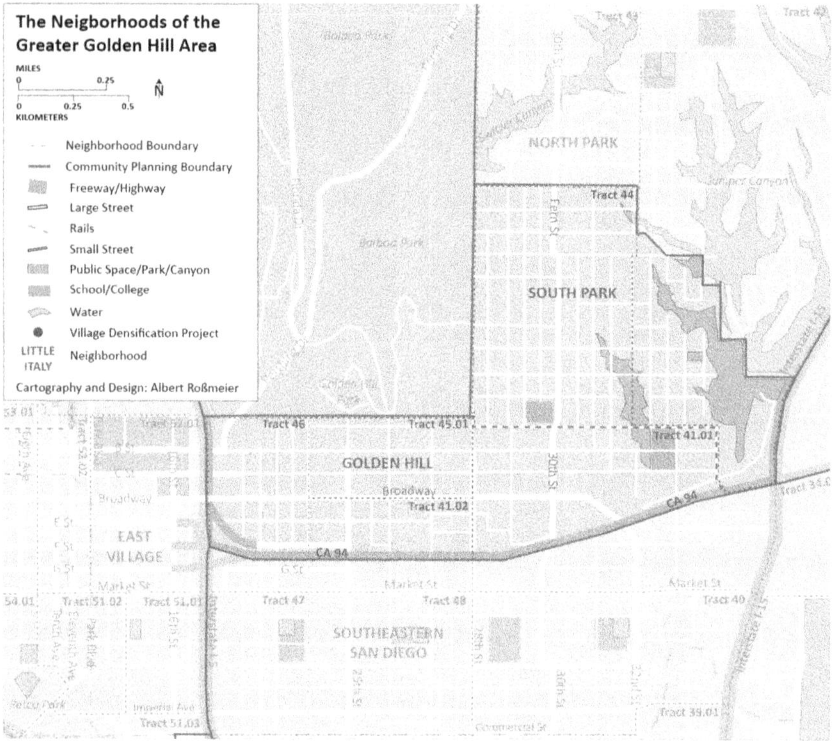

Figure 5.17 The Neighborhoods of the Greater Golden Hill Area. (Source: Own cartography and design)

and everyday experiences of neighborhood change in the course of San Diego's recent redevelopment trend will be illuminated for each neighborhood in order to continue the initial, quantitatively painted picture (section 5.3.2). Additionally, differences and peculiarities of the respective neighborhoods will be worked out. The chapter closes with an urban landscape and border-theoretical conclusion and interpretation of the observed phenomena (section 5.3.3).

5.3.1 The Historic Growth-Decline-Growth Sequence of the Neighborhoods of Golden Hill and South Park

With its history reaching back to the late 1860s and early '70s, Golden Hill marks the first residential community outside of Alonzo Horton's New Town. Due to its relatively large lots with views on San Diego Bay and the Pacific Ocean as well as its proximity to the newly dedicated Balboa Park—formerly known as City Park until 1910 (Amero 1990) –, the neighborhood quickly developed into a fashionable residence of wealthy San Diegans. In the following decades, the area has been developed with residential buildings reflecting different architectural styles, such as Victorian, Spanish Colonial Revival, Craftsman, and Farmhouse, which are still prevalent in today's mixed building stock. By the 1930s, after the population increase of the 1880s (see Figure 4.2) as well as the boom phase initiated by the Panama-California Exposition (Griffin and Weeks 2014, p. 66), the properties in the neighborhood were mostly built out. The early success of the neighborhood of Golden Hill was strongly linked to the horsecar (introduced in 1886 by the San Diego Street Car Company) and subsequently the Class 1 streetcar (operated by John D. Spreckel's San Diego Electric Railway Company until 1939) connection to downtown, running along Broadway in east-west direction. This transit connection elevated the community from a rather rural settlement to an early streetcar suburb (Ford 2014, p. 181). Additionally, easy access to Golden Hill Park (see Figure 5.17), the earliest recreational development in Balboa Park was provided in the northwestern corner of Golden Hill which constitutes an important infrastructural feature of the neighborhood to this today (CofSD 2013b, pp. HP181–HP182; cf. Montes 1977).

The official development of the neighborhood of South Park is dating back to the year of 1870, when a subdivision map has been filed for the section of Golden Hill east of today's Balboa Park. However, extensive development and settlement did not take place until the first decade of the 19th century, when parcels have been actively marketed, infrastructure, such as sidewalks and electricity, had been implemented, and a first electric streetcar connection to downtown San Diego had been introduced—all by the Bartlett Estate Company. Ultimately, the development efforts of the investment and real estate company helped South Park turn "from a rural community [...] into a developed residential district" (CofSD 2013b, p. HP182), while contractually bounded building restrictions ensured for the construction of high quality single-family homes (instead of low-cost apartment buildings) in the new, prestigious neighborhood (SOHO 1992, p. 2). Equivalently to the success of Golden Hill, in the 1910s, South Park experienced an early development surge due to its proximity to Balboa Park and the Exposition which led to the construction of numerous homes in the then prominent Craftsman and Spanish Colonial Revival styles (CofSD 2013b, p. HP182). Thus, in the 1920s, the development opportunities in South Park have as well reached capacity in terms of availability of free lots since the interest in the established neighborhood has increased even more with the continuing streetcar expansion north into the community area of North Park.

After the successful development in its first decades, during the Great Depression, the neighborhoods of Golden Hill and South Park recorded a significant decrease of interest by prospective homebuyers and thus also of new construction. By that time, "Golden Hill began to witness physical decline [...], falling victim to poor maintenance, physical distress and vandalism" (CofSD 2013b, p. HP183). Additionally, due to the vast population growth of the city during World War II and the concomitant housing shortage (as illustrated in detail in chapter 4), zoning laws have been adopted in several inner-ring neighborhoods of San Diego, including Golden Hill but not South Park, allowing for higher densities and larger building structures. Following these legal changes, the development paths of Golden Hill and South Park have drifted apart—at least in terms of housing development. Throughout and after World War II, South Park has largely maintained its physical structure of low-density single-family housing on large lots, even after the removal and decommission of the streetcars and their tracks (CofSD 2013b, p. HP184). However, in Golden Hill, starting in the 1940s, construction of moderate and large-scale apartment complexes set in in the form of infill development as well as replacement of single-family homes. Additionally, larger two-story homes were subdivided into multi-unit apartment buildings in order to make space for new San Diegans (see Figure 5.18). However, following the mass availability of private automobiles and the rapid production of 'cookie cutter suburbs' in North County, the white middle-class flight commenced, which entailed significant spatial reorientation of economic and planning attention away from the formerly prosperous inner ring. Thus, in the following decades, the demographic composition of the entire community area changed substantially. Especially the neighborhood of Golden Hill developed from an exclusive streetcar suburb of prestigious mansions and large properties with high homeowner rate to one of San Diego's most ethnically diverse and dense neighborhoods with a high rate of renters and working-class residents living in the area's aged building stock as well as in new but poorly maintained multi-unit apartment buildings.

However, starting in the 1980s, approximately 100 years after the successful foundation of Golden Hill and the subdivision of South Park, increases in the cost of suburban life (following the national oil crises of 1973 and 1979) as well as renewed economic and planning focus on downtown San Diego seemed

Figure 5.18 Subdivided Historic Residential Building in the Neighborhood of Golden Hill. (The picture shows a historic subdivided former single-family home in Golden Hill featuring four different addresses and entrances, which testifies to the historic zoning adaptions. Source: Picture Albert Roßmeier 2021)

to slowly redraw white middle-class interest to the outdated inner-ring neighborhoods. “Homeownership in Golden Hill steadily increased [and] a growing desire emerged among residents to eradicate blight, reduce density and restore the community’s historic character” (CofSD 2013b, p. HP184). Relatedly, an increased sense of community and place in the neighborhood was certainly facilitated by the construction of Interstate 5, 15 as well as State Route 94 in the late 1950s and ‘60s, which in addition to Balboa Park bounded the neighborhood of Golden Hill and South Park physically and severed Golden Hill from the industrialized neighborhood of East Village at the eastern downtown edge (Pryde and Stutz 2014, p. 199). Finally, throughout the 1980s and ‘90s, once again, a gradual replacement of residents set in which became increasingly evident in the following decades and is reflected both quantitatively in census data as well as qualitatively in subjective experiences and perceptions of several interviewees, as it will be discussed in further detail in the next subsection. First, however, quantitative data will be used to illuminate the processes of demographic and socio-economic neighborhood change in Golden Hill as well as South Park.

Officially, the neighborhood of Golden Hill is bounded by Balboa Park and A Street on the north (its boundary with South Park), 34th Street and 34th Street Canyon on the east, State Route 94 on the south (its boundary with Southeastern San Diego), and Interstate 5 on the west (the boundary with downtown). The neighborhood boundaries of South Park run along Balboa Park in the west, Juniper Street in the north (its boundary with the North Park community area), and parts of 32nd Street, Marlton Drive, as well as 34th Street canyon in the east. These neighborhood boundaries do not align exactly with census tract boundaries (see Figure 5.17) and therefore only allow for rough comparisons of different neighborhood portions. Three of the four local census tracts of Golden Hill include parts of the surrounding neighborhoods of the community area of Southeastern San Diego (Sherman Heights, Grant Hill, and Stockton), which over time became home to a large and relatively young, low-income Hispanic community (in 2012, over 84 percent of the residents were Hispanic; CofSD 2015, p. 1/12). Of the two census tracts covering the neighborhood of South Park, the bigger eastern tract also includes large parts of the community of North Park.

Consequently, below, 2000 and 2010 census data on the community level will be used to help trace change processes in the entire Greater Golden Hill area (SANDAG 2003p, 2016c). Subsequently, regardless of the inaccurate boundary overlaps, the following census data calculations will be contextualized with available results of the recent 2020 census on the tract level to cautiously work out spatial disparities of the respective sections in San Diego's inner ring (U.S. Census Bureau 2021a). Surprisingly, comparison of the census cohorts of 2000 and 2010 reveals that the overall population of the community planning area has decreased by almost 12 percent, from 17.989 to 15,848 residents. Additionally, also the total number as well as the share of residents below 18 years has decreased (from 4,486 persons or 24.9 percent of the total community population in 2000 to 2,863 or 17.9 percent in 2010). So did the total number of residents enrolled in school (including nursery/preschool, kindergarten, high school, college, and professional school; from 31 percent in 2000 to 26 percent in 2010). However, while the share of underage residents enrolled in school was 20.8 percent in 2000 and diminished to 14.6 percent in 2010, the share of residents inscribed in college or professional school has slightly increased from the year 2000 (9.6 percent in total) to the year 2010 (11.6 percent in total). Relatedly, not only the number of college or professional school students has increased, also the share of residents with educational attainment higher than high school degree has risen considerably—from 30.7 percent in 2000 to 41.8 percent in 2010. Considering the diminished share of underage residents and school kids in the community, the average household size has decreased as well, from 2.51 persons per household in 2000 to 2.23 in 2010. In addition to these changes, the population share of white residents has risen over the years (from 34.6 percent in 2000 to 44.5 percent in 2010), while the total number of residents of Hispanic origin (52.8 percent in 2000 and 43.5 percent in 2010) as well as of black or African American residents (6.8 percent in 2000 and 5 percent in 2010) has decreased. Summing up, in the years from 2000 to 2010, the demographics of Greater Golden Hill have developed from a community of predominantly Hispanic families with children to an area of mostly white singles and childless households with higher educational attainment.

Corresponding with the decrease of the total number of residents in the Greater Golden Hill area, also the total number of housing units has slightly declined, from 7,369 in 2000 to 7,285 in 2010. Contrastingly, the number of owner-occupied housing units has increased between 2000 and 2010. At the turn of the millennium, 1,671 units have been occupied by their owners (22.7 percent of the total number of housing units). By 2010, the number had risen to 1,849 owner-occupied units (25.4 percent of the total number of housing units). What is more striking, however, is the change in share of housing unit types in the community area: While single-family housing has made up 38 percent and the share of multi-family housing added up to 62 percent of the total number of units in 2000, the share of single-family units has risen to 48 percent in 2010, reducing the share of multi-family housing units to 52 percent. Additionally, also the development of the median housing value shows a stark contrast between the two cohorts: In 2000, the median housing value in the combined neighborhoods of Golden Hill and South Park has been $198,940, which then increased by roughly 156 percent within 10 years, reaching a median value of $509,879. Relatedly, also the median household income increased between the years 2000 and 2010, however only by 52 percent—from $30,478 to $46,476. In terms of income and poverty, comparing the cohorts of 2000 and 2010 reveals an overall decrease of the share of population below the official poverty status, declining from 26 percent in 2000 to 17 percent in 2010. In addition to the increase of median household income and the decrease of poverty in the combined neighborhoods of Golden Hill and South Park, the household availability of private vehicles developed positively as well. In 2000, 5,753 households out of 6,984 occupied housing units (82.4 percent) were in possession of at least one private vehicle, while in 2010, 6,278 households (90.7 percent of the total number of occupied housing units) had at least one private vehicle available.

In conclusion, the comparison of the 2000 and 2010 census data of the Greater Golden Hill area helps to start painting the picture of a changing inner-ring community and its population. In terms of demographics, the calculations have illustrated the decrease of the total number of residents in South Park and Golden Hill as well as the community's significant shift of racial composition. Concerning changes in housing, the community area has faced a minor decrease of the total number of housing units combined with a significant increase of the share of single-family housing between 2000 and 2010—in the course of the redevelopment processes, previously subdivided family homes and medium sized apartment buildings have been (re-)converted into single-family residences (Joassart-Marcelli 2021, p. 155). This can be tied to real estate cycles and the change of demand and supply, which has taken place in the area over the decades. Combined with a higher homeowner rate and the increased availability of private vehicles in the community area, these developments describe a change toward a wealthier version of the area with tendencies of social homogenization and de-densification in the sense of a 'suburbanization of the central city' (Frank 2018, 2020) rather than its presumed urbanization and diversification. Lastly, this testifies to the contradictive concurrency, rivalry, and hybridity of different development trajectories in the evolving neighborhoods of San Diego's inner ring.

In a next step, by means of first results of the 2020 census, it becomes possible to look briefly into spatial disparities of the different neighborhood sections of Golden Hill and South Park. However, due to limited availability at this point, recent data can only be retrieved for a small number of topics and for the tract level instead of the community level. As stated above, South Park is split into two tracts, one of which includes large parts of the Greater North Park area, while the tracts in Golden Hill include parts of the planning area of Southeastern San Diego. Nevertheless, looking deeper into 2020 data in terms of racial composition, interesting, stark contrasts between the different tracts become apparent: The tracts covering South Park (and southeastern North Park) feature a share of 65.7 percent (tract 44) and 66.8 percent (tract 42) of white residents. In Golden Hill, the highest share of white residents is recorded closest to downtown and Balboa Park (tract 46 with 55.1 percent), while tract 41.02—which covers the designated historic district of Golden Hill (see Figure 5.17) as well as parts of Sherman Heights and Grant Hill—features the lowest (34.8 percent). The two other tracts of Golden Hill show shares of white residents of 50.2 percent (tract 45.01) and 42.7 percent (tract 41.01). Additionally, in terms of Hispanic origin, South Park and the included parts of North Park record lower shares of population of Hispanic origin (tract 44 with 26.9 percent and tract 42 with 25.2 percent) compared to the tracts covering Golden Hill and parts of Southeastern San Diego.

Here, the tracts 45.01, 41.01, and 41.02 record higher percentages of Hispanic population (from 45,3 to 58 percent) than the tract bordering downtown, tract 46, which shows a share of residents of Hispanic origin of only 32,9 percent.

Another topic provided in the 2020 census map tool is the share of population of age 18 and older, which offers interesting insights when put in the elaborated context of neighborhood development. As stated in the quantitative map tool, the tracts closer to downtown record higher percentages of adult population compared to the tracts on the eastern community edge. Tract 46 on the western neighborhood corner of Golden Hill, adjoining downtown and Balboa Park features a share of 91.3 percent of residents of the age 18 and older. From west to east, from Golden Hill to South Park, the share drops continuously with the most eastern tract (42) comprising of only 80.8 percent of residents of the age of 18 and older. Accordingly, also in terms of population densities, the tracts closer to downtown and especially along Balboa Park record higher densities compared to the tracts on the eastern community edge, which are permeated by 32nd Street canyon and bordered by the wooded fingers of 34th Street canyon. Tract 45.01 along Balboa Park in Golden Hill shows the highest population density with 20,638 persons per square mile. The other western tracts of Golden Hill feature higher densities as well, tract 41.02 records 12,941 persons per square mile and tract 46 features a value of 12,801. However, the two most eastern tracts of South Park and Golden Hill record significantly lower densities: Tract 41.01, covering parts of Golden Hill and Stockton (in Southeastern San Diego) features 9,098 persons per square mile, while tract 42 records only 6,044.

In summary, the illustrated composition of 2020 census data suggests that the western neighborhood sections—which are the areas closer to downtown San Diego and Balboa Park—feature higher shares of white residents and concomitantly lower shares of Hispanic residents, higher densities, as well as higher shares of childless households than the eastern areas around the structure of 34th Street canyon and Interstate 15. In other words, the western neighborhood sections closer to the trendy and pedestrian-friendly locations of downtown and Balboa Park comprise of rather white and more urbanized households of singles and childless couples, while the southern and southeastern tracts—despite being zoned as medium-high density areas (CofSD 2013b, p. LU32)—testify to more Hispanic and typical suburban family households.

Lastly, the illustrated distribution of population densities across the two neighborhoods is mirrored in the underlying structure of housing densities, which differs strongly between Golden Hill and South Park. As described above, this is due to the zoning history of the respective neighborhoods. Both neighborhoods comprise largely of residential uses, however, as illustrated in the current community plan (CofSD 2013b, p. LU32), the predominantly white neighborhood of South Park features almost exclusively 'low' densities (of 1–9 Du/Ac), whereas the residential areas of Golden Hill lie predominantly in the 'medium' density range (of 16–29 Du/Ac) with two additional sections featuring 'low-medium' densities (10–15 Du/Ac). On top of that, in Golden Hill, zoning for 'medium-high' residential densities (ranging from 30 to 44 Du/Ac) has been realized in the southeastern neighborhood corner along Broadway eastward of 28th Street. This area (lying in tract 41.01) is currently characterized by low population density, however, features new apartment buildings and is planned to house more residents in the future. As stated above, the highest population density in Golden Hill is currently found along Balboa Park (in tract 45.01), but in terms of new infill development future growth will be directed more toward the predominantly Hispanic southeastern neighborhood corner (see Figure 5.19).

Figure 5.19 Construction of a New Multi-Unit Apartment Building in the Southeastern Section of the Neighborhood of Golden Hill. (The picture shows the implementation of density in the form of a construction site of a multi-unit apartment building in the less densely populated southeastern neighborhood corner of Golden Hill. Source: Picture Albert Roßmeier 2021)

Summing up, the quantitative and descriptive illustrations build a useful basis for the subsequent qualitative analysis of the change processes the community area of Golden Hill is undergoing. By means of the calculations using recent data from the 2020 census, it became possible to go beyond the initial quantitative illustration of the entire community and to differentiate the neighborhoods of Golden Hill and South Park. Thus, it has been shown that the various neighborhood sections differ in parts strongly in terms of socio-spatial and racial distribution as well as concerning population and housing densities. In the following, qualitative insights will be discussed in order to build a wider understanding of the resurgent inner-ring community and its various subjective (border) realities.

5.3.2 Postmodern Valorization of Historicity, Centrality, and Walkability in Greater Golden Hill—Qualitative Insights

Below, in the first paragraphs, individual interpretations and experiences concerning the redevelopment of the neighborhood of Golden Hill will be presented, followed by a discussion of recent change processes in South Park. The subsection closes with own, subjective insights and experiences from the conducted participant observations in the two neighborhoods.

In accordance with the presented quantitative data, a middle-aged resident from Ocean Beach, who grew up in various neighborhoods of central San Diego (R-12), shares that Golden Hill—similarly to the discussed 'gay gentrification' in Hillcrest—has been inhabited by an aged population until "young people started to discover the neighborhood in the 1990s." From their perspective, in the 1970s, the neighborhoods around Balboa Park—especially North Park, Golden Hill, and South Park—have been "unsafe" (R-12) communities of the local black and Hispanic population and were characterized by high crime rates and gang activity. Equivalently, interviewee EP-01 states that Golden Hill has previously been "a very dangerous area, a no-go area" of gangs, consisting of dilapidated and even abandoned houses. These depictions are in line with the experiences of interviewee R-11, a senior resident from East Village, who moved to San Diego in the 1980s. During that time, Golden Hill was "a run-down neighborhood, rents have been very low, and the houses were in bad shape" (R-11). In addition to these characteristics, in the same vein as interviewee R-12, the interview partner R-11 deplores the high crime rates of the late 20th century as well, which kept them from moving to the Greater Golden Hill area. But even "15 years ago" (R-11), in the early 21st century, when demographic change processes have already begun and Petco Park in the adjacent neighborhood of East Village had been built, crime rates have individually still been perceived as high. Accordingly, a friend

Figure 5.20 Renovated Historic Single-Family Home with Green Lawn and Spacious Front Yard in the Neighborhood of Golden Hill. (Source: Picture Albert Roßmeier 2021)

of interviewee R-11 felt the need to move out of Golden Hill and to a neighborhood they perceived as safe. In contrast to this, now, "the people in Golden Hill are friendly and it has really changed. All the historic homes are getting renovated, but it is becoming more expensive as well" (R-11). The interviewee states that the neighborhood of "Golden Hill has gentrified," thus describing the social and physical as well as reputational change of a stigmatized, ethnically diverse inner-ring neighborhood toward a trendy and walkable community of predominantly white and young urban professionals that are living in renovated historic single-family homes.

Summing up the development of Golden Hill, interviewee R-10, a young resident from Hillcrest, describes: The neighborhood of Golden Hill "is up and coming. Ten years ago, this was the ghetto, but now people are going down there." Additionally, the interviewee points out that Golden Hill still features more Mexican culture than other trendy inner-ring neighborhoods, such as Hillcrest and North Park, due to its proximity to Barrio Logan, which is interpreted by several interviewees as a "very hip" (R-10) neighborhood as well. In terms of

Mexican heritage, interviewee AR-08 perceives and criticizes a loss of culture: "Golden Hill used to be [called] 'Las Lomas,' which is Spanish for 'the Hills' and now it's called 'Golden Hill.' Oh, renaming is a big thing, [...] renaming for the purpose of redevelopment." Thus, the interviewee is deploring the loss of means of social and spatial everyday life identification, exemplified by the dwindling significance of the unofficial, gang-related[6] neighborhood name of 'Las Lomas,' which they view as negative side effect of the overarching redevelopment and gentrification processes in San Diego.

Sharing from a more analytical perspective compared to the emotional-experiential standpoints illustrated above, interviewee AR-10 explains that Golden Hill was a "low-income neighborhood lacking basic infrastructure, such as grocery stores and other amenities." Addressing the demographic changes and the upgraded historic building structure of the neighborhood, the interviewee states further that today Golden Hill "is like a new neighborhood." In a similar vein, interviewee R-07, a young mother living a 'binational lifestyle' in the San Diego/Tijuana region shares that the neighborhood of Golden Hill "was all working class and one of the cheapest places to live, but it got turned into one of the hipster areas as well, with old historic homes and high rents." This statement reflects the drastic increase of median housing values in the Greater Golden Hill area, as illustrated above (similarly perceived and discussed by the interviewees R-06 and R-11). In a similar vein, linking current trends with the neighborhood's past, interviewee R-13 from North County San Diego shares: "Golden Hill [...] is a fantastic neighborhood and it is not cheap. You could tell it wasn't always a nice place." The interviewee explains that the increase of housing values as well as increased home ownership in Golden Hill did not only affect the neighborhood physically in the form of renovations and other 'beautifications' but also socio-spatially through recurring efforts to expel homeless residents from the neighborhood:

> "It seems like if I were a homeless person 20 years ago, I had more places to go, I had more neighborhoods that I would be welcomed and ok and people would allow for me to be, and it seems like now there's fewer choices because places like, you know,

[6] The neighborhood name 'Las Lomas' has only been mentioned by one interviewee (AR-10). While the name might have (had) everyday life significance for a broad spectrum of residents, online search results only testify to the existence of the names 'Las Lomas' and 'Varrio Lomas' as means of spatial identification for the eponymous neighborhood gang 'Lomas 26' of Golden Hill, as discussed by Miller (2013). According to interviewee AR-10, the formation and existence of rivaling violent gangs in San Diego was related to the freeway construction in disinvested neighborhoods: "all the areas have their own gang and I feel like the freeway was the initiation of dividing the people and it worked."

> Golden Hill, they would never tolerate a homeless person on the street there anymore. The residents have higher expectations because they, you know, they're paying more for their residences. So, I don't necessarily know if the numbers have changed of the homeless, but it seems like it's just getting more concentrated in certain areas" (R-13).

Border-theoretically framed, interviewee R-13 is describing an ongoing rebordering process (cf. Liao et al. 2018) that restructures the socio-spatial organization of central San Diego. As argued by the interviewee, residents of Golden Hill are responding to increasing rents and real estate market prices in the area with higher demands on the neighborhood environment, both socially as well as aesthetically. Within this social development of neighborhood demands, individuals identified as non-conform, such as homeless residents, are continuously excluded from the neighborhood by police actions in order to construct or sustain a certain aesthetic and social neighborhood standard. As a result, for certain members of society, the *boundaries of accessible space* are simultaneously shifting and getting less permeable. Accordingly, during several participant observations (PO-01, PO-02, PO-05) as well as within statements of various interviewees, the former warehouse district of East Village appeared to serve as a place of refuge for the large homeless population in San Diego (Roßmeier 2019, 2020; Roßmeier and Weber 2021), which is bounded immaterially in less and less traversable ways in the course of the ongoing redevelopment processes of the neighborhoods around Balboa Park. Ultimately, Golden Hill, formerly a redlined neighborhood of a large low-income black and Hispanic population living in dilapidated building stock has been removed from the 'mental maps' (Lynch 1992) of accessible places of certain parts of San Diego's society as the immaterial boundaries around the homeless refuge of East Village have become ever tighter. In how far this socio-spatial development and political policy of 'homeless containment' is affecting East Village and its redevelopment will be discussed in detail at a later point (section 5.4).

In line with the present border-theoretical approach, it is of interest in what ways the neighborhood of Golden Hill is bounded spatially by the various interviewees and how or by what means the differences between Golden Hill and East Village or downtown are constructed. As it will be demonstrated by means of the following examples, different spatial interpretations can also be the result of different modes of individual spatial approaches, which are ranging from resident-specific subjective as well as socialized everyday life interpretations to approaches through the lens of special knowledge, which is typical for experts, such as planners, developers, or academics (the present methodical approach of combining different perspectives and modes of interpretation in the various interviews is presented in section 3.3.1; landscape socialization and different modes of constructing landscape are discussed in depth by Kühne 2019, pp. 61–64).

The interviewee EP-12, for instance, a young college graduate from Hillcrest, shares a rather experiential insight into their everyday life construction of Golden Hill and its spatial references to the adjacent communities of downtown San Diego: "Golden Hill is really named because it is a hill, and it starts going up a really steep hill out of downtown, so I would say that that's a signal that you're in Golden Hill, like, too steep to bike up, I would say." Here, topography is used for the subjective construction of Golden Hill and its boundaries to downtown San Diego. Arguing from a relational and non- or more-than-representational perspective, the described hill can be understood as a material foundation of the neighborhood of Golden Hill, which is connected to the subject, the interviewee, through the shared subjective everyday life (bordering) experiences they made on their bicycle (cf. Gergen 1999, p. 128; generally on bordering experiences see Szytniewski et al. 2020; Wille and Nienaber 2020). Combining this phenomenological reading with border-theoretical thinking, the exemplified experiential interaction of the social and the material dimension reveals the inseparable link between the subjective construction of space, meaning, and individual (or social) boundaries. Additionally, arguing from a landscape-theoretical point of view, the individual highlighting of material aspects (the hills of Golden Hill versus the flat topography of the downtown area) as part of the subjective process of constructing and differentiating the neighborhoods of Golden Hill and downtown resembles the social construction process of landscape, which is based on subjective selections of material elements that are assigned symbolic meaning. This bespeaks the strong theoretical as well as everyday life connection of the concepts of landscape and neighborhood, each underlying both subjective criteria (individual experiences and perceptions) and social conventions (common sense interpretations and trends).

In contrast to the example of everyday life experiences, interviewee EP-09, an urban planner from North Park, is addressing the adjacency of Golden Hill and downtown from a more analytical planning perspective by means of their expert knowledge. On the one hand, the interviewee points out that the freeways that are running through the inner city of San Diego function as relevant neighborhood borders: "There is a physical barrier, the freeway" (EP-09). On the other hand, however, the interviewee differentiates between various layers or dimensions of demarcation in a way that acknowledges fuzzy border and boundary situations wherein the various layers do not align perfectly on top of each other but are potentially always subject to shifts: "In the economic [dimension], or the economic change [...] is spilling over [from downtown] toward other neighborhoods. But physically, there is still a huge barrier for mobility and for getting over the freeway [Interstate I-5]. It's a design issue, I really, it's one of my personal goals to make that better" (EP-09). Border-theoretically framed, interviewee EP-09 pictures the outward movement of the economic boundary of downtown San Diego, the boundary of an exclusive real estate market serving young, professional singles and childless couples, which is drifting away from the physical frame of the city center into adjacent communities, here Golden Hill. In this sense, the interviewee is addressing a certain overlap of San Diego's downtown and the community of Golden Hill in the economic dimension. However, as further pointed out by the interviewee EP-09, the physical border of Interstate I-5 is restricting mobility, thus preventing a strong everyday life connection between downtown and Golden Hill—at least for the time being, for the interviewee shares personal motivation to mitigate the physical demarcation through planning operations in order to connect the neighborhoods of the inner city and the inner ring further. Time and again, nonprofit organizations in San Diego forge such ideas and create visions for so-called 'freeway caps' over Interstate I-5 and State Route CA 94, which are intended to "restitch the streets together" (AR-06, discussed as well by AR-02) through the realization of recreational space on top of the physical neighborhood border. However, only one of the envisioned 'freeway lids' has been realized yet due to insufficient funding (the realized cap covers a section of Interstate I-15 in City Heights).

In terms of the discussed expansion of downtown San Diego into the surrounding neighborhoods, interviewee EP-08 is arguing from a theoretical planning perspective as well, however, contrary to the interviewee EP-09, they point out significant differences and persistent contrasts between downtown and the first ring neighborhoods: "Logan Heights, Barrio Logan and Golden Hill, those areas are really special, wonderful, and vibrant, but these areas are definitely not downtown, [...] the zoning and the government structure is different" (EP-08). While interviewee EP-09 emphasizes a spillover of economic or real estate trends from downtown into the inner ring—metaphorically described by AR-11 as "a wave [that is] moving away from the city center" –, interviewee EP-08 is arguing for the dichotomy of the communities. Ultimately, the discussed statements reveal two pivotal points about neighborhood boundaries: Firstly, the statement of interviewee EP-09 presents (neighborhood) borders and boundaries as multilayered and as existing simultaneously in various dimensions. Secondly, it can be concluded that boundaries are subjectively drawn in diverse ways and thus potentially also at different locations.

Figure 5.21 Interstate 5—the Physical and Official Border between Downtown San Diego and Golden Hill. (The picture shows Interstate 5, which cuts through the neighborhoods of East Village and Golden Hill since its construction in the 1960's. Additionally, a homeless person entering this border zone is pictured. Source: Picture Albert Roßmeier 2021)

In the following, selected individual perceptions and interpretations of the interviewees concerning the redevelopment of South Park will be presented. According to several interviewees, the neighborhood of South Park underwent demographic change processes throughout the past decades, which resemble the demonstrated development of the adjacent neighborhood of Golden Hill. In the 1970s and '80s, according to biographical insights shared by interviewee R-12, South Park used to be an overcrowded area of family homes with barred windows and neglected front yards. Today, as pointed out by the interviewee, the front yards of the properties in South Park are well-kept lawns after "more and more white people moved to the neighborhood within the last 20 years and started buying and renovating run-down houses" (R-12; see also Figure 5.20). Thus, the interviewee is describing a subjectively perceived connection between the observed physical changes of the neighborhood and certain social or socio-economic developments (Figure 4.21).

In a similar vein, by referring to the historic development of the neighborhood of South Park, interviewee AR-04 is putting their experiences and interpretations in a larger framework. Starting in the 1950s, after "the streetcars were built out, the neighborhood declined for a while." However, according to the interviewee, real estate prices and rents in the neighborhood have increased significantly within the last two decades. The interviewee, a mother of two who is living in an academic two-income household, reports further: "In 2008, we rented a house in South Park but the rent increased and we couldn't afford it anymore." From their perspective, the redevelopment of the older inner-ring neighborhoods of Golden Hill and especially of South Park has already reached an advanced stage (similarly assessed by Joassart-Marcelli 2021; Joassart-Marcelli and Bosco 2020).

In contrast to these insights, interviewee R-10 as well as R-11 are not putting their descriptions into the historical context of the neighborhood but are describing the changes of the social composition of South Park as more recent. In comparison to other developing neighborhoods around Balboa Park, interviewee R-11 states that the processes of "gentrification in South Park started much later than the redevelopment of Hillcrest and North Park. [...] The neighborhood has only recently been gentrified. In the last 5 or 6 years people started buying and renovating old houses." The perceived change includes "the arrival of new businesses and really nice restaurants." Similarly, interviewee R-10 explains: "South Park is an up-and-coming neighborhood, it has an artsy vibe." The atmosphere would be "laid back" and the local residents are described as "easy going" (R-10). On top of that, the neighborhood is (still) viewed as "more affordable" (R-10).

However, in contrast to the two interpretation lines picturing both an earlier as well as a more recent neighborhood change of South Park, some interviewees did not experience significant redevelopment and gentrification processes in South Park or other communities southeast of Balboa Park at all. For instance, interviewee EP-06 states concisely: "South Park didn't really change." Another example for this line of interpretation is interviewee R-06, a resident from Hillcrest that claims that "South Park, Logan Heights and Barrio Logan are primarily residential, you know, there is nothing going on. You don't go there to meet people in a bar." In these examples, aspects such as increased median housing values, the changing socio-economic composition, or the opening of new restaurants and bars in South Park are not considered in the processes of subjective neighborhood construction. Ultimately, this demonstrates that neighborhood change as well as neighborhoods in general are perceived and experienced in very different and contingent ways. In line with Berger and Luckmann (1966) as well as Schütz and Luckmann (1973), this is due to the subjectivity of individual experiences in everyday life, which yield an infinite plurality of different realities of the redeveloping inner-ring neighborhoods of San Diego.

Continuing with individual perceptions, in the following paragraphs, two specific sections of an interviewee's interpretation of the physical structure and the neighborhood change processes in South Park will be used to discuss both border-related aspects and multilayered tendencies of urban/rural hybridization.

> "I didn't even know that, like, I don't think I have ever been to South Park until I moved back to San Diego a few years ago. I've been to North Park, like, at some point, yeah, at some point in time, just to speak of that. Now, it's [South Park] like big, there's a lot of breweries there, it's popular. But yeah, South Park was kind of off my radar until, like, a few years back. [...] They have a nice little main street area with cute shops and restaurants and bars and things like that. But it feels more like that small neighborhood. It's kind of isolated, like, it's a little bit, it takes a little bit longer to get to the freeway from other neighborhoods, so, I don't know, it has this small neighborhood vibe" (EP-12).

In the cited paragraph, interviewee EP-12 is referring to the perceived isolation of South Park from other communities, which stems from its limited freeway access (the closest accesses are found in the surrounding neighborhoods of Golden Hill and North Park). In this sense, the neighborhood lacks a certain reachability in the overarching, postmodern freeway net of connected communities and centers, which is—among other aspects—interpreted and framed by the interviewee as a "small neighborhood vibe." Illuminating the statement from a border-theoretical stance, the described 'seclusion' is becoming a key element in the interviewee's performance of neighborhood identity construction. On the basis of this subjective attribution, b/ordering and othering processes (van Houtum and van Naerssen 2002) take place through which the neighborhood is differentiated from other communities in closer proximity to the freeway net. Relatedly, as demonstrated above using the example of Golden Hill, here, the subjective process of spatial identity creation is ultimately connected to the imprecise drawing of individual neighborhood boundaries.

However, the interviewee's individual perception of the neighborhood boundaries of South Park and its described seclusion are not only based on infrastructural aspects (San Diego's freeway net) but also on the topographic structure and the physical surroundings of the community. Interviewee EP-12 specifically refers to the large park and canyon areas, which in addition also constitute the neighborhoods official boundaries (CofSD 2013b, p. IN2)—Balboa Park in the west, Juniper Canyon in the north, and the larger, finger-like structure of Juniper Canyon in the east, which extends southward along Interstate 15.

> "Yeah, so, there's that canyon [Switzer Canyon] and there's really nothing on either side of the canyon, so that's one factor. And then you have the actual park, on another side is the park [Balboa Park], there is no development going on because it's a park, so that's another factor. [...] Yeah, so then, there's other canyons, like, Juniper Canyon, if you're looking at a map, that's another one. And then it's sort of, kind of, bordered by the 805, and that's like, kind of, in little pockets, almost. Oh, and the slight hill, slightly more, and because it is surrounded by canyons, park, canyons, it feels a little more separated to me" (EP-12).

In this statement about the perception of South Park and its unique, separated location in San Diego's inner-ring, interviewee EP-12 is addressing a rather contradictory spatial aspect of the neighborhood, which has become apparent in one of the conducted participant observations as well (PO-05). In the field notes of the neighborhood walk from North Park over South Park and Golden Hill into downtown San Diego, the author referred to the unique, yet contradictory situation of South Park as 'green seclusion in central location.' This is due to the observation that the fingers of the larger network of Juniper Canyon constitute significant physical borders in South Park since this structure creates numerous dead ends in the community, the described "little pockets" (EP-12). The bordered neighborhood edges have been underlined visually by numerous smaller fences demarcating the developed single-family area from the undeveloped, wild canyon on the other side. Additionally, views from the 'small pockets' into rural eastern San Diego County and on the Laguna Mountain range in the background of the canyon fingers have enhanced the author's border experiences at—what feels like—the 'multiple eastern edges of San Diego's semi-urban landscape' (PO-05; see Figure 5.22).

In contrast to the description of the physical seclusion of South Park, both the statements of interviewee EP-12 as well as impressions from the phenomenological walk (PO-05) emphasize a developing centrality of South Park in terms of its emerging or renewed 'urban village' qualities. Interviewee EP-12 shares that they have been unaware of the neighborhood of South Park until a few years ago. Now, the neighborhood would be popular—South Park and Golden Hill "are growing. A lot of startups move down there" (EP-01) as well as "young people. It is definitely gentrifying" (EP-01). According to interviewee AR-02, the renewed interest in the neighborhood lies also in its proximity to the recreational area of Balboa Park, which—in the physical as well as the functional dimension—simultaneously demarcates and connects South Park to other communities. Additionally, the neighborhood features a certain degree of walkability: historically, the rather 'quiet neighborhood' (as experienced in PO-05) is structured by relatively small lots with single-family homes along tree avenues with

Figure 5.22 Green Seclusion in Central Location—South Parks 'Little Neighborhood Pockets' Between the Various Fingers of Juniper Canyon. (Source: Pictures Albert Roßmeier 2021)

shady sidewalks. Furthermore, the commercial main streets—30th Street and Fern Street—are lined with storefronts and businesses catering to a young and 'urban' clientele, comprising of record stores, various fitness and yoga studios, juice bars and coffee shops as well as small craft beer pubs and restaurants serving Italian or vegan dishes (PO-05; see also Figure 5.23).

Summing up, in South Park overarching economic influences, the surrounding physical structures, and current social trends blend and—in a contradictory way—create a physically secluded but at the same time central community, which is "attractive to single persons and childless couples who wanted a synthesis of suburban comfort and urban opportunities [...]. The small-town, somewhat 'alternative' flair of South Park is especially notable in the center it shares with Burlingame, where small owner-managed shops and cafés occupy historically interesting premises—with a corresponding rise in real-estate prices" (Kühne et al. 2016, p. 28). For instance, one example for the reinterpretation and valuation of historicity in the neighborhood is the small bar called 'Station Tavern,' which occupies the site of an old streetcar station from the first half of the 20th century and integrates several corresponding elements into their otherwise contemporary building and interior design (PO-05). Thus, the gastronomic business is an emblematic element of the different hybridization tendencies found in the neighborhood of South Park. Here, various contradictions and opposites are unified and combined, conferring the historic neighborhood a 'new feel' while presenting it, on the hand, as 'small,' suburban, and secluded by green canyons and parks but, on the other hand, also as growing, socially diversifying, and centrally connected. In line with this, the neighborhood is characterized by physical and social hybridization tendencies that are associated with the overarching process of urban/rural hybridization (Kühne 2019, p. 124). Hybrid lifestyles both emerge and fit in the described 'urban landscape' context of South Park, which draws simultaneously on historic and contemporary elements and combines suburban or rural influences with urban centrality and diversity in a pastiche-like manner. Ultimately, it can be stated that both the ambiguously framed landscape concept within the interpretive category of urban/rural hybridity as well as border-theoretical thinking help to illuminate the various contradictions and reinterpretations of the case example of South Park. In conclusion, summing up the described developments concisely, interviewee EP-06 states: The neighborhoods around Balboa Park, "Golden Hill, South Park and North Park are changing, they are becoming more mixed" (equivalently shared EP-08).

In the following, another observed phenomenon will be discussed against the background of the elaborated hybrid context in order to further underline the multilayered neighborhood change that is taking place in South Park. During the

Figure 5.23 Cultural 'Appropriation' and Hybridization in Greater Golden Hill: The Popular Seafood and Vegan Taco Trucks on a Parking Lot in South Park. (The picture shows two food trucks on the parking lot of a large retail store in South Park. Source: Picture Albert Roßmeier 2019)

participant observation in the neighborhoods east and south of Balboa Park (PO-05) as well as during several recreational trips to San Diego, two taco trucks operating regularly from a parking lot of a large retail store in South Park caught the author's attention (see Figure 5.23). The observed situation at the commercial neighborhood center stood out not only because of the discovered olfactory stimuli during the phenomenological walk, but because the author identified the pictured vegan taco truck as a pastiche-like reinterpretation of Latino culture in a hybrid, recombined way (for the critical illumination of the social construction and the relationship of food, ethnicity, and gentrification in the case context of central San Diego see Joassart-Marcelli 2021; Joassart-Marcelli and Bosco 2020). While one of the trucks offers various Mexican street food-style seafood dishes and caters predominantly to the local Latino community, the second food truck in the photograph offers 'vegan tacos,' a supposedly healthier version of Latin-American street food, which is addressed to the newer, young residents of South Park. Similar to the hybrid cuisine of 'Tex-Mex,' which is combing common U.S.-American taste preferences with ingredients believed to be typically Mexican, vegan tacos can be interpreted as a phenomenon of so-called 'cultural appropriation,' which draws on Latino street food culture but caters to a white clientele with larger stocks of economic capital and cosmopolitan taste preferences. Interviewee AR-07, a young Mexican American graduate student comments the cultural reinterpretation of tacos as follows: "Vegan tacos are a big thing now. Never in my life, I had a vegan taco. [...] It's the capitalization of culture, basically" (AR-07). In terms of the development of street food and its popularity in San Diego, the interviewee describes further:

> "The food trucks specifically, that's such a weird thing to me because growing up, we have our own flea markets here. We call them [unintelligible], and they are called 'swap meets.' There is one in National City, a swap meet, we used to go there literally like every Sunday. A lot of Mexican, like, very, very Mexican. [...] And at that swap meet they would have food trucks. And it wasn't like these little fancy food trucks, it was literally like tacos and nachos and that kind of stuff, or fresh fruit. And those were the food trucks I kinda like grew up with. And then, I would occasionally see them parked liked on cheap little market parking lots where they would sell fish tacos and stuff like that. And that was, like, even 10 years ago wasn't even a thing and now food trucks are totally popping up. [...] For example, Balboa Park, Food Truck Fridays" (AR-07).

In this statement, interviewee AR-07 is addressing the observed change of consumption preferences and food aesthetics using the examples of food trucks and Mexican food in general in San Diego. The addressed, overarching social trend is particularly reflected in the neighborhood of South Park in the form of the discussed vegan taco truck, which ultimately constitutes a pastiche of the everyday necessities of people with lower stocks of economic capital in Latin-American countries that is now trimmed toward postmodern, individualist and aesthetic-focused consumption demands of young white professionals in the described urban/rural context of South Park.

As a last part of this section, in the following, certain differences between the neighborhoods of South Park and Golden Hill will be elaborated by means of experienced phenomena and situations during the participant observation (PO-05). Thus, it becomes possible to continue the quantitatively as well as qualitatively painted picture of the redevelopment trend in central San Diego, which is taking place in various ways and intensities in the different communities. The following discussion illuminates how the neighborhoods of South Park and Golden Hill relate to each other as well as to adjacent communities in terms of overlaps, similarities, and differences. In comparison to South Park, the discussed social trends and demands related to food, consumption, and leisure activities are reflected in less pronounced ways in Golden Hill (PO-05). In contrast to the observed business composition of South Park with several fitness studios and coffee shops in a pedestrian-friendly atmosphere, Kühne and Schönwald (2015a, p. 164) identify the majority of the businesses along larger intersections of Golden Hill as car-centered and consisting predominantly of "mini malls, fast food restaurants, liquor stores and gas stations. [The businesses in Golden Hill are] typical indicators of suburban communities with small stocks of symbolic capital."[7] Ultimately, this description coincides with the experiences made during the neighborhood walk. The majority of the local businesses in Golden Hill do not represent contemporary aesthetics and symbols found in other trendy neighborhoods of San Diego (such as Hillcrest) and cater to a clientele with medium and lower stocks of economic capital. Additionally, the topographic structure of the neighborhood diminishes the possibilities for pedestrians and bicyclists to move around the area comfortably—due to the sloping hills and the partly sun-exposed sidewalks, the neighborhood has been assessed as area of low walkability (PO-05, cf. interviewee EP-12).

In contrast to the business structure, the observations of Kühne and Schönwald (2015a, p. 164) on private properties and front yard landscaping no longer seem to correspond to current characteristics. The researchers state: "Here [Golden Hill], front yards with unconventional design are found regularly"[8] (Kühne and Schönwald 2015, p. 164), which they interpret as signs of a generally low level of sanctions against unconventional behavior in the neighborhood. However, this stands in contrast to the statement of interviewee R-13 who describes the increasing demand of the residents of Golden Hill for social and aesthetic conformity and integrity concerning signs of homelessness in the neighborhood. Additionally, as shown above, interviewee R-12 pictures the improvement of front yards in the Greater Golden Hill community from unkempt dirt areas to green lawns throughout the last decades. Lastly, also during the neighborhood walk (PO-05), only few of the front yards in Golden Hill have been identified as unconventional or neglected. Finally, this underlines the dynamic of the elaborated socio-economic and aesthetic neighborhood change, which has been taking place in the Golden Hill area in recent years. However, despite the increase of the median housing value and the urge of official planning instances to maximize "opportunities to re-use the existing housing stock" (CofSD 2013b, p. LU16), Golden Hill still features a higher number of unrenovated homes and empty lots compared to the trendy community areas of Uptown, North Park, or the neighborhood of South Park (PO-04, PO-05). This is especially apparent in the eastern section of the neighborhood, which has been quantitatively identified as a low-density area with a higher share of Hispanic population compared to the western neighborhood sections adjacent to downtown and along Balboa Park.

---

[7] „Minimalls, Schnellrestaurants, Liquor Stores und Tankstellen. Typische Indikatoren für suburbane Siedlungen mit geringer Ausstattung symbolischen Kapitals."

[8] „Unkonventionelle Vorgartengestaltungen sind hier regelmäßig anzutreffen"

In addition to the differences in the business composition and the appearance or condition of private homes, South Park and Golden Hill vary strongly in terms of population density and building sizes. Historically, Golden Hill features higher densities than South Park and is determined to accommodate more residents in the future through the construction of new apartment buildings in the southeastern part around the intersection of the prominent 30th street and Broadway (CofSD 2013b, p. LU20). Currently, this neighborhood section features several older as well as newer multi-apartment buildings. However, the eastern neighborhood part is also interspersed with single-family homes and undeveloped lots, especially eastward of 30th street where the population density seems to drop significantly and—once again—views into the sparsely developed eastern San Diego County and on the Laguna Mountain range enhance the impression of reaching the borders of 'semi-urban' central San Diego (PO-05, see Figure 5.19 and Figure 5.22). Accordingly, the eastern part of Golden Hill consists of a hybrid mixture of heterogeneous structural elements—new apartment complexes and construction sites are found in close vicinity to dilapidated single-family homes and empty lots – , particularly when compared to the properties and the building structures of South Park as well as of the western neighborhood parts of Golden Hill, which appear rather uniformly. In other words, physical change processes and aspects of a *physical-structural hybridization of urban and suburban/rural influences* are especially noticeable in the eastern parts of Golden Hill. The neighborhood of South Park and the areas adjacent to downtown and along Balboa Park in Golden Hill, in turn, testify more toward *social hybridization* and the crossing of *urban and suburban/rural lifestyles* or the everyday life routines of new and long-term residents (cf. Kühne 2019, p. 124). In line with these observations, it can be concluded that the Greater Golden Hill area is characterized by tendencies of urban/rural hybridization in varying intensities and dynamics in the different corners of the community.

In addition to these insights, it also becomes possible to draw border-related conclusions for the Greater Golden Hill area. Generally, in the overall phenomenon of neighborhood change and urban/rural hybridization in San Diego's inner ring, the discussed redevelopment trend is spreading from trendy, gentrified neighborhoods to adjacent, low-income areas, which are thus socially rebranded and reinterpreted from "'oh, this is a dangerous area' to 'oh, it's just a disinvested area, an area of opportunity,'" as interviewee EP-11 explains. Thus, overlaps of neighborhoods emerge in various dimensions through processes of assimilation, hybridization, and gentrification. This kind of progression has been addressed by interviewee EP-11, an urban planner from the planning department of San Diego, who witnessed the penetration of development dynamics from several trending neighborhoods in central San Diego into adjacent 'less interesting' communities throughout the last decade. For instance, the interviewee mentions spill-over processes from the neighborhood of Point Loma into the Midway District: "Over there, people call parts of it [Midway District], like, Point Loma or kind of trying to extend that Point Loma concept into the Midway district, which has been disinvested for a long time" (EP-11). Relatedly, concerning central San Diego, interviewee EP-11 addresses the multilayered expansion of North Park into the community of City Heights as well as of South Park into Golden Hill. Interviewee EP-11 states that "South Park [is], kind of, being extended in what used to be called Golden Hill."

The statement of interviewee EP-11 is, on the one hand, indicating that the neighborhood of South Park underwent significant change processes before the overarching redevelopment trend has reached the area of Golden Hill. This goes along with the statements of other interviewees (such as R-11) and is also indicated in a short portray of Golden Hill's redevelopment and rebranding by The San Diego Union-Tribune[9]: "Diverse and eclectic, the area [Golden Hill] is undergoing a renaissance, as its neighbor, South Park, did a few years ago" (Schimitschek 2017, n.pag.). Thus, Golden Hill is communicated as a neighborhood that has been reached at a later point in the overall redevelopment and reinterpretation process of San Diego's inner ring—after the 'development wave' went from Hillcrest over North Park into South Park, from where it is now progressing even further south and east into traditional Hispanic low-income communities, such as Barrio Logan and Southeastern San Diego (cf. Joassart-Marcelli 2021, p. 153). This illustration matches the impressions from the phenomenological neighborhood walk in which contrasts and disparities, in other words, aspects of physical and social hybridization have been observed more frequently in Golden Hill compared to other communities. This has led to the conclusion that Golden Hill is still in an earlier stage of redevelopment than, for instance, the somewhat more advanced neighborhood of South Park (PO-05).

On the other hand, the presented section of the interview with the urban planner EP-11 can be interpreted from a border-theoretical perspective. The interviewee is describing how the neighborhood of South Park is gradually overlapping with areas that were formerly known or perceived as Golden Hill. In this process, the individually and socially constructed neighborhood concept of South Park as a walkable and trendy place to be is spreading into the adjacent, less

[9] The San Diego Union-Tribune is—according to the company itself—"San Diego's largest media company" (2021). It "features coverage of daily news, sports, shopping, entertainment and is the second largest newspaper in southern California, preceded only by the Los Angeles Times" (San Diego State University 2021).

'fancy' area of Golden Hill and catalyzes a spatial reassessment. Following the argumentation of the interviewee, areas that have been known as Golden Hill in everyday life are now perceived as parts of the neighborhood of South Park due to redevelopment and subsequent reinterpretation processes. Thus, the interviewee is describing the shift of perceived neighborhood boundaries in the course of the spatially progressing redevelopment trend. This is ultimately underlining the social constructivist and post-structuralist understanding of neighborhoods as open and contingent spatial ideas instead of fixed territorial units (cf. Coulton 2012; B. A. Lee et al. 1994; Suttles 1972).

In addition to the remarks on the relations of eastern Golden Hill and South Park, the connections and demarcations between the northern part of the downtown neighborhood East Village and western Golden Hill will be illuminated in the following. In this section, the respective border situation has already been discussed on the basis of different interpretations of various interviewees, which will now be linked with subjective experiences and observations from the neighborhood walks (PO-01, PO-05). Initially, during the walks in East Village and Golden Hill as well as along Interstate 5, which constitutes the physical border of the two communities (see Figure 5.17 and Figure 5.24), strong differences became apparent. In the physical dimension, the demarcation between the two neighborhoods seems most drastic, on the one hand, because of the freeway, which cuts the previously connected communities in two individual areas and, on the other hand, due to the existing building structure. As illustrated above, the western part of Golden Hill consists predominantly of older single-family homes, while East Village has a large number of new multi-use building complexes. This structural disparity is the result of different zoning and land use regulations, as addressed by interviewee EP-08 (see Figure 5.24).

In the symbolic and socio-economic dimensions, strong differences between the two sides of the freeway are visible in terms of development expenses but also in respect to homelessness. In East Village, new luxury apartments buildings and multi-use developments in contemporary urban design, such as the so-called 'I.D.E.A. District' and the 'Makers Quarter' project, represent the urban policy shift toward reinvestment and the prestigious repurposing of downtown San Diego. This structural-symbolic difference between Golden Hill and East Village runs exactly along the freeway and describes a precise border between the dense and new inner city and the less dense former streetcar suburb. Additionally, as pictured by the interviewee R-13 and others, Interstate 5 describes a strong boundary in terms of signs of homelessness and social tolerance toward San Diego's homeless population. When crossing the freeway bridge from Golden Hill into East Village (PO-05), the high concentration of homeless residents in

Figure 5.24 The Multi-Layered Border between Golden Hill and East Village along Interstate 5. (The picture displays the difference of the building structure of Golden Hill and East Village. The foreground shows a wide street with multiple parking spots in front of low-density suburban-style housing. In the background, high-density apartment towers are visible. Source: Picture Albert Roßmeier 2021)

the former warehouse district became immediately apparent. However, the strong presence of homeless persons as well as of signs of homelessness—tents, drug paraphernalia, and trash on the sidewalks as well as permanently parked cars filled with large amounts of personal belongings—is not limited to East Village but starts already on the bridge and especially the two underpasses between the two neighborhoods. It became obvious that the San Diegans in need make use of the physical border—the infrastructure around the freeway that is the bridges, the underpasses, and the embankments—as a certain place of refuge (see Figure 5.21 and Figure 5.33), which adds to the complexity of the various everyday roles of urban borders. Finally, in the structural, symbolic, and socio-economic dimension, the disparity between Golden Hill and East Village is very pronounced and resembles some of the visible differences between Bankers Hill and East Village.

However, when looking closer at the two physically divided communities, certain connections and dynamics become visible. As argued by interviewee EP-09, from an economic or real estate viewpoint, the redevelopment efforts in downtown have ignited a trend of neighborhood restructuring,'place making,' and gentrification, which is spreading into the inner-ring neighborhoods—from East Village into Golden Hill and other communities east and southeast of downtown. As already demonstrated, this redevelopment trend has been moving around Balboa Park and has thus also reached Golden Hill from the adjacent neighborhood of South Park in recent years (cf. Joassart-Marcelli 2021, pp. 160–161; Joassart-Marcelli and Bosco 2020). As a consequence of the development of Golden Hill, socio-economic similarities with the northern corner of East Village become traceable—for instance in terms of population density or racial composition (U.S. Census Bureau 2021a). Taking a closer look at the share of Hispanic or Latino population of the census tracts in Golden Hill and East Village, a significant decrease from east to west and thus a gradient across the strong physical border of Interstate 5 becomes visible. In other words, the Hispanic influence is increasing continuously with greater distance from downtown and East Village to the more suburban, eastern end of Golden Hill instead of increasing abruptly at the freeway. In conclusion, even though Interstate 5 constitutes a significant physical border in everyday life, which is restricting mobility and other connections, the previously connected neighborhoods feature certain relations and 'cross-border' developments in the course of the overarching redevelopment trend in central San Diego, as illustrated in the digital neighborhood sketch (see Figure 5.25) in the following section. In the future, Golden Hill's "relationships with Balboa Park, Downtown, and other pre-World War II neighborhoods [...] will be enhanced [even more] through land use refinements, mobility improvements, and better urban design practices" (CofSD 2013b, p. LU16), as determined in the recent community plan for the Greater Golden Hill area.

5.3.3 Urban Landscape and Border-Theoretical Conclusions for the Greater Golden Hill Area—Interconnected and Central Despite its Natural and Asphalted Borders

In the following, a summary of the key results and insights into the individual interpretations of the neighborhood change in the Greater Golden Hill area will be given. On top of that, this subsection serves the purpose of elaborating urban landscape and border-theoretical conclusions—in written as well as in graphic form. The provided sketch of the neighborhoods of South Park and Golden Hill (see Figure 5.25) constitutes a postmodern experimental approach to cartography, which exemplifies different flows, networks, relations, as well as physical cutoffs and serves the purpose of facilitating the synthesis of urban landscape and border theory with empirical insights (on creative, contemporary approaches to landscape visualization and mapping see Edler et al. 2020).

The community area of Golden Hill is officially split into the two neighborhoods of South Park and Golden Hill, which stretch along the eastern and southern edges of Balboa Park. The communities' edges on the south and east are defined by State route CA 94, Interstate 15, and a complex structure of finger-like canyons, as shown in the overview map (see Figure 5.17). In the north, South Park borders the neighborhood of Burlingame in the community area of Greater North Park. Developed in the late 1800s through the emerging streetcar network around downtown and Balboa Park, Golden Hill as well as the subsequent addition of South Park have been successfully established as prestigious residential suburbs. Home to influential families and personalities, the community of Greater Golden Hill constituted a popular and desirable residential area in San Diego throughout the early 1900s and the Exposition years due to its proximity to Balboa Park and the scenic views on the San Diego Bay. However, following the two boom phases of the city, both Golden Hill and South Park began to fall from favor and experienced physical decay in the course of the Great Depression. The beginning distress has been accompanied as well as catalyzed by the white middle-class flight into the newly emerging suburbs in North County San Diego, which has caused the socio-economic composition of the entire inner ring to change (on the historic settlement development see chapter 4). Consequently, throughout the following decades, both South Park and especially the neighborhood of Golden Hill have changed from exclusive neighborhoods of the white affluent parts of San Diego's society to ethnically diverse, densely populated areas characterized by low stocks of economic and symbolic capital.

Throughout the 1980s and '90s, however, following the development efforts in downtown and the 'gay gentrification' of Hillcrest, the Greater Golden Hill community area once again started to attract white middle-class interest (as discussed by the interviewees EP-01, R-10, R-12). As a result of the redevelopment trend, between 2000 and 2010, the socio-economic structure of the overall community has gradually developed from predominantly Hispanic low-income families (in the year 2000, 52.8 percent of the residents were Hispanic compared to 43.5 percent in 2010) to mostly white single and childless households with higher numbers of college students as well as higher shares of educational attainments (in the year 2000, 34.6 percent of the residents were White compared to 44.5 percent in 2010, additionally, in 2000, 30.7 percent of the residents had educational attainments higher than high school degree compared to 41.8 percent in 2010; SANDAG 2003p, 2016c). A closer look at the various census tracts of the area (using 2020 census data) revealed that the northern sections of the community—the neighborhood of South Park—as well as the adjacent parts of Burlingame and North Park are recording higher shares of white residents than the tracts covering

Golden Hill and the northern parts of Southeastern San Diego. In the year 2020, the ethnic composition of South Park consisted of more than 65 percent white residents and the western parts of Golden Hill along East Village and Balboa Park of more than 55 percent, compared to the eastern section of Golden Hill (tract 41.01), which featured a share of less than 43 percent of white residents (U.S. Census Bureau 2021a). In conclusion, the quantitative data suggests a gradient in the share of white residents from north to south and from west to east, or in other words, a higher share of Hispanic population with greater distance to the already gentrified areas of East Village, Balboa Park, and North Park.

As illustrated in the sketch, not only the ethnic composition of the two neighborhoods follows the spatial pattern of a gradient from downtown and the Balboa Park area to the eastern neighborhood section of Golden Hill, also the share of adult population as well as the population density in the community area decline significantly from west to east. The western sections of Golden Hill and South Park feature higher shares of residents at the age of 18 and older compared to the eastern parts of the two neighborhoods that stretch along Interstate 15 and the rugged canyon structure. While the areas along downtown and Balboa Park comprise of more than 91 percent of adult residents, indicating a relatively high number of childless households, the lower share in the eastern community areas (less than 81 percent of the residents are adults) testifies to a comparably higher number of family households with children. In addition to this disparity, also population densities vary strongly between the western and the eastern neighborhood sections. While the western neighborhood parts feature high 'urban' densities of up to 20,638 persons per square mile, the eastern sections of Golden Hill and South Park are considerably less densely populated with only 6,044 persons per square mile (U.S. Census Bureau 2021a). Summing up, the neighborhood sections along East Village and Balboa Park constitute of rather urban environments in terms of density with a higher share of white residents living predominantly in childless households. In contrast to this, the eastern neighborhood parts testify to rather suburban household structures—due to a higher number of underage residents and a low population density—and feature a higher share of Hispanic population.

This quantitatively illustrated gradient resembles the spatial pattern of the qualitatively elaborated redevelopment wave, which is moving from the 'trendy' North Park area into South Park and from there into Golden Hill as well as from the gentrified East Village district into western Golden Hill and even further into Barrio Logan and Southeastern San Diego (outlined in the following section). As described in several interviews, this overarching development process is shifting the boundaries of affordable and accessible space in central San

Diego more south and east (AR-02, AR-04, AR-08, AR-11). Similarly, in the symbolic dimension, social reinterpretation and reassessment processes as well as image branding efforts are expanding the boundaries of San Diego's trending neighborhoods into disinvested and previously underserved communities—in this respect, interviewee EP-11 has discussed the expansion of (the concept of) South Park into areas formerly referred to as Golden Hill.

Simultaneously, however, Greater Golden Hill features strong physical borders and official demarcations to its adjacent neighborhoods that are of significant everyday life relevance. As argued by the interviewees EP-08, EP-09, and EP-12, in the physical and (planning-)political dimensions, the borders between the downtown area and the former inner-ring suburb of Golden Hill—constituted through Interstate 5, State Route CA 94, and the unique topography—remain stable and continue to divide the dense inner city from the less dense neighborhoods around it in terms of everyday life mobility restrictions and different underlying planning guidelines (zoning etc.). Additionally, as observed in the neighborhood walks (PO-04, PO-05), East Village and Golden Hill vary strongly in terms of their building structure: While the immediate area west of Interstate 5 consists mostly of new multi-unit apartment and mixed-use buildings, Golden Hill on the east side of the freeway features predominantly older single-family homes. The various viewpoints from the hilly streets in Golden Hill into the East Village district facilitate strong b/ordering and othering processes due to the visible differences of the physical structure (the built environment) but also because of the wide asphalted border (Interstate 5) between the neighborhoods (see Figure 5.24). Concerning South Park, its location between the canyon structures on the east (which are creating numerous dead ends and neighborhood pockets) and Balboa Park on the west as well as the absence of freeway accesses in the neighborhood are facilitating impressions of a 'natural' or 'green seclusion' (see Figure 5.22). In the participant observations (PO-04, PO-05), the strong physical borders, the surrounding freeways, the hilly terrain in Golden Hill, and the multiple fingers of the canyons in South Park have created various bordering experiences. Nevertheless, in the course of the described redevelopment wave, symbolic, social, and economic overlaps or transitions (EP-11, R-07) emerge in San Diego's inner ring, which connect North Park to South Park and East Village to Golden Hill in different ways and are causing social and physical hybridization processes in the form of new building developments, new trendy gastronomic offers, such as vegan taco trucks (see Figure 5.23), as well as the combination of typically urban and suburban lifestyles. Thus, despite the natural and asphalted borders surrounding the two neighborhoods and the described "small neighborhood vibe" (EP-12), significant multi-layered connections between the Greater Golden Hill community and the adjacent redeveloped neighborhoods have been established over the years, which confer the community a certain centrality.

Summing up the theoretical approach of this section, on the one hand, urban border theory helps to bring attention to the discussed neighborhood changes in terms of aspects of inclusion/exclusion and their everyday life relevance for different parts of San Diego's society. Landscape theory, on the other hand, is suitable to look closer at the reinterpretation processes of the inner-ring neighborhoods. Of particular interest here is the social constructivist understanding of landscape as a practice and as 'a way of seeing' (Cosgrove 1985, 1998 [1984]; Cosgrove and Daniels 1988). The aim of the application of this perspective is *not* necessarily to label San Diego's inner ring as 'landscape' but to make use of the specific way of thinking that has evolved in the field of landscape research. The approach to landscape as 'a way of seeing' allows to understand neighborhood change in Golden Hill and South Park as the result of the social attribution of positively assessed characteristics, such as authenticity and vibrancy, to these areas. Generally, in such social and individual reinterpretation or socialization processes, a more or less defined area is interpreted in a new and contingent but specific way that resembles the social discovery of landscape in a 'meaningless world' (Greider and Garkovich 1994, p. 2). In conclusion, from this perspective, authenticity, vibrancy, and "artsy vibe[s]" (IP-02) for instance, are a subjective way of seeing and feeling (in a phenomenological-experiential way) rather than essences of neighborhoods (cf. Burr 2005, p. 6). Lastly, this perspective seems especially beneficial for the analysis of neighborhoods such as Barrio Logan and Southeastern San Diego, which are in earlier stages of redevelopment, gentrification, and social reinterpretation, as it will be discussed in detail in the following section.

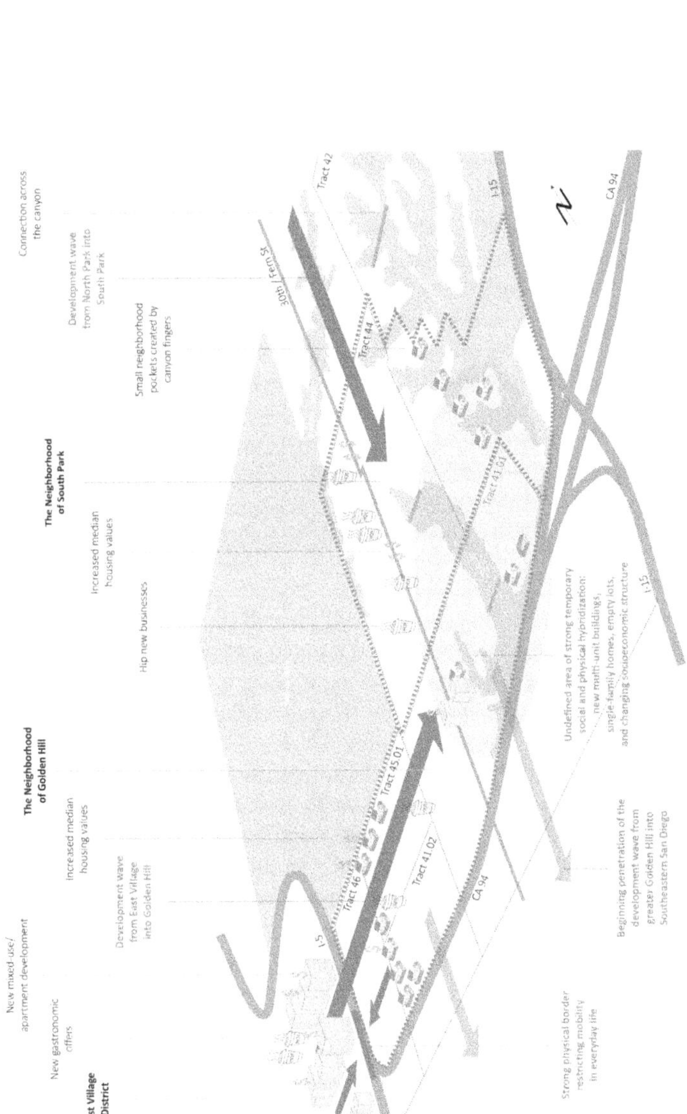

Figure 5.25 Digital 3D Sketch—Networks, Flows, and Demarcations within and around the Community Planning Area of Greater Golden Hill. (General information is provided in black, border-related conclusions are given in red. The red arrows are used to display the direction of the elaborated development wave, the transparent arrows indicate the direction of its commencing current progression. Along Interstate 5, two arrows pointing toward each other are used to display the physical demarcation and its high everyday relevance for mobility and in terms of b/ordering and othering processes. Social and physical hybridization processes became especially apparent in the eastern community section which are highlighted by the transparent red circle. Source: Own cartography and design)

5.4 Downtown San Diego's Multidimensional Expansion—The Redevelopment of East Village and its Ambiguous Connections to the Hispanic Communities of Barrio Logan and Southeastern San Diego

The following section constitutes the final component of the present empirical analysis and focuses primarily on processes of neighborhood change in the East Village area at the official (south)eastern edge of downtown San Diego (see Figure 5.26). The initial goal is to elaborate quantitative insights into the comprehensive downtown redevelopment that has taken place in the core since the 1970s and has gained new momentum in the East Village area in the last two decades. Subsequently, a qualitative analysis will illuminate the social perceptions and experiences of the changes in the community. However, the recent development efforts are not only affecting the former warehouse district of East Village, which lies within the official boundaries of San Diego's downtown, but also adjacent areas, such as the lower-income Hispanic communities of Southeastern San Diego and Barrio Logan, are increasingly facing redevelopment and gentrification pressures. Consequently, in addition to the focus on East Village, the following section is tracing the changing individual realities in the neighborhoods beyond the official eastern downtown edge.

First, the historic trajectory of downtown San Diego and the formerly connected areas of East Village, Southeastern San Diego, and Barrio Logan will be presented (section 5.4.1). In a next step, recent quantitative developments in East Village and its neighborhoods to the east and southeast will be illuminated (section 5.4.2). Building on the historically and quantitatively painted picture, qualitative insights into the recent redevelopment processes in East Village will be given (in section 5.4.3). Subsequently, the impacts of the development trend on the unique border and boundary situation between East Village on the one side and Barrio Logan and Southeastern San Diego on the other side will be elaborated (section 5.4.4). A summary of the key results and a written as well as cartographic synthesis of the quantitative and qualitative insights with the underlying urban landscape and border theories will conclude the chapter (section 5.4.5).

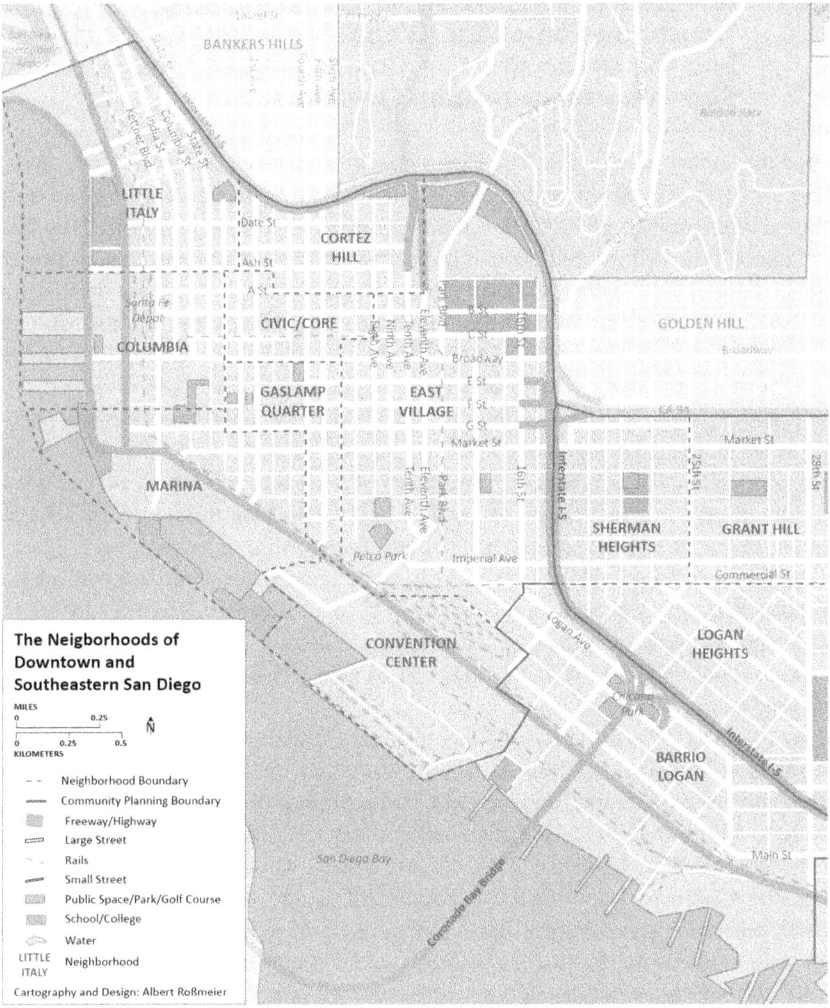

Figure 5.26 The Neighborhoods of Downtown and Southeastern San Diego. (Source: Own cartography and design)

5.4.1 A Historical Perspective on the (Attempted) Transformation of San Diego's Warehouse District into the Urban Creative Quarter of East Village and its Sprawling Gentrification Dynamics

Since the street grid of 'New San Diego' has been mapped out in the mid 1800s, the history of downtown San Diego is pervaded by a series of boom-and-bust phases. After first attempts to develop the area in the 1850s and early '60s had not gained broad social and economic interest, the real estate speculator Alonzo Erastus Horton had successfully initiated a development boost in the 1870s and '80s, which entailed the relocation of the county government and all other official authorities from the settlement in the Mission area to New Town. Paired with the activity of other local boosters and the long-awaited connection of the city to the intercontinental railroad system, in 1885 had come what "San Diegans today still refer to as the 'first Boom Time'" (Comer-Schultz 2011, p. 52)—a short era characterized by the arrival of large numbers of tourists, eager investors, and other curious groups that had begun to settle in and around the thriving New Tow as well as in the emerging community of Southeastern San Diego (see Figure 4.2). In 1888, however, the onset of a severe real estate crisis had ended the boom phase abruptly and led to a significant slowdown of population growth until the early 20th century. In addition to the demographic effects, the downturn has caused an infrastructural and functional restructuring of New Town with consequences still relevant today.

In the late 19th and early 20th century, the settled commerce in San Diego began to move north of Market Street, leaving behind an abandoned area in the south, which became the red-light and gambling district Stingaree, known today as the touristic Gaslamp Quarter. At that time, Broadway had started to serve "as a de facto boundary between two distinct worlds" (Ervin 2007, p. 194) with the CBD in the north and a mixture of liquor stores, bars, adult bookstores, and other low-cost businesses and their clientele in the south. Adjacent to this neighborhood in the east, switching yards and numerous warehouses, which "still give character to [these] parts of downtown today" (CCDC 2015, p. 1.7), have been constructed amidst existing residential properties in the course of the commencing industrialization and militarization of the waterfront. These developments have created the community of East Village as a noisy and noxious place of work and residence for the industrial workforce. With certain equivalence to today, this area has "throughout much of the city's history [...] been geographically amorphous" (Kayzar 2008b, p. 138). Historically, it has stretched eastward from Fifth Avenue, including areas of what are today Sherman Heights, Logan

Heights, and Barrio Logan. These communities share a similar history of ambiguous boundaries, which have "grown and divided over time" (Norris 1983, n.pag.). Consequently, today's neighborhood of Barrio Logan had carried various names throughout its history—until 1905, it was referred to as East End and afterwards as Logan Heights until the separation of the large area east of 17th street from today's downtown and its fragmentation into smaller neighborhoods (summed up as Southeastern San Diego today) due to the freeway constructions of the 1960s.

In comparison to East Village or Logan Heights, residential and commercial development in Barrio Logan has been delayed. "Available records suggest that as late as 1904, only a few, scattered businesses were in operation. Most were small grocery stores, while others, such as feed and seed stores, reflected the semi-rural atmosphere" (Norris 1983, n.pag.). But this atmosphere gave way in the years that followed; while East Village experienced a further shift from residential to commercial and industrial uses in the 1920s (first through the arrival of car dealerships and later on through the high military production needs; Eddy 1993), the local workers and the African American community were relocated to the east, into Barrio Logan and Southeastern San Diego (CCDC 2004, p. II-9) where the construction of apartment buildings and workers' housing had begun to change the suburban settlement structure of Victorian-style mansions (see Figure 5.27). "Industrial growth since the 1920s has differed fundamentally from earlier growth in that newer economic enterprises, by and large, have taken over land formerly used for residential purposes. Residential displacement by industry has resulted in the further erosion of residential land values" (Norris 1983, n.pag.) and the image of the large, undefined eastern outskirts of downtown San Diego. In the 1930s, zoning updates in Logan Heights made it possible for industrial uses to move into direct vicinity of the already decaying residential buildings. Ultimately, "as a result of the many recent land-use changes [...], Logan Heights today offers the visitor little of a sense of unity or of historical continuity" (Norris 1983, n.pag.). In other words, the area east of downtown San Diego—Southeastern San Diego and Barrio Logan but also East Village—are historically characterized by the direct juxtaposition of various uses, cultures, and groups.

In addition to the zoning changes, Japanese immigrants working in the tuna industry, Mexican refugees fleeing the revolution, and other new residents brought to the city by the military buildup during World War I had begun to settle in the densifying neighborhoods east of downtown where they replaced the affluent white families that left for new, trending neighborhoods north of Balboa Park (cf. Joassart-Marcelli 2021, p. 47). Especially the Panama-California Exposition, the emergence of new architectural styles (such as the various revival and the Craftsman styles), and the quickly developing streetcar network around the park have sparked the interest in emerging neighborhoods such as Mission Hills, North Park, and Kensington, while at the same time older communities with Victorian architecture, such as Southeastern San Diego and South Park, gradually lost their prestige. Additionally, with rising incomes and the growing mass availability of private vehicles in the 1920s and '30s, "places that previously required a high expenditure of time to reach, such as La Jolla, but were considered scenic, now became attractive as residential locations especially for people with a high stock of symbolic capital" (Kühne and Schönwald 2015, pp. 141–142). However, as a consequence of the use of discriminative zoning ordinances, restrictive covenants, and other segregative methods in the new and remote neighborhoods, the older residential areas east of downtown, as well as the city center itself, became part of the few available areas in San Diego for the non-white population and persons with a low stock of symbolic capital as early as the mid-1920s (cf. Ford and Griffin 1979; Killory 1993, n.pag.). Thus, "Southeastern San Diego today remains one of the most ethnically diverse and historic communities in San Diego" (CofSD 2015, p. 1.5).

In the World War II and the post-war era, the City of San Diego has recorded a continued influx of new residents (see Figure 4.2), which has boosted the Southern Californian trend for suburban—and thus socio-spatially as well as socio-economically segregated—living even more. Prompted by the drastic need for housing throughout the region and the financial interests of local developers, the booming building industry has transformed large areas of undeveloped land in the countryside and along the shore into low-density suburban neighborhoods. During this time, numerous new freeways have been constructed, which ensured the continuation of sprawling residential development in the periphery and allowed for the establishment of significant employment and consumption centers. For the white middle-class population outside the urban area of San Diego, the primary effect of the new transportation infrastructure was the further connection of suburbia. In the disinvested and neglected inner city and the community to its east, however, the construction of freeways and bridges entailed destructive effects. In the 1950s and '60s, Interstate 5, State Route CA 94, and the Coronado Bay Bridge have cut the spatially rather undefined community of Logan Heights into smaller neighborhood fragments, thus creating the communities of East Village (formerly referred to as Centre City East), Golden Hill, Southeastern San Diego, and Barrio Logan in the east as well as the neighborhood of Little Italy (formerly known as Harborview) in the west. On the one hand, the constructions of these physical borders caused the displacement and loss of residence of a certain number of the local population and thus the further reduction of residential use in the ambiguous area. On the other hand, however,

Figure 5.27 The Victorian, Queen Anna Style Villa Montezuma in the Neighborhood of Sherman Heights, a National Register Historical Landmark Erected in 1887 by the Writer and Musician Jesse Shepard. (Source: Picture Albert Roßmeier 2020)

the division of the large community of Logan Heights has not only taken place physically but also socially and symbolically through the fragmentation of its common identity (cf. E. E. Delgado and Swanson 2019; K. Delgado 1998).

Subsequently, in the year of 1970, the anger and frustration of local residents and activists over multiple undesired planning decisions (such as the freeway constructions, the widening of Harbor Drive, and the settlement of numerous junk yards in the area) and the beginning construction of a new police station under the pillars of Interstate 5 and the Coronado Bay Bridge culminated in the occupation of the construction site and its designation as 'Chicano Park.' Since its occupation, the park has not only served as a recreational-cultural but also as a symbolic space for the local and national Chicano culture, representing the fight for spatial participation, equality, and justice in the disinvested community and beyond (K. Delgado 1998; Le Texier 2007). Due to the large and carefully restored collection of murals displaying a mix of mythological, historical, and social aspects relevant to the Chicanos and their hybrid culture, the park has been designated a National Historic Landmark in 2017 (see Figure 5.28). In a broader sense, "the fight for Chicano Park, and the creation of the Barrio Logan concept [...] must be seen as attempts to re-establish a sense of neighborhood in an otherwise fragmented area" (Norris 1983, n.pag.), which has struggled to maintain and create a shared identity in the course of the historic (material and immaterial) bordering processes (cf. Kühne et al. 2022).

Figure 5.28 Restauration of Large Wall Murals on one of the Freeway Pillars in Barrio Logan's Chicano Park. (Source: Picture Albert Roßmeier 2021)

In the 1960s and particularly the 1970s, not only the newly formed communities of Southeastern San Diego and Barrio Logan were facing the consequences of the regional segregation and the 'white flight' phenomenon, but also downtown San Diego has been left behind in the success of the suburban residential and commercial development. Over the years, "San Diego's historical core had degenerated into a skid row and vice district" (Erie et al. 2011, p. 49). Former residents but also large retail and department stores had gradually left the decaying, dense city center for the orderly suburban malls and neighborhoods (Ervin

2008, p. 420). As a reaction to this growing loss of importance of the once booming downtown area but also in light of the striking revenue crisis of the city in the 1970s, the City Council has gradually conveyed private development a larger role in hope for the increase of tax revenues and the overall correction of the city center's trajectory. During this time, the planning ideology and approach in downtown San Diego has shifted from the further construction of office towers toward "entertainment destination development" (Kayzar 2008a, p. 449). A symbolic catalyst for this new era was the approval of a new multi-level retail shopping center in the heart of downtown San Diego, the Horton Plaza mall, in 1974. This project was embedded in a large urban design plan that included the overall redevelopment of the historic Horton Plaza and its park into a 6.5 block retail and entertainment center as well as the rebranding of 15 blocks of the adjacent Stingaree area into the touristic Gaslamp Quarter. Two years later, in 1976, the Marina District and the Columbia Office Zone have been added to the eager redevelopment project list. Following a modernist planning approach, the overarching goal of these redevelopment projects was the removal of blight through demolition and subsequent new constructions and thus the satisfaction of the "middle- and upper-class desires for a sanitized urban experience" (Ervin 2008, pp. 422–423; cf. Eddy 1995). For these purposes, in 1975, the non-profit corporation Centre City Development Corporation (CCDC), a division of the city's redevelopment agency, was founded to autonomously oversee the redevelopment process in the Horton Plaza area. Despite several obstacles and setbacks—including funding difficulties, concerns about historic preservation, and the struggle to convince major department stores to relocate downtown—one decade after the formation of the CCDC, Horton Plaza mall successfully opened in 1985 (Eddy 1995). In line with the shift of the city's planning ideology and its urge to create tax revenues, this project has furthered "exclusiveness, an orderly environment, and consumerism. Over time, the reinvented downtown became a place to explore and spend money—an urban Disneyland" (Ervin 2008, pp. 422–423).

Failing to comply with the new designation of the city center and its rising rents, in the 1980s, former residents and the local transient and homeless population have been relocated to the eastern section of downtown, the large Centre City East area, which constituted of underutilized warehouses, empty lots, and other industrial structures (Rumpf 2016, p. 15). This transition was supported by the CCDC and the City of San Diego, which had begun to grant temporary conditional use permits in the neglected areas east of the redevelopment zone in order to facilitate the 'removal' of deviant citizens and other non-conforming uses (Ervin 2007, p. 194; Kayzar 2006, p. 153). An important objective of this push was vacating buildings and properties in order to accommodate higher paying tenants and uses in the costly renovated Horton Plaza and Gaslamp area. On the one hand, this relocation and gentrification process has pushed out low-income households, business owners, seniors, the local homeless population, and a community of artists that have previously resided in the old Stingaree neighborhood and the city center. For the displaced artists and creatives, the gentrification of the Gaslamp Quarter created the (forced) opportunity to realize live/work lofts in vacant commercial and industrial buildings in the adjacent Centre City East area—today's East Village—where they initiated a temporary form of an unofficial arts district.

On the other hand, also specific uses and organizations, such as social service providers, have been pushed out of the city center and have instead been granted use permits in the eastern downtown portion. Additionally, at the same time, federal spending on social welfare has been significantly reduced, which promoted the foundation of private social service organizations throughout the country, thus creating the task for local municipalities to accommodate them (Kayzar 2008b, p. 141). In San Diego, low-income housing options, such as SRO rooms and social services in general, have traditionally only been welcomed in the East Village area—"zoning within most of the city and county, combined with strong NIMBY objections from local [suburban] communities, interacted with the city's political networks to shape the decision to make the east end a primary destination for service providers" (Kayzar 2006, p. 156). As a result, in the 1980s and '90s, the redevelopment of the city center has promoted the concentration of artists, homeless residents, and other people in need in the eastern corner of downtown San Diego—an area that has over time developed into a spatially amorphous and undefined "zone of discard" (Kayzar 2008b, p. 139) for the entire region.

However, this area has not stayed untouched by redevelopment efforts in the course of the (attempted) renaissance of downtown San Diego. In 1989, for instance, San Diego's convention center has opened between the Marina District and the switching yards of the city's industrial harbor just south of East Village. The facility underwent a first major expansion in 2001 and is currently again subject to additional construction measures. Symbolically, the opening and the expansion of the convention center both reflect a continuation of the entertainment-focused development ethos that has been prevalent in downtown San Diego since the 1970s. Physically, the development can be seen as an addition to the constructed frame of the eastern downtown area in which over time several redevelopment projects have been realized with the intention to enhance spatial order and definition and to create a (new) sense of place. Another milestone of redevelopment in East Village was the construction of the new main branch of the San Diego Public Library and its opening in 2013. As a new facility with up-to-date technology and unique architectural design features, the controversial and costly project was hoped to increase the community's attraction and strengthen its reputation as an important and official part of downtown (Erie et al. 2011, pp. 153–158).

However, the culmination of the efforts to transform the former Centre City East area was the approval of a "mammoth redevelopment project" (Erie et al. 2011, p. 176) in 1998 by the San Diego voters that ultimately led to the rebranding of the eastern portion of downtown San Diego into East Village. The plan included the $454 million construction of the ballpark Petco Park for the San Diego Padres, the city's baseball team, which opened in 2004, as well as ancillary private development of hotels and businesses on a combined area of 26 blocks at the community's southwestern edge. "Most of the stadium's costs were borne by the public sector under a redevelopment P3 [public-private partnership] among the City of San Diego; the Port District; CCDC; and JMI Realty, the real estate arm of Padres owner John Moores" (Erie et al. 2011, p. 184). What makes the redevelopment project particularly unique is its "extensive use of public-private redevelopment partnerships that delegate[d] substantial public resources and powers to the private sector. Indeed, San Diego has used P3s more extensively and, with Petco Park, on a larger scale than is typical of cities elsewhere" (Erie et al. 2011, p. 177). From the public point of view, "linking the ballpark to private-sector investment [...] provided a political solution for how to sell the expensive project to local voters" (Erie et al. 2011, p. 187). Moores, the owner of the San Diego Padres, instead benefited from framing and presenting the new stadium as a redevelopment project, which allowed him to obtain financial subsidy from the public sector for the new ballpark. Additionally, the project was paired with the slogan 'more than a ballpark,' which has referred to promised financial and infrastructural revitalization outcomes for the entire downtown area. These promises ultimately led the San Diego voters to approve the large plan and its financing model, which was based on tax revenues.

The ballpark project and the ancillary development in the surrounding area significantly changed the physical structure but also the social composition of the East Village community, as it will be explained by means of quantitative and qualitative data in the next two sections. However, in the years following the ballpark's opening, the promised redevelopment dynamics have not proven to be as impactful as previously anticipated by city officials and local investors. On the one hand, this is due to the downturn of the housing market in the late 2000s, which has put "many large-scale center city residential construction projects [...]

on hold [while] submissions of new project proposals to the redevelopment corporation have diminished substantially" (Kayzar 2008b, p. 136). On the other hand, however, not only market trends are to blame for the lack of revitalization outcomes in East Village. Kayzar (2008b, p. 137) is pointing out "several elements of the live, work, and play planning goal for downtown that [were] failing to meet the expectations of center city investors and proponents" in the 2000s and early 2010s: A significant aspect was the ratio of housing to employment opportunities throughout the downtown area and particularly in East Village. The lack of jobs in East Village had created the need for residents to commute out of the community and the city center to the peripheral employment centers of the region. This had an evident effect on the street life, the daily activation of infrastructure, and the demand for local (gastronomic) businesses. Furthermore, the local businesses and the limited recreational space are rather serving the temporarily appearing masses of baseball fans and their entertainment-oriented demands instead of fulfilling the everyday needs of residents. In addition to that, a large number of the housing stock in East Village is occupied as vacation and second homes. Thus, even though downtown tourism destinations such as the Gaslamp Quarter and Little Italy are "extremely successful and lively" (Ford et al. 2008, p. 86) with "street life sufficient to support restaurants and retailers[,] [...] holes in the urban fabric remain throughout downtown, and investment activity is inconsistent. Nowhere is the contrast between revitalization success and failure more evident than in the East Village, and in this sense the community remains far behind the even partial successes of other center city communities," as Kayzar (2008b, p. 137) states. Until today, the community is strongly characterized by the contradictive juxtaposition of new and eager development projects and the "disproportionate share of many of the region's burdens, and these burdens often mitigate community successes at a rate unknown in other areas of downtown" (Kayzar 2008b, p. 137).

Ultimately, it becomes obvious that not only the urban fabric in East Village, which consists of new residential and old industrial buildings, is entangled in an unsorted way but also social confrontations are taking place where newcomers and their demands and lifestyles are potentially coming into conflict with existing residents and the community's homeless population. In the following, a detailed quantitative and qualitative analysis is aiming to generate a wider understanding of the recent change processes in as well as around East Village, its socio-spatial consequences, and their potentially conflicting individual interpretations. Lastly, as it will be discussed below, even though the success of the redevelopment efforts in East Village is rather contentious, neighboring communities, such as Southeastern San Diego and Barrio Logan, are increasingly affected by rising rents and property values, shifting neighborhood images, and the changing perspective of civic leaders, planners, and investors on the future potential of the southeastern downtown edge. This has complex consequences for the relations of the different communities to each other and for their border and boundary situations, which will be elaborated in detail below.

5.4.2 Recent Demographic and Housing-Related Developments in East Village in Comparison to the Communities of Southeastern San Diego and Barrio Logan

In the following section, census data is used to trace the quantitative development of the historically connected communities on the southeastern downtown edge between the years 2000 and 2020 in order to help interpret and understand the recent change processes in the respective neighborhoods. Due to limited availability of 2020 census data on the community level, the illustration of demographic and housing-related developments is split into two parts. Firstly, a more general overview and comparison of the community planning areas of Centre City (downtown San Diego), Barrio Logan, and Southeastern San Diego for the years 2000 and 2010 will be presented (see Table 5.4). In order to put the community developments into a larger context, census data for the entire City of San Diego will be added to the comparison. Secondly, four adjacent census tracts along the official boundary lines of the various communities (tracts 47, 49, 50, and 51, see Figure 5.29) will be contrasted, using selected data from the 2000, 2010, as well as the 2020 census in order to underpin and differentiate the initial comparison on the community level with small-scale data (see Table 5.5).

The four tracts are covering large parts of the neighborhoods of Sherman Heights (tract 47) and Logan Heights (tract 49) in Southeastern San Diego as well as areas of Barrio Logan (tract 50) and East Village (tract 51). The selection of these census tracts is reflecting the focus area of the present section, which is dealing particularly with the community edges along Interstate 5 where they adjoin each other. Due to the thematic interest in the multi-dimensional expansion of the downtown area into the adjacent communities, the eastern ends of both Barrio Logan and the large Southeastern San Diego community will not be discussed in detail due to their distance from downtown and East Village and their seemingly persistent low-density residential character (as assessed during the participant observation PO-02). In contrast to the eastern community areas, the recent development in East Village and northwestern Barrio Logan seems to differ. For in the 2020 census, the census tract 51—an area stretching over the southeastern parts of East Village and the most northern section of Barrio Logan—has been split (into three distinct tracts: 51.01, 51.02, and 51.03), which generally indicates disproportionate population growth. As a last remark, it is beneficial to repeat (see section 3.3.4) that the following quantitative illustration is not serving the positivist purpose of explaining the 'inevitable reality' in the communities and tracts on the southeastern downtown edge but is used instead as a means to complement the subsequent discussion and interpretation of the changing individual border(ing) realities and experiences in this transitioning part of the city.

Prior to the presentation of the census comparison, it is necessary to outline the physical borders and official boundaries of the three selected communities (see Figure 5.29) in order to locate the various official neighborhoods, their sub-districts, and the chosen sections for the analysis. The downtown area—officially and in planning terms referred to as Centre City community planning area—is physically demarcated by Interstate 5 in the north and east and the San Diego Bay in the south and the west. The section of the southern boundary between Interstate 5 and the bay is officially running along Commercial Street and the switching yards of the harbor. Between the bay and Interstate 5 in the north, the boundary is described by Laurel Street. Thus, the large East Village area, encompassing the official eastern and southeastern end of downtown San Diego, is as well bordered by Interstate 5 in the north and the east and is adjoining Balboa Park and the communities of Golden Hill and Southeastern San Diego. In the south, the large neighborhood is bordering the community of Barrio Logan and the convention center area. The western boundary of East Village has been drawn along Sixth and Eighth Avenue where the community borders the Gaslamp Quarter, the Civic/Core district, and the neighborhood of Cortez Hill until its boundary merges with Interstate 5 in the north. Additionally, due to the large size of the East Village area, it is officially divided into four different subdistricts, three of which, however, are yet almost nameless or unbranded. The districts are officially referred to as Northwest, Northeast, Ballpark and Southeast and especially the two southern areas are "poised to experience some of the most significant changes under the [current] Community Plan" (CCDC 2015, p. 3.36). Accordingly, the focus of the present section is directed on the southeastern community area, which intersects with Barrio Logan, Sherman Heights, and Logan Heights.

Southeastern San Diego lies to the east of Interstate 5 and constitutes of the official neighborhoods of Sherman Heights, Grant Hill, Stockton, and Mt. Hope in the north and Logan Heights, Mountain View, Southcrest, and Shelltown in the south. The spread-out community is—due to segregative historic urban and

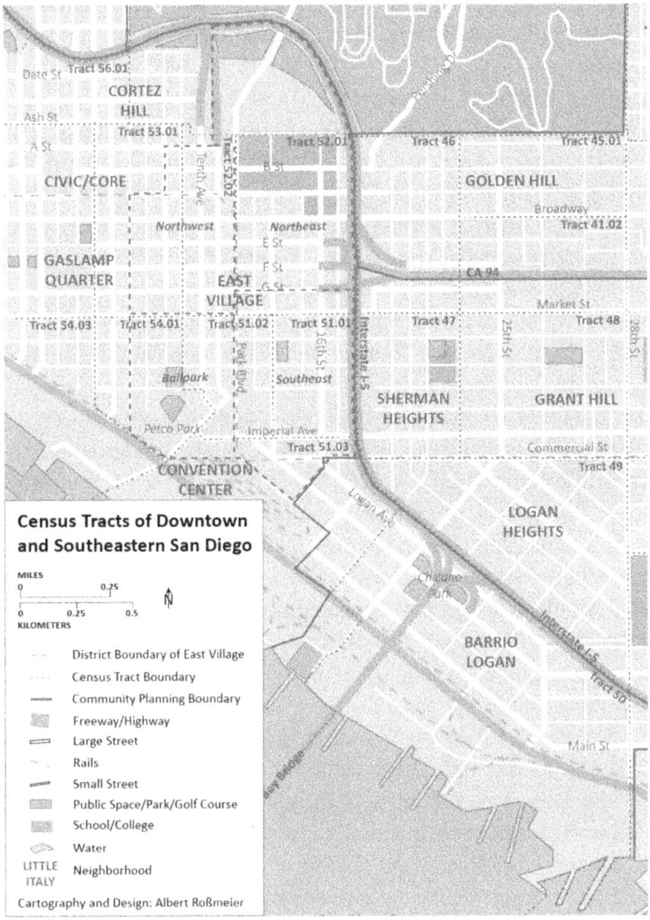

Figure 5.29 Selected Census Tract and Special Focus Area. (Source: Own cartography and design)

traffic planning—almost entirely bordered by freeways with State Highway 94 in the north and Interstate 805 in the east. Division Street and the jurisdiction of National City are describing the southeastern boundary of the community. Lastly, the smaller community of Barrio Logan is nestled between San Diego Bay to its south, Division Street and National City to its southeast, Interstate 5 and Southeastern San Diego to its north and northeast, and East Village to its northwest. Historically and also due to its unique location, the physical structure of the community of Barrio Logan is also determined by the establishment of a conflictual mix of land uses and physical structures. These include the emission-rich industrial port and its switching yards in the north, the large United States Naval Base in the south, and expanded traffic infrastructure, such as Harbor Drive, the Coronado Bay Bridge, and Interstate 5. These aspects will be outlined further in the subsequent qualitative section by means of the gained individual insights. Beforehand, the discussion of census data is presented in the next paragraphs.

When comparing the population development of the three selected communities between 2000 and 2010 and putting it in contrast with the development of the City of San Diego, the most drastic percentual population growth has taken place in the Centre City community planning area with an increase of 80 percent, from 17,513 to 31,494 residents. In tract 51—combining the southeastern corner of East Village and the northern part of Barrio Logan—this growth rate has even been exceeded. Between 2000 to 2010, the population in this area grew by 113 percent and in the overall timeframe from 2000 to 2020 by 253 percent, from 3,360 to 11,869 (cumulated value for the split tracts of 51.01, 51.02, and 51.03). In contrast, the entire community of Barrio Logan underwent a less drastic population growth of only 34 percent, which equals an increase from 3,636 residents in 2000 to 4,865 in 2010. Tract 50, the central part of the community, however, has recorded a population decrease. There, the population declined by eight percent between 2000 and 2010, from 2,430 residents to 2,227, and then rose again slightly to 2,312 in the following decade, which means an overall decline of five percent compared to the year 2000. Similarly, the community of Southeastern San Diego—the largest community of the three selected areas in terms of the total population—experienced a minor negative growth as well. From 2000 to 2010, the community recorded a population decline of 1 percent, from 57,571 to 56,757 residents. In tract 47, the boundaries of which are roughly describing the neighborhood of Sherman Heights, the population decline has been even more pronounced. While tract 49 (Logan Heights) declined by only 1 percent in the two decades from 2000 to 2020, tract 47 recorded a decrease of the total population of 26 percent from 2000 to 2010 and an overall decline of 41 percent from 2000 (2,521 residents) to 2020 (1,493 residents). Putting these numbers into the regional context, the jurisdiction of San Diego has recorded a population increase of six percent from 2000 to 2010. In summary, the immediate downtown edge, tract 51, is facing a strong population increase, while the central part of Barrio Logan and the neighborhoods to the east of Interstate 5 have recorded a significant outflow of residents.

Similar to the outlined population development of the various communities and tracts, the total number of family households is following the depicted trend of each community and tract in the last two decades. While the downtown area is recording a disproportionate increase of 161 percent between 2000 and 2010, from 1,628 family households (which equals 19 percent of the total number of households) to 4,254 (and thus an increase to 24 percent of the total number of households), the communities of Barrio Logan and Southeastern San Diego have faced small decreases. In Barrio Logan, the total number of family households has decreased by nine percent, from a share of 75 to 58 percent of the total number of households, and Southeastern San Diego has recorded a decrease of seven percent between the years 2000 and 2010, which equals a decrease of the share of family households from 82 to 75 percent. In accordance with the overall population increase, the City of San Diego, in turn, has recorded a slight growth of the number of family households by 4 percent in this time frame.

In addition to the development of family households, it is beneficial for the present illustration to take the development of the total number of underage residents in the selected tracts into consideration. In each tract, the total number of population under the age of 18 mirrors the development of the total number of family households. This means that while tract 51 is recording a significant increase of underage residents between 2000 and 2020 (a growth of 56 percent in the first decade and an overall increase of 107 percent in the two decades), the tracts 50, 47, and 49 have faced decreases. Especially the neighborhood of Sherman Heights (tract 47) has recorded a significant decrease of 75 percent of its population under the age of 18 in the two decades, from 935 and a share of 37 percent of the total population to 230 underage residents, which equals a share of only 15 percent. However, even though tract 51 is the only tract of the selected four in which the total number of underage population has increased, the share of residents under the age of 18 has decreased as well (from 18 to 11 percent). In conclusion, the share of underage residents is continuously decreasing throughout downtown and the neighborhoods on its edge—regardless of whether the population in the respective area is growing or decreasing.

Furthermore, the ethnic composition of the neighborhoods on the southeastern downtown edge and accordingly also the spoken languages in the local households have changed significantly in the two decades from 2000 to 2020. At the community level, Barrio Logan has recorded the most drastic changes in terms of the total number of white population and the population (above the age of five) speaking only English at home. Here, the share of white residents has increased from 6 to 16 percent (from 203 to 795 residents) from the year 2000 to 2010, which equals a growth of 292 percent. Additionally, the population of

Barrio Logan that is speaking English as the only language in their household has increased by 167 percent, from 642 persons (which equals 20 percent of the total number of residents above the age of 5) to 1,717 (38 percent). In the Centre City community, the ethnic composition has shifted as well slightly, from 55 to 57 percent white residents, which means a growth of 86 percent (from 9,677 to 18,046 residents). In terms of the language spoken at home, the population speaking exclusively English has increased by 93 percent, raising their share from 69 to 74 percent. Thus, in the cohorts of 2000 and 2010, the majority of downtown residents are English-speaking and white, while the community of Southeastern San Diego, for instance, is home to only 3 percent of white residents in 2000 and 4 percent in 2010 (which equals an increase of 29 percent, from 1,799 to 2,329 residents). Here, the total number of exclusively English-speaking persons (above the age of 5) has decreased by 13 percent, reducing the share from 26 to 22 percent. In comparison to the three communities, between 2000 and 2010, the City of San Diego has faced a decrease of white residents (by 3 percent) as well as of the share of white population, declining from 49 to 45 percent. Although the number of exclusively English-speaking households living in the jurisdiction of San Diego has increased, their share has slightly decreased (from 63 to 61 percent).

The tract level, however, is allowing for a more detailed, small-scale insight into the development of the ethnic composition and is pointing toward certain spatial concentrations. Even though all of the three communities have recorded significant increases of the total number as well as the share of white population in the years between 2000 and 2010, the four selected, adjoining tracts within these communities have experienced even higher growth rates—especially in the decade from 2010 to 2020. Despite a slight decrease of the total number of population, the most drastic increase of white residents is found in the tract covering Logan Heights (tract 49) with a growth of 506 percent between the years 2000 and 2020 from 121 to 722 residents. In terms of the share of white population, however, tract 49 recorded the lowest increase (from 2 to 15 percent) over the two decades. Additionally, with only 15 percent of white residents, tract 49 is characterized by the lowest share of white population in comparison to the other three tracts and especially to tract 51 (covering parts of East Village and Barrio Logan), which comprises of a majority of white residents. In turn, the strongest increase of the share of the local white population is found in tract 47, starting at 10 percent or 240 residents in 2000 and reaching 37 percent or 551 white residents in 2020. Even though this growth is a significant increase of 130 percent (in the timeframe from 2000 to 2020) of the total number of white population, this is the weakest total growth of all the four tracts.

In comparison, the tracts 50 (central Barrio Logan) and 51 (southeastern East Village and northern Barrio Logan) have similar growth rates. While the increase of white residents in tract 50 has been rather marginal in the first decade (26 percent, from 96 to 121 white residents), the area has recorded a strong growth between 2010 and 2020 of 310 percent (from 121 to 496 white residents, which equals an increase of 417 percent from 2000 to 2020). In tract 51, the white population has increased by 459 percent between 2000 and 2020 (from 1,126 to cumulated 6,297), which has made white residents the predominant ethnic group in the area. Accordingly, compared to the other tracts, tract 51 also features the highest share of white population, which developed from 34 percent in 2000 to 53 percent (cumulated) in 2020. In terms of the three split tracts of 51.01, 51.02, and 51.03, the highest share of white residents (63 percent of the total population) is found in the most western part of the former tract 51 (now 51.02), which is the tract that borders the ballpark Petco Park. In conclusion, while downtown and its southeastern edge are growing significantly with a predominantly white, English-speaking population, the community of Southeastern San Diego in the east of Interstate 5 is facing a population decrease and comprises of only 4 percent of white residents and a majority of Spanish-speaking households (32 percent of these households are speaking English "not well" or "not at all", SANDAG 2016d).

Comparing the three selected communities in terms of the educational attainment of the local population (25 years and above), the strongest percentual increase is found in Barrio Logan with a growth of 444 percent from 2000 to 2010, from 114 (6 percent of the population above the age of 25) to 506 (17 percent) persons with college degrees. However, the highest total number of adults with college degrees is found within the central city area where the group of residents with college degrees has increased from 4,4430 (a share of 30 percent) to 12,771 persons (a share of 49 percent) between the years 2000 and 2010. As a result, in 2010, the downtown area is featuring a disproportionately high share of college-educated residents compared to the overall city of San Diego (45 percent). The lowest share of residents with college degree, in turn, is found in the community of Southeastern San Diego (1,944 residents or 7 percent of the total population above 25) in the year 2000, which increased to 2,873 persons (9 percent) in 2010. In comparison to this development, in the community of Barrio Logan the group of college-educated residents has recorded a higher increase between 2000 and 2010, from 114 residents with college degree (6 percent) to 507 (17 percent).

In addition to the educational attainment also the development of the median household incomes in the three communities is worth considering. Between 2000 and 2010, the citywide median household income has increased by 38 percent, from $45,825 to $63,198. In comparison to this, the communities of Southeastern San Diego and Barrio Logan are characterized by relatively low median incomes and have experienced below-average increases. On the one hand, the growth rates are only 30 percent in Barrio Logan and 33 percent in Southeastern San Diego and, on the other hand, the total median household income in 2010 grew to values less than half of the citywide income ($26,761 in Barrio Logan and $31,414 in Southeastern San Diego). The central city area, in turn, has recorded an above-average increase of the median household income between 2000 and 2010 with a growth of 115 percent, from $22,888 to $49,296. However, despite the stark increase, this value is still below the citywide average.

Summarizing the quantitative discussion of demographic developments presented above, the Centre City community experienced a strong population growth between 2000 and 2010 and is—compared to the City of San Diego—characterized by a lower share of family households and lower median household incomes but features higher shares of white, exclusively English-speaking, and college-educated residents. Barrio Logan, in turn, is characterized by low shares of white residents, persons with college degrees, exclusively English-speaking households, and particularly low median household incomes. However, the beginning influx of higher educated, white, and English-speaking residents, the stark increase of the median housing value, and the decrease of family households and units is particularly conceivable, which can be interpreted as an assimilation to or expansion of the socio-spatial trends in the adjoining East Village area. Accordingly, Joassart-Marcelli (2021, p. 154) states that "Barrio Logan, City Heights, and Southeastern San Diego began gentrifying more recently and remain in transition, with several tracts experiencing little change. Among these neighborhoods, Barrio Logan is at a more advanced stage of gentrification, with property value increasing by 67 percent in the past seven years alone." In equivalence with these findings, Southeastern San Diego appears to follow central San Diego's urbanization and redevelopment trend in a less pronounced way—the community records the highest share of family households and units, it has not experienced significant growth in terms of the total number of housing units or the median household income, and it is characterized by the lowest shares of white, exclusively English-speaking, and college-educated residents. In the following section, these quantitatively painted neighborhood change processes will be discussed in detail by means of qualitative insights.

Beforehand, however, the development of housing-related aspects in the three communities and selected tracts will be discussed. In comparison to the development of the median household income, the consideration of the median housing values in the respective areas offers a very different picture. Even though the median household income in Barrio Logan and Southeastern San Diego has only risen by less than 35 percent between 2000 and 2010, the median housing value has increased by 286 percent in Barrio Logan (from $111,207 to $428,906) and by 268 percent in Southeastern San Diego ($95,629 to $351,456). These two growth rates are disproportionately higher than the citywide increase of 120 percent (from $230,076 to $506,107). The highest median housing value is found in the Centre City community area with $530,361 in 2010 after a relatively small increase of 52 percent.

In terms of the total number of housing units, the downtown area has faced a disproportionate increase in the years from 2000 to 2010 compared to the citywide development. While the total number of housing units in the City of San Diego has increased by 10 percent between 2000 and 2010, the Centre City planning area has faced a growth of 151 percent, from 9,454 to 23,689 units. In this timeframe, the growth rate of downtown San Diego has been exceeded by the increase recorded at the southeastern edge of downtown, in the border area of East Village and Barrio Logan (tract 51). In tract 51, 2,997 new housing units have been constructed between the years 2000 and 2010, which equals a growth rate of 368 percent. In the following decade, 2,795 new additional units have been realized in the tracts 51.01, 51.02, and 51.03, which equals an increase of 712 percent in the overall timeframe from 2000 to 2020. Of the 6,606 cumulated total housing units found in these three tracts in 2020, the majority (3,745) is located in the tract 51.02, which lies in the immediate east of Petco Park and as shown above also features the highest share of white residents and the lowest share of population below the age of 18 compared to all the other considered tracts. In comparison, tract 51.01 contains 1,991 units, while only a relatively smaller number of 870 housing units are located in the ambiguously structured tract 51.03, which includes the convention center, parts of the harbor area and its switching yards, as well as the southern parts of East Village, and the northern section of Barrio Logan (see Figure 5.29).

In comparison to the high growth rates of the downtown area, the community planning area of Barrio Logan recorded an increase of 18 percent of the total number of housing units between 2000 and 2010, from 1,051 to 1,244 units. However, in the same timeframe, the central portion of the community (tract 50) has faced a slight decline of 4 percent, from 664 to 640 units. Thus, the decline of the total number of housing units in the central section of the community has been offset by the growth that has taken place on the northern community edge, in tract 51, which is the redeveloping transition area to downtown. However, in the following decade, tract 50 has recorded an increase of its housing stock as well, from 640 to 714 units, which equals a growth of 8 percent between the years 2000 and 2020 and bespeaks the community's commencing transformation toward redevelopment. Similar to Barrio Logan, also the community of Southeastern San Diego has experienced an increase of the total number of housing units. In the decade from 2000 to 2010, the community gained 228 housing units, which constitutes a slight increase of 2 percent. Considering the development in the two selected tracts of the community, tract 47 (Sherman Heights) in the immediate east of downtown has recorded a decrease of the total number of housing units by 8 percent, while tract 49 (Logan Heights) has faced a growth of 17 percent in the two decades between 2000 and 2020. The reason for the decline of the total number of housing units remains unknown. However, insights from various interviewees on the development of Golden Hill, for instance, are pointing toward the reconversion of formerly subdivided single-family homes, which can be interpreted as physical manifestations of new residents with higher stocks of economic capital and hybrid urban/rural lifestyles or demands for spacious and independent living in urban settings.

Accordingly, also the development of single-family housing units in the selected communities will be discussed. In all of the three planning areas, the number of single-family units has declined between 2000 and 2010. The most drastic decrease as well as the lowest share has been recorded in the downtown area. The total number of single-family homes in the Centre City area has decreased by 57 percent (from a share of 5 to 1 percent). In Barrio Logan, family housing units have decreased by 35 percent from 2000 to 2010. This development is remarkable because, at the turn of the millennium, 75 percent of the housing stock in Barrio Logan has constituted of single-family units, but within one decade this share has decreased to only 39 percent, which means that the community is now providing a more mixed stock of housing options. Southeastern San Diego has experienced a slight decline of the total number of single-family units by 4 percent between 2000 and 2010. Thus, the share in Southeastern San Diego has decreased from 82 to 73 percent, which means that—in contrast to the other two communities—single-family housing is still the predominant housing form throughout the planning area.

Lastly, the development of vacancy rates in the selected tracts will be considered. In the tracts 50, 47, and 49, the total number of unoccupied housing has changed only marginally between 2000 and 2020, resulting in no percentual change of the local vacancy rate. In both cohorts, a vacancy rate of 8 percent has been recorded in tract 50 as well as in tract 47, while tract 49 features a lower rate of only 6 percent. This is different for the tracts 51.01, 51.02, and 51.03. In these areas, extensive housing construction has taken place since the early 2000s.

However, the median housing value is disproportionately high and, as it will be discussed in detail below, the reception and the overall success of the East Village redevelopment and rebranding efforts are highly ambiguous (cf. Kayzar 2008b, p. 137), which is mirrored in relatively high vacancy rates, ranging from 4 to 26 percent in the three split tracts. Even though the highest total number of unoccupied housing units (706 units) of the three split tracts is found in tract 51.02—the tract with the largest housing stock –, the highest vacancy rate is found in tract 51.01 at the eastern end of East Village along Interstate 5 (26 percent and a total of 510 unoccupied units), which also features the lowest total number of residents. The relatively high vacancy rate of 26 percent in tract 51.01 is only exceeded by the vacancy recorded in tract 54.01, which contains Petco Park and parts of the tourism and nightlife district Gaslamp Quarter. In 2020, 29 percent of the total housing units in this tract are unoccupied. In turn, the lowest vacancy rate of all the selected tracts is found in tract 51.03 (4 percent). This portion is covering the transition area between East Village and Barrio Logan and is—in comparison to the tracts 51.01 and 51.02—characterized by the lowest number of housing units (due to the inclusion of the convention center, the harbor area, and other commercial structures), the lowest share of white residents, but the highest share of underage residents, and a relatively high total population. Thus, this border area can be interpreted as a *transition zone* between Barrio Logan, the mixed-use neighborhood of predominantly lower-income, Hispanic family households with a higher stock of single-family housing on the one side and the predominantly white, relatively unaffordable East Village community on the other side, which constitutes an ambiguous area itself due to the conflicting and contradictory mix of new and old buildings, land uses, and residents.

In the following section, the recent change processes that are taking place in the highlighted communities and tracts will be discussed by means of individual insights of residents, planners, developers, and other specialists, such as local academics. This will complement the quantitative perspective on the complex urban landscape changes with various individual interpretations in order to take account of the processuality and infinite variety of subjective realities and (bordering) experiences in this ambiguous part of the city.

Table 5.4 Demographic Neighborhood Developments between 2000 and 2010 in the Central City, Barrio Logan, and Southeastern San Diego Community Planning Areas.

		Total Population	Family Households	Total White Population	Population Speaking only English	Population 25 years and above with College Degree	Median Household Income (in $)	Median Housing Value (in $)	Total Housing Units	Single-Family Units
Central City	**2000**	17,513	1,628 (19%)	9,677 (55%)	11,907 (69%)	4,440 (30%)	$22,888	$348,397	9,454	456 (5%)
	2010	31,494	4,254 (24%)	18,046 (57%)	22,848 (74%)	12,771 (49%)	$49,296	$530,361	23,689	196 (1%)
	Change	**+80%**	**+161%**	**+86%**	**+92%**	**+188%**	**+115%**	**+52%**	**+151%**	**−57%**
Barrio Logan	**2000**	3,636	734 (75%)	203 (6%)	642 (20%)	114 (6%)	$20,604	$111,207	1,051	734 (75%)
	2010	4,865	666 (58%)	795 (16%)	1,717 (38%)	506 (17%)	$26,761	$428,906	1,244	480 (39%)
	Change	**+34%**	**−9%**	**+292%**	**+167%**	**+444%**	**+30%**	**+286%**	**+18%**	**−35%**
Southeastern San Diego	**2000**	57,571	11,627 (82%)	1,799 (3%)	13,178 (26%)	1,944 (7%)	$23,553	$95,629	14,954	11,627 (82%)
	2010	56,757	10,808 (75%)	2,329 (4%)	11,473 (22%)	2,873 (9%)	$31,414	$351,456	15,182	11,153 (73%)
	Change	**−1%**	**−7%**	**+29%**	**−13%**	**+148%**	**+33%**	**+268%**	**+2%**	**−4%**

(continued)

Table 5.4 (continued)

		Total Population	Family Households	Total White Population	Population Speaking only English	Population 25 years and above with College Degree	Median Household Income (in $)	Median Housing Value (in $)	Total Housing Units	Single-Family Units
Jurisdiction of San Diego	**2000**	1,223,400	271,398 (60%)	603,892 (49%)	714,429 (63%)	305,953 (39%)	$45,825	$230,076	469,689	264,933 (56%)
	2010	1,301,617	281,704 (58%)	586,799 (45%)	745,499 (61%)	385,460 (45%)	$63,198	$506,107	515,412	278,965 (54%)
	Change	**+6%**	**+4%**	**−3%**	**+4%**	**+26%**	**+38%**	**+120%**	**+10%**	**+5%**

Source: Own composition and calculation based on compiled census data (SANDAG 2003a, n, o, q, 2016a, b, f, d).

Table 5.5 Quantitative Neighborhood Change between 2000 and 2020 in four adjacent Census Tracts of East Village, Barrio Logan, and Southeastern San Diego.

			Total Population	**Total Population under the Age of 18**	**Total White Population**	**Total Housing Units**	**Vacant Housing Units**
Tract 51: East Village	**2000**		3,360	608 (18%)	1,126 (34%)	814	96 (12%)
	2010		7,140	950 (13%)	2,996 (42%)	3,811	790 (21%)
	Change		**+113%**	**+56%**	**+166%**	**+368%**	**+723%**
	2020	Tract 51.01	3,199	391 (12%)	1,548 (48%)	1,991	510 (26%)
		Tract 51.02	4,428	166 (4%)	2,778 (63%)	3,745	706 (19%)
		Tract 51.03	4,242	702 (17%)	1,971 (46%)	870	37 (4%)
		Cumulated	11,869	1,259 (11%)	6,297 (53%)	6,606	1,253 (19%)
	Change (In Relation to 2000)		**+253%**	**+107%**	**+459%**	**+712%**	**+ 1,205%**
Tract 50: Barrio Logan	**2000**		2,430	966 (40%)	96 (4%)	664	51 (8%)
	2010		2,227	765 (34%)	121 (5%)	640	57 (9%)
	Change		**−8%**	**−21%**	**+26%**	**−4%**	**+12%**
	2020		2,312	569 (25%)	496 (21%)	714	60 (8%)
	Change (In Relation to 2000)		**−5%**	**−41%**	**+417%**	**+8%**	**+18%**
Tract 47: Sherman Heights	**2000**		2,521	935 (37%)	240 (10%)	755	63 (8%)
	2010		1,858	412 (22%)	363 (20%)	763	120 (16%)
	Change		**−26%**	**−56%**	**+51%**	**+1%**	**+90%**
	2020		1,493	230 (15%)	551 (37%)	697	59 (8%)
	Change (In Relation to 2000)		**−41%**	**−75%**	**+130%**	**−8%**	**−12%**

(continued)

Table 5.5 (continued)

		Total Population	Total Population under the Age of 18	Total White Population	Total Housing Units	Vacant Housing Units
Tract 49: Logan Heights	**2000**	5,014	1,878 (37%)	121 (2%)	1,411	85 (6%)
	2010	5,028	1,636 (33%)	171 (3%)	1,368	38 (3%)
	Change	**0%**	**−13%**	**+41%**	**−3%**	**−55%**
	2020	4,978	1,260 (25%)	733 (15%)	1,646	101 (6%)
	Change (In Relation to 2000)	**−1%**	**−33%**	**+506%**	**+17%**	**+19%**

Source: Own composition based on compiled census data (SANDAG 2003g, h, i, j, 2016l, m, n, o; U.S. Census Bureau 2021a). The subdivision of the communities into the various census tracts is shown in Figure 5.29. Tract 51, covering both parts of East Village and of Barrio Logan has been split into three distinct tracts for the 2020 census, which typically indicates population growth in the respective area. 2020 census data has not been available for the level of the community planning areas.

5.4.3 Redevelopment Efforts, Image Branding, and Gentrification at the Southeastern Downtown Edge—East Village between Success and Failure

The present section is dealing with the redevelopment processes in downtown San Diego and especially in East Village. The focus is directed on the transformation and the re-branding of the city's former industrial area Centre City East into the East Village entertainment and creative district, which has taken place since the early 2000s. By means of a discussion of individual experiences and perceptions of the redevelopment in East Village, its current state, as well as its long-term vision, different interpretation threads will be worked out. For this purpose, statements of interviewees from the three ideal-typical perspective groups of the present study will be contrasted. The subsequent section will deal with the adjacent communities of Barrio Logan and Southeastern San Diego and the multi-dimensional bordering processes that are taking place at the southeastern downtown edge. That section is focused on the various ways in which the redevelopment is perceived to progress from the inner-city to its edges and on how this progression forms a transition(al) space between the inner city and the traditionally Hispanic communities of the inner ring.

As outlined above, the planning and construction of the Horton Plaza mall in the 1970s and 80s is emblematic of the onset of the decadelong political and economic efforts to transform San Diego's city center from a neglected scene of decay and poverty to a prospering hub of tourism, upscale housing, and creative jobs. Concerning the past of downtown, an investor from Bankers Hill states: "Downtown was nothing, Horton Plaza was nothing but homeless [people] when I arrived here [in San Diego] 30 years ago" (EP-06). In a similar vein, the environmental planner EP-12 from Hillcrest states that "people worked downtown, but nobody was living downtown. Downtown used to not be a place that you, like, downtown was not as nice as it is now. [...] I did not go downtown, downtown has changed a lot. It was kind of, like, grimy and not the best so that's a big change." Reflecting on the time before the construction of the prestigious mall at Horton Plaza, the retired Ocean Beach resident R-12 shares that the area around Horton Plaza was frequented by "prostitutes, it was only porn theatres, gentlemen clubs, and lot of Chinese import stores." In accordance with this statement, also the planner and urban research scholar AR-02 describes: "When I first came here, there wasn't a lot going on in downtown. Downtown was suffering in the late '60s and '70s [...]. People were leaving, but there were certain efforts to not let this happen." The interviewee is referring to the construction and the opening of the Horton Plaza mall in 1985 and to the redevelopment of the former redlight district Stingaree into the touristic nightlife area Gaslamp Quarter. Concerning the transformation of the Stingaree district, which had previously served as an enclave for Chinese immigrants and a refuge for artists and other transients, the geography graduate AR-08 shares: "Before that [the redevelopment], I guess like further back, in the '70s, it was like, this area was not too, nobody, like, goes there because there's the poverty, it was very unmaintained. But after this development, then you have a hotspot, pretty much. And everybody in San Diego is like 'Oh, the Gaslamp, the Gaslamp,' so it became a popular area."

As discussed above and described in various interviews, the attempts to reinstate the area's prestige have taken place on the basis of the displacement of low-income and immigrant households as well as of homeless residents, on the one hand, and through the attraction of white suburbanites to the newly designed shopping center on the other hand. In the course of the demolition and redevelopment project, "homeless people were pushed from Horton Plaza to Balboa Park and then to Petco Park until this one was developed," as interviewee EP-06 from Bankers Hill recalls. According to the downtown place-maker and entrepreneur EP-08, the efforts to displace homeless people and other residents in need have been enhanced by the relocations of several service providers, such as the San Diego Rescue Mission. "The Rescue Mission in the Gaslamp Quarter was right there, but when they did the redevelopment, they just shut the Rescue Mission and then a lot of homeless people moved to East Village" (EP-08). However, despite the removal and containment of 'undesirable' facilities and their users in the eastern end of downtown—the spatially rather undefined Centre City East –, San Diego's city center had still not experienced the success it was promised. Referring to the years after the opening of the Horton Plaza mall, the urban research scholar AR-11 described:

> "Many people in San Diego, even in you know, 10 years and certainly when I moved here in 1993, even in City Heights, which is about 6 miles from downtown, there were many people in my neighborhood that said: 'Why would you go downtown, what for? Like, to buy heroin?' Like, what other reason would you have to go downtown other than maybe to go to the county building to, you know, register or pay your taxes or something. You know, people really, there just was really no sense of what the role is."

By the end of the 20th century, this dilemma has created the need for city leaders, developers, and other stakeholders in downtown to forge a new role for the city center that would guarantee persistent success and prosperity. For the megaproject Petco Park, the city and its redevelopment arm, the Centre City Development Corporation (CCDC), partnered with the owner of the San Diego Padres baseball team and pursued the costly idea to rededicate the Centre City East area into an urban sports event area with supporting commercial and upscale residential development surrounding it. The developer EP-07 recalls: In the first years after its construction, "Petco Park kicked off the development significantly" in the newly renamed East Village area. East Village "has an enormous potential" (EP-07). According to the scholar AR-11, this major physical and infrastructural transformation is a new "recreational presentation of the city, you know, a reframing of it as a place where you hang out, which […] creates a new relationship between the inner-city and the suburbs that had eroded." In terms of this new presentation and framing, the resident R-12 states that "10 years ago, no one called it Cortez Hill or East Village, you know, this area used to be really rundown."

Arguing from a theoretical and methodical standpoint, these statements about the transformation of the East Village area are representing three different modes of interpretation that can be ascribed to the three defined ideal-typical perspective groups of the present study. While the developer EP-07 is concerned with the revitalization outcomes of the large transformation, the urban scholar is reflecting on socio-spatial aspects and the historical meaning of the city center in the sprawling region. The interviewee from the perspective group 'resident,' in turn, is reporting about the sphere of everyday life, about individual impressions and the shared social idea San Diegans have of the former Centre City East area. With reference to landscape theory, these examples resemble the subjective construction of reality and the transformation of physical space into "symbolic meanings that emanate from the values by which people define themselves" (Greider and Garkovich 1994, p. 1) and the special knowledge they gained. For San Diego and the urban landscape of East Village, this comparison allows for the conclusion that the interpretation of physical spaces, neighborhoods, or borders is always taking place within various modes that highlight certain aspects and leave out others entirely.

In order to further emphasize this subjective selectivity within social perception and interpretation, various statements of interviewees of the three ideal-typical perspective groups will be contrasted and discussed in the following. As in the example above, the interviewees from the perspective group 'residents' have particularly shared biographical insights, common sense interpretations, and personal experiences of the redevelopment efforts in today's East Village. One example is R-01, a college graduate and family father that has lived in East Village in the early 2010s. The interviewee describes general observations concerning land use changes and the overall appearance of the district and recalls subjective experiences. According to R-01, ten years ago, the majority of the building stock constituted of one- and two-story warehouse buildings, most of which were officially vacant but utilized by the local homeless population. Referring to the perceived desolate state of the community, the interviewee R-01 describes: "When I lived down there, I remember times where I had to walk out to my car late at night, and sometimes I would be scared because people were fighting and screaming, you know, I saw some violent and aggressive things." However, the interviewee describes further that the redevelopment efforts evoked a change of the local zoning laws that allowed for the increase of density and thus facilitated the construction of taller buildings in the area. "Now, when I drive down there, you don't see such things anymore, you see more the evidence of gentrification, you know, upscale apartments, entertainment, luxury restaurants, those kind of things. It's hard for me to believe what I'm seeing, you know, the whole shape has changed, the cityscape" (R-01). The interviewee shares further that today's housing stock in East Village is consisting mostly of upscale condominiums and desperately lacks affordable space. Equivalently, also interviewee R-12 describes the change of the community from a neglected industrial warehouse district to an exclusive neighborhood with high rents. The interviewee states that downtown and East Village used to be affordable until the eastern end has experienced significant investment in the last 10 to 15 years (R-12).

Interviewee R-11, a long-term resident from the downtown neighborhood Cortez Hill is sharing from a biographic-experiential perspective as well. According to the interviewee, the East Village community was the last district in downtown to redevelop, it suffered from a bad reputation due to a large homeless population and the decade-long accumulation of service providers. However, interviewee R-11 deplores that the recent development in the area is overwhelming and "insensitive, it's all about the money." The residents—including themselves—are left behind in this process that is solely centered around development profits, as the interviewee describes. R-11 states further that even though the public discourse in East Village is focused on the need to construct low-income housing, the housing options that are officially considered low-income are not affordable for households with below-average incomes. Additionally, concerning the extent of the redevelopment processes in the area and its effects for long-term residents, the interviewee R-11 refers to a specific bar in East Village, which has been operating since before the construction of Petco Park and the rededication of the district. From the interviewee's perspective, this bar has been a staple for residents and locals in the times prior to the development. However, in addition to the renovations and high-rise constructions, the redevelopment has caused various businesses to adapt to the different group of incoming users and new residents. As R-11 deplores, this trend has also caused the respective bar to attract mostly younger crowds with its large selection of craft beers and loud dance music. Additionally, also during the participant observation in East Village (PO-02), this bar, its design, the noise level, and the customers in front of it have been recorded as attributes that match the idea of the downtown edge as an entertainment district with a trendy, young community rather than its previous version as an underutilized warehouse area.

Summing up, the interviewee R-11 is sharing personal impressions and experiences that criticize the development as 'insensitive' gentrification and highlight the negative aspects of the redevelopment. However, among others, the interviewee also discusses effects of the community change in East Village that they view as desirable, such as the increased cleanliness of public spaces. Another example for this interpretation approach is the interviewee R-02, that views gentrification as a process that particularly entails positive aspects for the community, such as the diversification of recreational uses and overall improvement of security in a specific area (similarly interpreted also by R-06). Furthermore, interviewee R-07, a young mother who is generally opposing the gentrification of central San Diego, is the opinion that new residents can have a positive impact on neighborhoods if they actively participate in the local community, for instance through efforts to improve local public schools. Within these interpretations, neighborhood change and the influx of new residents are rather framed as a chance to enhance and improve the local community of an area than as an inevitable threat. Ultimately, this bespeaks the subjectivity and variety but also the complexity of the ways in which urban landscape and neighborhood change is interpreted and experienced in San Diego.

Special knowledge—here in the form of education and experience in the fields of urban and regional planning, geography, or border studies—is adding another layer or lens to the interpretative contingency, making interpretation processes even more complex. In comparison to the ideal-typical perspective group of 'residents,' interviewees from the group termed 'academia/research' are putting their statements in a larger framework of national or global urban development trends, they present 'zoomed-out' observations that are discussed against the background of common themes of the respective academic fields, and they are concerned about long-term effects of developments and visions for districts. However, just as the interviewees from the perspective group 'residents,' also this group is sporadically sharing personal anecdotes and experiences since the majority of these interviewees is living in the cross-border metropolis of San Diego-Tijuana. Concerning East Village and its long-term vision, the various interviewed scholars are generally of the opinion that "East Village is, you know, a part of downtown, so, East Village will definitely urbanize and is already urbanized," as interviewee AR-09 sums up. Interviewee AR-06 states that due to an update of the local zoning laws, "you see new construction [in East Village], it is all over the place." In terms of current development plans, interviewee AR-02 reports that the idea of local developers and city leaders is "to really turn East Village into what is called an innovation district, it all started with Petco Park," to attract more "startups, designers, and innovators to East Village" (AR-02). As a result, "downtown San Diego is becoming more urban, with more shared economy, more shared workspace" (AR-06). Kühne et al. (2016) are interpreting this as a process of 'urfsurbanization' in which the formerly less densely populated East Village area is developing toward a stereotypical version of an urban downtown.

However, concerning the long-term vision of East Village and the entire downtown area, several interviewees of the group 'academia/research' voice concerns. According to AR-06, "it is time to look at the vision, East Village doesn't have one." The interviewee is convinced that the future development in and around downtown San Diego should be "less freeway oriented" and that it needs to "enhance the neighborhoods but without creating a tremendous juxtaposition because this creates enclaves. We want balanced societies, not that juxtaposition." Thus, the interviewee is criticizing the current situation of separate and physically divided autonomous communities, which are existing next to each other and without relevant infrastructural links. A possible way to tackle this separation is to "reform the zoning on the corridor level to end the juxtaposition of neighborhoods and help them merge together" (AR-06). The goal is "to make those connections" beyond the freeway lines and to introduce zoning laws that "allow for mixed-use urbanism" along corridors, as the interviewee explains. In addition to AR-06, also the interviewee AR-09 is addressing the future vision for East Village and downtown. In an ambivalent way, the interviewee states: "I think East Village is a good model because they are trying to, a lot of jobs are located down there, a lot of high-tech jobs and young people wanna live there. But the big challenge in East Village is, [...] I think downtown is kind of struggling because the city hasn't decided what they want downtown to be." This critique is addressing the city's approach to create future visions of cultural vibrancy and economic prosperity for formerly—more or less—neglected districts solely through costly and exclusive megaprojects, such as the Horton Plaza redevelopment or the Petco Park transformation.

Concerning the exclusiveness of these projects and the state of homelessness in East Village, AR-02 states: "In some aspects we have been very successful because we have a quite vibrant downtown, but in that process, there is a lot of social issues, which have been ignored, which were out of site but now they are coming to the forefront." In this respect, arguing from a more critical perspective, interviewee AR-04 is forecasting that "East Village is going to be a wealthy part of San Diego. The homeless are being pushed out, first into Barrio Logan and then more south and east." Equivalently, also the young bi-national scholar AR-01 is deploring that the development in East Village is "pushing out people and away from services." Thus, the interviewees are describing how the redevelopment takes effect in the form of a bordering process that regulates and updates access to and participation in the inner city—through the privatization of public space, the increase of median housing values, but also the displacement of service providers (as described by AR-11). From their perspective, the *boundary of accessibility* between the exclusive, high-rent inner city and the lower-income communities on the downtown edge is being shifted outward to the southeast and is thus integrating more and more parts of the East Village area. These aspects are mirrored quantitatively in the illustration above (see Table 5.4 and Table 5.5) and will be discussed in detail in the following section (5.4.4).

Bringing the two discussions about the future potential of the area and its produced exclusiveness together, interviewee AR-09 sums up:

> "It has a lot of potential, I think it has a lot of positive qualities: It's close to downtown, it's near the bay, there are already a lot of jobs that are located there, high-tech jobs, and young people wanna live there, you know, downtown. So there's a lot of good entrepreneurship and a lot of good energy. [...] Young people, millennials, and other folks who are moving downtown and doing live/work are part of the scene down there. So it seems like it has a lot of potential, but I think the cost of housing is still a concern and the city needs to do more to encourage affordable housing. And that's a big debate in the city right now."

In order to further underline the variety of interpretation threads that result from various perspective modes, a discussion of selected statements of and discourses shared by planners and developers in San Diego is following. In comparison to the two presented modes, the interviewed developers and planners are concerned with concrete revitalization outcomes as well as the branding and the themes of districts. The East Village developer EP-07, for instance, is highlighting the success of their development efforts. "Nine years ago there was nothing there, from the user perspective it was not interesting [...]. There was no place to eat, and yesterday I ate at the place that was voted the best Taco place in San Diego. [...] Now there are parents with strollers, and nine years ago there were only homeless people." For the interviewee, the success of the redevelopment is measured by the return of young families to downtown and East Village and the diversification of gastronomic infrastructure. The addressed population growth has been shown above by means of census data: Between 2000 and 2020, the tract covering the southeastern part of East Village and the northern section of Barrio Logan (tract 51) has been the only tract on the southeastern downtown edge that has recorded a significant increase of the total number of underage residents (see Table 5.5). Furthermore, between 2000 and 2010, the number of family households has also increased in the entire downtown area, from 19 to 24 percent, whereas the share has decreased throughout the surrounding communities in the same time frame (see Table 5.4).

However, despite these outcomes, the interviewee emphasized that "there is still a lot of work to do" (EP-07). For according to several other interviewees, the East Village area is lacking employment options (cf. Kayzar 2008b, p. 137). In reflection of San Diego's planning approach to convey responsibility and powers to the private sector, the interviewee EP-07 reports that their team of private developers was "trying to figure out how we could create something in downtown that is creating jobs." This motivation was spurred by the large number of housing units that have been realized throughout downtown (see Figure 5.30) but have unbalanced the ratio of uses, specifically of housing to employment, as criticized by various interviewees. In line with the illustration of census data above (see

Table 5.5), the city center has faced "a dramatic increase in housing and that was great, but we wanted downtown to be vibrant and sustainable, we wanted downtown to be a magnet for talent, for young, educated, and creative people [...]. These people want to be downtown and not in the suburbs anymore, it's a trend" (EP-07). Thus, the interviewee describes their efforts to convey a certain image to the former industrial district. In the same vein, also EP-08, a place-maker and entrepreneur in downtown is envisioning East Village as a "really recognizable neighborhood, a funky makers quarter space where we celebrate community and walkability, it's really awesome." In order to realize this vision, as interviewee EP-07 continues, "we need to start to focus on job creation and we did that through the lens of creative place making, we wanted to highlight the importance to create that environment [...]. For future prosperity it is necessary to create a great city." The interviewee is convinced that "companies are going where the talent is [...], they are constantly searching where the talent is" (EP-07).

Referring to the efforts of the private developer and their theme for East Village, the urban and policy planner of the City of San Diego, EP-09, is outlining their idea further: "The vision there is to really better link the universities that are there, the City College, the New School of Architecture, several universities that are kind of in a linear, you know, that are in a line, and to better marry those with more creative kinds of business, if you want. [...] So, it's really, I think the vision there then is more to create an education hub with research and creative thinking." Accordingly, the nonprofit development organization Downtown San Diego Partnership is framing East Village as "downtown San Diego's eclectic hipster neighborhood and another hotspot for tech and innovation startups" (DSDP 2016, p. 9). On top of that, with its broad infrastructure of "studios, artis lofts, galleries, restaurants and shops" (DSDP 2016, p. 9), East Village is also envisioned as "the new arts district," as the young graduate and place-maker EP-05 highlights. In summary, the presented statements demonstrate the continuation of intertwined public and private efforts to redevelop the city's eastern end, which have begun with the construction of Petco Park in the early 2000s and are carried on with the current vision of a creative work-life district at what is referred to as the 'innovation economy's next frontier' (DSDP 2016). However, the confidence of various public and private stakeholders concerning the positive redevelopment outcomes and the prosperous and creative future of the East Village area is contrasted by a variety of interviewees (from the other two perspective groups) that frame the efforts as conflictual or only partly successful. These doubts are tied to a number of aspects that will be presented in the following.

Figure 5.30 New Loft- and Industrial-Style Residential Buildings in East Village in between Construction Sites and Fences. (New multi-unit residential building in industrial aesthetic and installed fence on the public sidewalk intended to prevent homeless encampments (left). High-rise and multi-unit buildings of the 'IDEA District' development behind construction site (right). Source: Pictures Albert Roßmeier 2022)

One of the discussed obstacles for the development success in East Village is the size of the community. Due to its large size, as argued by various interviewees of the different perspective groups, the community is lacking definition and form and—as a consequence—also identity. As outlined by interviewee AR-11 using the example of the rather undefined City Heights community, ambitions to name and bound certain areas or neighborhoods in San Diego have historically been tied to the prestige and the value of the land and its real estate. Due to the early onset of the phase of decay and industrialization of the large eastern portion of downtown, which was one with Logan Heights before the freeway constructions, the area has historically become rather amorphous (Norris 1983). But with the construction of Petco Park and the efforts to convince voters to approve the costly redevelopment, the area and the ballpark itself have been paired with an extensive branding concept that has promoted the area as 'more than a ballpark.' Concurrently, civic leaders introduced the name East Village "in order to attract or to get rid of the stigma behind the inner-city neighborhood and to give it a new branding, pretty much. And that branding also comes, I believe, the signs, the neighborhood signs, Barrio Logan is one of them, City Heights, you know, all throughout San Diego there are these neighborhood signs and I think that was part of the movement to rebrand, right, to attract more people and make it a nice area" (AR-08).

However, as discussed by various interviewees, the continued branding efforts have not led to a persistent, successful change of the area's social perception and identity. Compared to East Village, "Little Italy and the Gaslamp have a very distinct identity. [...] Little Italy has its own sign and is more compact, you know, this makes it remarkable. But East Village has a lot of blocks and you can't really overview it, its borders, it's just not a connected district" (EP-05). Also the development specialist EP-08 deplores: "The issue is that it's 130 blocks big, so it might be promoted as different parts and districts in the future." As a result, "walkability is definitely a challenge" (EP-05) in East Village. One the one hand, this is because "it's definitely too big" (AR-06) and, on the other hand, "there's no main street, there's no, like, where would you, like, just everything is scattered all over," as the nonprofit stakeholder EP-10 describes. Reflecting further on the dimension and the spatial organization of the community, EP-10 continues: "It's huge, [...] maybe it's just too big and you can kind of divide it, so this is your ballpark district, this is your entertain or education and entertainment, you know, whatever. But I think it's not a bad idea to think about doing that, and then you can get each of these areas some identity and focus because it's just, it's tough to figure it out." Phrasing the perceived absence of an "iconic image" (EP-05) more directly, the interviewee EP-05 states that "two years ago, East Village had zero identity and no significance within the city. [...] It didn't mean anything before, and it still doesn't have an identity." The interviewee hopes that "it will evolve in a different direction" (EP-05) through the efforts of young residents to activate available space in East Village and claims that "you need to have the community on your side" (EP-05).

In contrast to this perspective, interviewee R-07 is critical of the creative vision for East Village and interprets the transformation efforts as ironic. In their view, the endeavor to build on the creative aspects of the community's early manufacturing history and creating a 'warehouse aesthetic' is furthering the demolition and erasure of the area's respective structures and social networks instead of their preservation and growth. In this respect, the interviewee is referring to the artist community that has settled in East Village after the redevelopment of Horton Plaza and the Gaslamp Quarter (cf. Kayzar 2006, pp. 148–149). Interviewee R-07 claims: "For decades, San Diego's arts scene was downtown. [...] The ballpark was the end of East Village as the artistic center, the artistic culture was completely destroyed." In conclusion, the two interviewees EP-05 and R-07 are sharing contrasting statements about art and artists in East Village. While EP-05 is emphasizing the potential of new creative, young residents and is convinced that East Village is San Diego's new arts district, interviewee R-07 points out that the redevelopment of East Village has displaced and destroyed the 'underground' community of artists, the galleries, and the lofts that have existed in the area's officially vacant structures since the 1990s.

In accordance with the interviewee's illustration of displacement in East Village, several other interviewees have discussed issues of affordability that have arisen in the course of the redevelopment processes. The developer EP-07, for instance, shares that "there are people who work downtown, but they can't afford to live downtown." An example for this is the interviewee EP-05, a place-maker who works in East Village but lives in Normal Heights due to the area's lower rents. Concerning their situation and the housing market in East Village, the interviewee states: "the cost for living is definitely a challenge" (EP-05). More drastically, the urban scholar AR-02 describes the housing situation in downtown as "a battle in concern of displacement [...] we have a housing crisis" in San Diego. In the same vein, also interviewee AR-09 shares concerns: "The cost of rental units is really, really high. So, are you excluding, like, 70 or 80 percent of the population can't afford to live downtown and that's not a good thing for a city." Thus, also the developer EP-07 acknowledges that "we have to build low-cost housing because there is this huge income disparity" and that it is necessary to "think to of ways which are not that exclusive." According to the urban sociologist AR-03, downtown is currently facing a lack of low-cost housing options due to high construction costs for affordable housing. The interviewee states that "you can't afford to build a building in downtown unless it is luxury housing" (AR-03). As a consequence, "a lot of social diversity has been driven out" (AR-03).

In addition to the struggles of defining a functional structure for the amorphous area, providing a suitable mix of housing options, as well as maintaining or creating a creative community, several interviewees have discussed high vacancy rates in East Village as an indicator for the area's 'failed revitalization.' As illustrated above by means of census data, the vacancy rate in East Village is disproportionately high compared to all the other communities and tracts analyzed in the present study (29 percent in tract 54.01 in the year 2020). In this respect, interviewee AR-09 states: "they are building a lot of ghost towers, investment portfolio pieces [...], towers with 60 percent of the apartments being empty." According to the interviewee AR-11, the area's vacancy rate is disproportionately high because "a minority of people would want, who really embrace that urban living. And the thing is, when the condo towers started going up in downtown San Diego, nobody would move into them. The investors were way ahead of the market." Similarly, the place-maker EP-05 states that "the growth is exceeding the demand [...]. Many residents are not working here or enjoying what the neighborhood offers, you know, they commute out." In a similar vein, the developer EP-07 deplores: "Most of the residents of downtown are not working downtown, most of the times you have this hallowed out feeling of downtown, you don't have people activating the space" in the ways the place-maker EP-05 is envisioning it. This is also due to the reason that the real estate in downtown is "primarily residential and even some of the residential is even like, people are Airbnb-ing it, so it's not, you know, it's hard to gage how many people are actually living in some of those high-rises. But it's primarily residential, there's some, like, creative offices, they are trying to start that" (EP-10). Bringing the struggles to 'activate' East Village and the strategy to establish creative office space together, the interviewee AR-11 shares: "There is a, I think, a real effort to build a tech economy into East Village now so that there's some reason, it's not just a big agglomeration for the region, it's a hub for something. Right now, it's essentially a bedroom community, I mean, you go there at the daytime and there's nobody there. Just like the suburbs, everybody... And where do people drive, they drive out to the office parks in the suburbs where there's just the offices" (AR-11). Similarly, also the interviewee EP-03, a real estate and land use specialist connects the residential situation in downtown to a regional imbalance of San Diego's job and housing market: "There are a lot of jobs in North County, but at the moment the most affordable area to live is in South Bay, which is a 1.5 hour commute." However, in hope for future development in East Village and the positive reception of the outcomes, the developer EP-07 points out that "we haven't run out of space yet."

The young developer EP-04 from Barrio Logan, in turn, is painting a more pessimistic picture for the future of East Village. Even though local developers, such as EP-07, "are fighting pretty hard to attract the tenants that they want" (R-07) in order to "create a certain community" (R-07) in East Village, the interviewee EP-04 is emphasizing that the predominant lifestyle in San Diego is not combinable with the circumstances of a dense and transit-oriented life in a downtown apartment without a designated parking space. The interviewee mentions that their boss is convinced that "San Diego is the city of the future." However, as EP-04 continues, "here it is all about Flip-Flops and fish tacos and not about bike lanes, it is about being able to live in the heart of the city and being able to have two cars. [...] People are too lazy to ride their bikes." Reflecting on the common lifestyle in San Diego, the interviewee R-13 from North County San Diego states that "everybody still drives, nobody is using public transportation here." For this reason, as argued by EP-04, the vacancy rates in the newly built condominium buildings of East Village—especially on Market Street—are particularly high. In a similar vein, interviewee R-10 points out that downtown is only of interest for tourists and not for the local San Diegans. The Hillcrest resident claims that even though newcomers would initially move to the city center, they are eventually relocating to other neighborhoods of the region. Similarly, also R-13 is describing an aversion of local residents to East Village and the downtown area. The interviewee themselves is rarely crossing Interstate 8, which constitutes a symbolic and physical demarcation line between the white, low-density parts of San Diego in the north and the Hispanic-influenced, higher density areas of the city in the south, as argued by R-13 as well as the border scholar AR-05. The interviewee R-13 is convinced that "people like that, when they are in North Park they are on the other side of Balboa Park from downtown [...]. There is just a buffer there. You know, people don't wanna live downtown, they wanna life near it" (R-13).

One reason for the described aversion is the "huge homeless population in East Village, it is a big issue here" (EP-03). The place-maker EP-05 deplores that "there is homeless people living everywhere in East Village and that is overshadowing what is happening here." Despite the efforts to create a new, creative atmosphere in East Village, residents such as R-03, R-07, and R-09 are commenting on the "really tragic homeless problem in downtown and East Village" (R-07). According to R-11 and R-13, particularly the area around the new Central Library as well as the facility itself are a display of the "increasing homeless population" (R-03) in San Diego. The interviewee R-13 reports: "The library is amazing, they put in a massive library there, beautiful building, it's like eight stories and nobody would, it's basically a homeless shelter. It's become a really high concentration of homeless [people], I would never go there and I would never take my kids there. I don't know anybody that would, it's really amazing. I don't mind, I think it's great if they have a place to go, I don't mind it. But personally, you know, it's been quite a concentration." According to AR-11, this concentration exists due to the historic role of East Village as a "sacrifice zone for homeless services to preserve the whole rest of the city." In line with R-13, also the place-maker and entrepreneur EP-08 is criticizing the concentration of homeless residents in East Village and views it as a restriction for development outcomes. Therefore, as the interviewee argues, the service providers "should be spread out more throughout downtown" (EP-08).

In addition to the discussed negative impacts of homelessness on the image of East Village, the community is commonly put into competition with the redeveloped neighborhood of Little Italy, which is widely perceived as vibrant and economically successful. In comparison to the Italian-themed community on the

Figure 5.31 Homeless Encampments in the Border Area of East Village and Barrio Logan. (Top: Large homeless encampment along an empty lot at the intersection of National Avenue and 16th Street. Bottom: Homeless residents, litter, and scattered tents along an old one-story building at the intersection of Imperial Avenue and 16th Street. Source: Picture Albert Roßmeier 2022 (top) and 2021 (bottom))

northwestern downtown end, "East Village really hasn't been a big success, nothing like Little Italy. Right when East Village was coming out then everybody, with the new baseball stadium, 'this is gonna be the thing'—Little Italy stole all the thunder. I mean, East Village is a far second choice behind Little Italy for people who wanna live in fantastic urban, I mean, it's a whole category below, it's just nothing like it" (AR-11). Reflecting further on the image and the branding success for the eastern downtown area, the interviewee AR-11 states that the East Village concept has not been adopted by San Diegans in the way civic leaders and other stakeholders had intended to. As a consequence, the interviewee reports that they "don't hear people even say the phrase East Village anymore. It's just really a kind of, for a few years when it was the new thing coming it was on everybody's tongue: 'East Village, it's the thing, it's a thing.' And it's almost, like, it's already this negative brand identity. And so people are talking about the ballpark area, you know, they are kind of rephrasing because East Village is like: 'Oh, that place that kinda didn't work out.'"

In accordance with this statement, also interviewee R-13, who is involved in a "grassroots homeless outreach" (R-13), appears to be critical of the success of the East Village redevelopment efforts. Their doubts are particularly tied to the presence of the homeless population and perceived high crime rates in East Village (similarly perceived also by the Golden Hill resident EP-06). The interviewee shares:

> "In my opinion, East Village is still a question mark about if it's an experiment that will work. I don't know if it will take hold or not because the crime and drugs and homelessness has to go somewhere, and at the moment it's concentrated in East Village. [...] I think that there's less, you know, with this improvement of areas around downtown, it's leaving them [the homeless population] fewer place to go, I think, because it seems like it's getting more and more concentrated in the East Village area" (R-13).

Thus, the interviewee is connecting the current redevelopment and 'beautification' dynamics of the inner-ring neighborhoods to the concentration of homeless residents in East Village, which is, according to the interviewee, particularly prevalent in the community's subdistrict 'Southeast' (see Figure 5.29) due to the presence of the large service provider 'Father Joe's Village.' Interestingly, when describing the area that the interviewee regards as the primary location of the local homeless population, R-13 delineates almost exactly the boundaries of the subdistrict Southeast: "There's a box for, if you were to take or go from, you know, Market at the top, Imperial at the bottom, the 5 freeway at the right, and like 11th on the left, that's, like, the homeless capital of San Diego. [...] That's its own pocket [...] it's a tent city and it's only gotten more and more crowded" (R-13; see also Figure 5.31). Thus, interviewee R-13 is describing social boundaries of everyday life *within* the larger East Village area, which ultimately resembles the idea of both the development stakeholder EP-10 and the current downtown community plan to divide East Village into smaller subdistricts that follow different themes. Furthermore, the statement of R-13 also bespeaks individually perceived *external boundaries*. According to the interviewee, the described containment of San Diego's homeless residents in the southeastern East Village 'pocket' is exacerbated by the onset of redevelopment and reinvestment dynamics in the inner ring, which impose new boundaries of inaccessibility between East Village and the neighborhoods around Balboa Park (particularly discussed for the Golden Hill area, see section 5.3.2). In order to elaborate more on b/ordering and othering processes and practices in and around East Village, in the following section the thematic focus will be shifted on the communities of Barrio Logan and Southeastern San Diego, which adjoin the southeastern downtown edge.

Summing up the presented discussion of the redevelopment of the City Centre East area into the East Village district, it can be stated that the imagined renaissance of this formerly industrialized community has not been as straightforward as anticipated by civic leaders, planners, and other private stakeholders. Correspondingly, also the social perception and interpretation of the costly megaproject Petco Park and its socio-spatial consequences appear to be highly ambiguous. While certain interviewees are strong believers of the initial entertainment and the subsequent creative theme for East Village, others are of the opinion that the development efforts have failed and exacerbated existing social issues in central San Diego, such as homelessness, the shortage of housing and job opportunities, as well as gentrification. Thus, the redevelopment efforts are simultaneously interpreted as successful but also as conflictual, somehow revitalizing but at the same time destructive. Ultimately, this very divergence creates East Village as an ambiguous and hybrid urban landscape that evades clear and simple interpretations in a pastiche-like manner.

5.4.4 How Far Reaches Downtown? The Multi-Dimensional Ambivalence of San Diego's Inner City Boundaries and Their Renegotiation at the Southeastern End

In the following section, it is of interest how the commencing spillover of redevelopment dynamics from the East Village area into the communities of Barrio Logan and Southeastern San Diego is individually perceived by a variety of voices. In light of the current developments in the formerly underserved neighborhoods, it will be worked out how social and spatial identities are constructed and assigned to each of the communities and how these interpretations change over time. For this purpose, the focus is shifted on multi-dimensional bordering processes and individual practices, experiences, and interpretations, which help to illuminate the ambiguous urban landscape at the southeastern downtown edge and its manifold perceptions further. Thus, it becomes possible to take immaterial-symbolic boundaries in account, which seem to be of particular everyday life relevance in this transforming and dissolving border area where white, college-educated inner-city dwellers reside in the dense west and Hispanic low-income families and artists live in the east and southeast in the inner-ring communities of Barrio Logan and Southeastern San Diego.

In terms of Barrio Logan and its current development, several interviewees are emphasizing that the community is currently in a visible state of change. Barrio Logan is perceived by a variety of interviewees to evolve from a neglected neighborhood suffering from high crime rates, gang activity, and obnoxious land uses to a safe(r) area for cultural, leisure, and dining experiences. Recurring on childhood memories, the business owner EP-01 from East Village states that Barrio Logan has been "scary and dangerous when I was a kid." According to EP-06 from Bankers Hill, the area "was definitely not safe." In terms of the industrialization of the neighborhood, the planner from Tijuana describes: "Yeah, I mean, I've seen Barrio Logan grow. 12 or 14 years ago [...] the area was highly contaminated. Water contamination, there was air contamination in a lot of that area because there was a lot of big semi-trucks coming in and out all the time and then you had the Mexican American, the working-class community, living there. So it was a really bad place to be, [...] there was a lot of gangs et cetera et cetera" (EP-TJ). For historically, not only East Village but also "Barrio Logan has been the dumping ground for unwanted land uses" (AR-02) in other areas of the region due to high emission rates. Accordingly, 52 percent of the community's land area are occupied by the Port of San Diego and the United States Naval Station (CofSD 2013a, p. v). In addition to this, ancillary service facilities of the port and the naval base are scattered throughout the community, all of which have historically contributed to the industrialization of the former Centre City East area and thus the formation of an 'industrial belt,' which stretches from East Village along the ocean front of Barrio Logan up to National City in the south.

The described heavy industrialization is regarded as particularly problematic because, one the one hand, the air contamination in Barrio Logan has caused disproportionately high asthma rates, as discussed by the interviewees AR-04, AR-08, R-07, and EP-TJ. On the other hand, the conflicting mix of residential and industrial land uses and especially reports about crime and gang activity have caused the communities on the southeastern downtown edge to gain a bad reputation. As a result, "people wouldn't come here a few years ago, they were afraid" (EP-04). The interviewee EP-04 continues and shares that their brother visited Barrio Logan regularly in the 1990s when "people used to be really scared of Barrio Logan because of their perception of what was going on. The thing which changed that is the influence of breweries and restaurants. The bars and restaurants that came here helped Barrio Logan to open up for other people." Similarly arguing for the positive evolution of Barrio Logan, the creative place-maker EP-02 reports that "the area had been a little sketchy, but now there are a lot of reasons to go down there." They point out the increase of small businesses and cultural events in the community, which began to attract people from outside the neighborhood. "Barrio Logan is a great example [...] they are putting up events, art related galleries, cruising culture of lowrider cars and that is drawing a lot of interest to that area [...]. There is only a handful of restaurants, but they are busy" (EP-02). However, not only the community south of Interstate 5 but the entire southeastern area is perceived to undergo new development—the neighborhoods of "Sherman Heights, Logan Heights, and Barrio Logan are changing at the moment" (AR-02), and "there is a lot of development happening" (R-03).

From a border-theoretical perspective, the discussed interview statements are describing a commencing dissolution of the strong symbolic boundary that has historically separated the safe and orderly neighborhoods throughout San Diego from the crime-burdened, contaminated Hispanic communities at the southeastern downtown edge. Historically, this symbolic-immaterial boundary has been erected in the course of the suburbanization and the simultaneous decay of the inner ring. On the one hand, this has taken place individually and socially on the basis of prejudices, subjective experiences, or urban landscape socialization—the process of internalizing common behaviors, interpretations, and ideas about a given area (Berger and Luckmann 1966). On the other hand, Barrio Logan has historically been demarcated and marked as 'lower quality neighborhood' in redlining maps, which lead to discriminative bordering and segregation practices. However, today, due to the expansion of the local infrastructure—the opening of new restaurants, cafés, and art galleries—several interviewees are reporting that "more people are trying to access it [Barrio Logan], and people become more comfortable with that kind of crossing, I guess, of that border from downtown to Barrio Logan" (EP-11). In line with Newman (2006, p. 177), this statement demonstrates that symbolic boundaries do "not necessarily exist in the form of a fence or a wall, but [...] can be as sealed as the strongest of inter-state borders separating two belligerent countries from each other. [...] The 'here–there' and 'us–them' cut-off points [...] may be as invisible as they are tangible and, equally, as perceived as they are real."

However, as discussed by several interviewees, the perforation of the symbolic boundary around Barrio Logan and Southeastern San Diego is only slowly taking part. The historic aversion and negative perception of the communities seem to persist for some interviewees and groups, while other 'outsiders' begin to appreciate and visit the newly opened businesses, art shows, and events. According to the developer EP-04 from Barrio Logan, now, the area "is cool to explore," however, "people don't really live here, they commute into here" (EP-04). Interviewee R-01 discusses the selective frequentation of Barrio Logan as well and explains that "there are only parts of cultures that are appealing to certain persons." Describing their observation further, the interviewee states that even though certain culinary aspects of the Mexican culture are widely accepted and appreciated among non-Hispanics in San Diego, the society that is perceived to practice the respective culture—the Central Americans—are not part of this fascination and trend. Interviewee AR-05 from Mexico is phrasing this concisely: "the Mexican culture is trendy but not the Mexican people. They take the Mexican culture as a product; it is a process of dehumanizing and romanticizing Mexico." Thus, the interviewees R-01 and AR-05 are addressing a twofold conflict of highly selective interpretation and acceptance. On the one hand, specific aesthetics and culinary components, such as colorful murals, tacos, and Mexican-influenced beverages, receive affirmation, partly also due to the postmodern appreciation of "presumed authenticity and sense of community" (Joassart-Marcelli and Bosco 2020, p. 40) in Barrio Logan and Southeastern San Diego. On the other hand, however, a rejection of humans identified as Hispanic is taking place, based on (perceived) socio-economic differences. Socially, this can be interpreted as an alienation that is the result of ongoing ordering and othering processes. Spatially, these b/ordering processes are reflected in the form of the symbolic demarcation of Barrio Logan and Southeastern San Diego as communities that offer only limited attraction and are not suitable to live in.

Phrasing this alienation and its spatiality more drastically, interviewee R-07 claims: "Everything around downtown is pretty much gentrified except of the southeast community, you know, Barrio Logan is still quite affordable due to straight up racism." Similarly interpreting, interviewee AR-11 states that "Americans are not comfortable in Barrio Logan, Anglo-Americans, by and large. Some of them from the Southern California region have become comfortable, they've become part of the regional culture even though they are not part of Mexican-American culture." In addition to this described general division and local cultural habituation or hybridization process, interviewee AR-11 mentions Chicano Park as an example for the contrasting interpretation and appreciation of recent Hispanic or Chicano history within the local population: "Chicano Park, I mean, that's a world significant sight to me, and most of white San Diegans wouldn't sat foot there." The interviewee continues and claims that the displayed political murals on the concrete pillars are regularly misunderstood by San Diegans that are not fluent in Spanish. The prominent mural stating "¡VARRIO SI, YONKES NO!" from 1977, which roughly translates into 'yes to the neighborhood, no to junkyards,' is, according to both AR-11 and R-01, commonly interpreted as a radical slogan against (white) U.S.-Americans due to the incorrect translation of the Spanish word 'Yonkes' into the English term 'Yankees.' Besides this common misinterpretation of the displayed art, the place perceptions of Chicano Park are strongly contrasting: While the interviewee R-06 from Hillcrest perceives the park as a run-down area suffering from vandalism, littering, and the presence of homeless residents, AR-11 honors Chicano Park as a national historic landmark that represents the Hispanic fight for spatial participation in San Diego and throughout the United States. Ultimately, these two spatial interpretations are exemplary for the contingency of urban landscape perception and construction and thus of the subjective valuation of space. Urban landscape and its perception are highly subjective and connected to individual memories and experiences. This means that certain places and physical structures, such as Chicano Park, can be of value for an individuum or a group, while they are not even recognized by others.

Besides the aversion to Chicano Park, interviewee R-06 negates the (new) local restaurants and bars in the communities and shares that the neighborhoods of Logan Heights and Barrio Logan are primarily residential, "you don't go there to meet someone in a bar" (R-06). Furthermore, also the Carmel Valley resident R-13 voices criticism against the neighborhood of Barrio Logan, although the interviewee's persistent negative perception seems to be particularly tied to the industrial burden of the community:

> "I haven't really had a reason to go there. I do know more and more people I know end up going there [...]. I know it's had a couple of cafés and things pop up. I think, in my opinion, the area would struggle. It's not as desirable of an area, it's waterfront, which is cool, but there's all the Navy ship docks there, there is the bridges over to Coronado, the 5 freeway. [...] Yeah, and you got the trains going through there all the time, and they are freight trains, you know, the long freight trains. And the traffic patterns are really screwed there with the Navy base. If I were an investor, I would not invest there" (R-13).

Reflecting further on the social image of Barrio Logan and Sherman Heights in comparison to the trendy community of North Park, the interviewee AR-11 shares that the majority of non-Hispanic Americans experience intimidation when entering the Chicano-influenced neighborhoods or—border-theoretically framed—when traversing the strong symbolic-aesthetic and social boundary between North Park and Southeastern San Diego. According to the interviewee, the persistent negative perception of Hispanic culture throughout San Diego is limiting the transformation and redevelopment success of the communities of Barrio Logan and Southeastern San Diego:

> "In Barrio Logan, there's no 1,600$ micro apartments, there's nothing even close. It's just, it's still, so there's all these neighborhoods in San Diego, Barrio Logan on the top of the list, Sherman Heights being second, but since 1995 people have been telling me: 'If I could just invest, these places are ready to go off, they're just gonna explode'– 'no they don't'—'but North Park did.' The difference in my opinion is, [...] North Park is extremely white space in San Diego and so all of our young hipsters that wanna drink craft beer and eat tacos, they feel very comfortable in North Park. There's no underlying Latino identity—at all. Whereas in Barrio Logan people will tell you: 'Hey, we didn't cross the border, the border crossed us—in other words, you're standing in our country right now.' You know, that doesn't happen in North Park. North Park is thoroughly claimed by, you know, charter culture" (AR-11).

In addition to described social perception and intimidation, several interviewees discuss further aspects that are perceived to halt and limit the development of the southeastern downtown edge. Interviewee EP-09, for instance, explains that the redevelopment processes in Barrio Logan are delayed due to an array of financial aspects and zoning regimentations:

> "Developers are still developing a lot—downtown, East Village, we have seen a boom. You can get more height, right? You can also get higher rents. Right. So you get more return on your investment, whereas if you do that in Barrio Logan, it is not going to happen. So we are not seeing a lot of development still in those areas. [...] Barrio Logan is slower to catch on because of this. One, they have smaller lots, to a large degree, but also because of perception, of people not wanting to move there. But all of that is starting to change now, too" (EP-09).

Thus, even though the social perception of Barrio Logan and its Mexican-influenced bars and restaurants is slowly beginning to change, as argued by EP-09, Barrio Logan's "zoning and government structure is different" (EP-08) from the regulations in downtown. For, as discussed by interviewee R-01 and AR-06, the zoning in East Village has been updated in the course of the Petco Park development, which then allowed for the construction of high-rise towers and the replacement of the former urban fabric of one- and two-story warehouses. In addition to the current low-density zoning regulations, which limit development options in Barrio Logan and Southeastern San Diego, the developer EP-07 states that the community of "Barrio Logan has zero interest in development, so there is nothing going to happen there. [...] Nobody is going there because it just takes too long and it is too hard, the residents are using their power to prevent certain developments" or to significantly delay projects, as the interviewee continues further. Concerning the self-perception and the common identity in Barrio Logan, the border scholar AR-05 describes: "Barrio Logan is not another regular neighborhood, Barrio Logan is a highly, highly organized community, also very politicized." Additionally, also interviewee R-01 emphasizes that "the neighborhood is protesting their culture and is very well organized." The described solidarity and cooperation in the community has been particularly demonstrated in 2016 in the course of the efforts to avert the publicly as well as economically driven plans to implement a national football league stadium in the neighborhood—"Barrio Logan was really fighting the football stadium, which they successfully did" (R-07; similarly argued by the artist R-05 from Barrio Logan). Underlining the social cohesion and the importance of community in Barrio Logan further, interviewee AR-05 states: "they [the residents of Barrio Logan] fight for that, they keep their community, but with the gentrification they are losing their property and their lands in Barrio Logan. [...] It's a fight, a community struggle" (AR-05; see also Figure 5.32).

Another aspect that is perceived by various interviewees to delay the redevelopment of the neighborhoods of Barrio Logan and Sherman Heights is the displacement of homeless residents from East Village and their arrival in the adjoining neighborhoods. Due to the high demand for developable land with higher density zoning in downtown and East Village, "the containment area is again shrinking" (Kayzar 2016, p. 42). For instance, interviewee EP-09 argues that "a year or two ago, the city started basically to force them out and now we definitely see more homeless in those surrounding, that first ring neighborhoods—Barrio Logan, North Park, South Park, Balboa Park" (EP-09). Equivalently, also the developer EP-04 from Barrio Logan states that "the homeless people are going to Barrio Logan and Logan Heights now." The investor EP-06 from Bankers Hill explains similarly: "When they cleaned up Petco Park and built the new library everything went to Barrio Logan." In line with interviewee EP-09, also the former East Village resident R-01 is reporting about the displacement of homeless residents from East Village and is convinced that the redevelopment trend is pushing them into the neighborhoods east of downtown as well as into "the wild landscape" (R-01) of Balboa Park, its canyons, and less visible or patrolable areas. However, the relocating homeless people are not only entering canyons and parks but are also gaining access to state-owned freeways (as seen in Figure 5.21 and Figure 5.33) in order to escape and hide from police patrols. This can be understood as an appropriation of the physical border lines between the neighborhoods,

Figure 5.32 The Conflicting Mix of Land Uses and Evidence of Social Opposition and Activism in Barrio Logan. (Top: Junk yard, single-family home, and the industrial port in close proximity. Bottom left: Caps for sale at a pop-up market on Logan Avenue in Barrio Logan with various slogans oposing police and ICE (U.S. Immigration and Customs Enforcement) as well as addressing the complex hyridity of the local Chicano community ("Ni de Aquí Ni de Allá"). Bottom right: Anti-investment activism in Barrio Logan with reference to the Atzlan mythology, which is based on the negation of the present (location of the) U.S.-American-Mexican border and the legitimization of the ancient boundary of the Aztec rome that has stretched over Mexico and several southwestern states of today's United States. Source: Pictures Albert Roßmeier 2022)

which serve as last refuges for groups that are not tolerated anymore on the sidewalks of the redeveloping areas of the city. Thus, the border line, the freeway with its unmaintained weeds and bushes, becomes even more of a socio-spatially relevant space in the city through the addition of this new meaning—it stretches and it transforms into an area that is utilized in an alternative way beyond its traditional use as a traffic infrastructure.

In terms of the discussed displacement, interviewee R-12 is referring to effects on the neighborhood of Sherman Heights, which has started to develop and gentrify recently. However, the arrival of homeless residents from East Village (see Figure 5.33) has halted the redevelopment in this area and as a consequence also causes the local real estate prices to decline (R-12). This claimed trend concerning the local real estate, however, is not supported by the presented census data above (see Table 5.4 and Table 5.5). Nevertheless, referring to the development dynamics, also interviewee AR-11 states that Sherman Heights "kind of got stuck. And even through an extent Barrio Logan, although Barrio Logan now got a downtown advantage where it's close to the downtown core, which has become stylish and attractive again. So I think it's experiencing a significant change." As the interviewee continues, they describe how the homeless residents in East Village "have been cast out, so they are part of that wave, you know, the wave we're talking about, from the center. The homeless people are in the wave, they are getting carried out by it" (AR-11). Thus, the interviewee is illustrating metaphorically how the redevelopment dynamics—and as a result also the homeless residents—of the central city are spreading and moving from the downtown area into the neighborhoods of the inner ring. From a border-theoretical perspective, the described 'development wave' can be interpreted as a gradual rebordering process and as an outward shift of the immaterial boundaries of affordable and accessible space from the inner city into the inner ring and from there—as the wave rolls—even further outward. In other words, the socio-economic boundaries of the downtown area are pushing across its previous location at the physical border of Interstate 5 (in the sense of a thinning border; Haselsberger 2014) and expand the area of an increasingly exclusive real estate market into the inner ring. As a consequence, economically and socially less affluent residents are being pushed into the (yet) less invested communities of Barrio Logan and Southeastern San Diego and into the immobile physical border spaces, the wooded inner-city freeway edges (see Figure 5.33).

Despite the voiced impressions about the delay and the limiting factors of the development and gentrification processes, their onset in Barrio Logan and Sherman Heights is largely undenied among the interviewed individuals. As articulated by various interviewees (AR-02, AR-08, R-07), the commence of the neighborhood development is connected to the relocation of San Diego's arts scene to Barrio Logan, which "got pushed more east as the investment in the East Village increased" (R-10). According to interviewee AR-02, the arrival of artists in Barrio Logan and Sherman Heights "has caused these neighborhoods to

Figure 5.33 Utilizing the Border—Overgrown and Littered Slopes along State Highway 94 between Sherman Heights and Golden Hill as Refuge for San Diego's Homeless Residents. (Source: Picture Albert Roßmeier 2021)

change." They perceive the influx of displaced residents from East Village and the investment into the eastern downtown edge as catalysts for the current development and the gentrification processes in the adjoining neighborhoods. AR-02 states further that "there have been some major political battles [in Barrio Logan], the people who have been there, they wanted to stay." Similarly, also Interviewee AR-08 is connecting the displacement of artists to the discussed gentrification of Barrio Logan and is doing so with a strong reference to urban bordering processes: "the artist were a tool used to expand that boundary, [...] they expanded the boundary into Barrio Logan, even though Barrio Logan was already an artistic hub, right? [...] And again, I'm supportive of the arts, but the way they were used, I would say that they were used to expand that boundary and allow access for developers to come in and make the area marketable for potential buyers" (AR-08). Thus, interviewee AR-08 is describing how the perception of Barrio Logan changed for investors and developers from a historically uninteresting and unprofitable neighborhood to an up-and-coming area of financial opportunity due to its 'artistic renaissance.' Accordingly, also the Barrio Logan community plan is highlighting the "evolving arts district along Logan Avenue" (CofSD 2013a, p. LU13). As argued by the planner EP-TJ from Tijuana, this reinterpretation process has also been enhanced by the looming scarcity and the rising prices of developable land in East Village:

> "Downtown started to grow really quick, brought in the stadium, and did a lot of big projects. Now, that raised the price of land huge and it also, there was a limited amount of land through that, so what was the place to continue that sort of growth? Well, it was Barrio Logan because it's right next to downtown, so a lot of people were looking now at Barrio Logan. And so new apartments were starting to kind of go there and being built there" (EP-TJ).

Summing up, the interviewees AR-08 and EP-TJ describe a new consideration and interpretation of Barrio Logan that stands in contrast to the statements of the interviewees EP-07, EP-08, and EP-09, who are arguing for the advantage of East Village over Barrio Logan as the primary area of developer activity. Within a critical focus on socio-economic processes, the urban scholar AR-08 is highlighting displacement and gentrification in Barrio Logan, while the planner EP-TJ is explaining the commencing development processes through an emphasis on the increased value and limited availability of developable land in East Village. Equivalent to EP-TJ also interviewee EP-03 states that Barrio Logan has become "an area for a lot of new development. It is one of the most creative neighborhoods in San Diego, I think. They do art shows and there is a lot of public art, but the neighborhood is about to turn in terms of gentrification."

According to the developer EP-04 from Barrio Logan, the onset of a gentrification process and the "increase of property values" (EP-04) in Barrio Logan are not only connected to the described artistic representation of the area but also to the unique neighborhood identity, which the interviewee views as a strong advantage in comparison to East Village. Interviewee EP-04 is of the opinion that "East Village doesn't have a sense of community, but Barrio Logan does have that, people care about the community, it is still a neighborhood feel versus being just another urban street, you just see giant cement boxes, it loses its warmth. [...] East Village is cold" (EP-04). On top of this, Barrio Logan is not only viewed as a community with a strong neighborhood identity and political solidarity but also as an area that is characterized by authenticity and provides unadulterated, 'real' experiences. According to interviewee AR-05, this new valuation is "part of a fashion and food trend" (AR-05) in which "aesthetics convey authenticity while allowing for creativity" (Joassart-Marcelli 2021, p. 178). Interviewee EP-TJ compares the reinterpretation of Barrio Logan to recent trends regarding the changing perception of Tijuana and its local food scene:

> "Very similar to Tijuana where the brewing, you know, the beer culture started to happen, the galleries. So now, instead of Barrio Logan being that place, that dangerous place, as they would say, it started to be that kind of cultural backyard of downtown San Diego. But, you know, it was only a matter to get people, I think, used to it and say 'look, this place, it's nice, it's fine, it's safe. Look, the Mexicans are ok and look there's Chicano Park and there's murals and there's all this,' right? [...] So it's now starting, I think, to be gentrified" (EP-TJ).

In line with this interpretation of the changing social attitude from aversion to appreciation or curiosity, also interviewee AR-11 shares that now "people are, like, 'hey, tequila and tattoos, those are really cool, so let's go to Barrio Logan and drink tequila and have tattoos and tacos.' These three things, all things with a 't,' that have become very acceptable, very transferable commodities" (AR-11). However, as perceived by interviewee AR-08, the onset of the trend to explore and visit the neighborhood for recreational use is furthering the commodification of the community in the sense of a 'cultural sellout.' The interviewee is connecting this negatively associated development to the changing ways in which the neighborhood is socially discussed: New residents are framing the neighborhood, "like, 'Oh, look at this neighborhood, it's really nice, it's really cultural, come and check it out' and little by little the cultural aspect, I believe, is marketed and then sold. And therefore, the people who practice that culture have been sold out. And so now they have to go wherever they could afford" (AR-08). Interviewee AR-04 is critical of this development as well and is referring to it as the "re-branding of Barrio Logan." The interviewee shares further: "As gentrification is coming in, there is more taxes coming in and the area gets more cleaned up" (AR-04). As a result, "everything is getting more expensive down there" (EP-01; equivalently argued by AR-10).

Connecting the onset of gentrification and 'beautification' with the opening of breweries in Barrio Logan (as discussed by interviewee EP-TJ for Tijuana), the young graduate student AR-07 describes: "It started recently, like, to gentrify, I'd say within the last 10 years because there definitely weren't breweries before. And I think it's being, they have, like, a lot of art shows there as well. And they've done some new affordable housing construction there, and I think that definitely ties into the whole urban redevelopment, more aesthetic beautification and along with that you see a lot of prices go up and residents being displaced" (AR-07). Similarly, also the planner EP-11 argues that the opened breweries in Barrio Logan and their "Mexican-influenced beers" (EP-02) are "kind of participating in that process of changing the neighborhood perception" (EP-11). This statement corresponds with a specific observation from the neighborhood walk through Barrio Logan (PO-02) in which a large bus brought the employees of a company on an organized brewery tour for non-residents to different culinary destinations in Barrio Logan. In line with the discussed change of perceptions, this tour reflects the recreational rebranding and commodification of the industrialized residential neighborhood and its ethnic aesthetics.

However, not only critical perspectives on the neighborhood change processes have been found. Several interviewees paint a rather optimistic picture of the redevelopment and highlight arising economic opportunities. While interviewee AR-04, for instance, deplores processes of "green gentrification" in Barrio Logan, AR-03 frames the recent development in East Village and Barrio Logan positively as "upgrades." In line with this interpretation, interviewee AR-09 emphasizes that even though "Barrio Logan has started to gentrify, [...] the neighborhood definitely became safer, there is more going on there, they have this thing on the weekends called 'Barrio Art Crawl,' they have a, I think, twice a month, they have it there, very nice, they have all those art galleries that opened up, all the restaurants, and it's a very nice fest there, a street fair kinda atmosphere" (AR-09). In addition to the discussed infrastructural improvements and the increase of safety in the neighborhood, interviewee AR-09 is of the opinion that the recent neighborhood development and rebranding is especially carried out by the local population. This perspective is particularly contrasting with the previously discussed interpretations of 'forceful' and unwanted development in Barrio Logan led by non-local, white developers and investors. The interviewee states: "Actually a lot of the redevelopment in Barrio Logan and that area is grassroots, it's not so much the city, it's individual entrepreneurs and local people just starting their, you know, 'the city is not doing anything, so we're going to do it ourselves,' which is good, I think. And then finally the city has stepped up and helped out" (AR-09). In the same vein, also the bi-national interviewee AR-01 emphasizes that "much of the gentrification is done by third or fourth generation Latinos purchasing property and pushing the people out there, [...] it's not all done by white people as it's always said." Thus, the interviewee EP-05 sums up that in Barrio Logan local "landowners invested in the ghetto." As a result of this investment, the community now experiences "more walkability and safety," as argued by AR-01. The interviewee AR-04 is referring to this process as *gentefication*[10] and is thus as well of the opinion that the local development is—at least partly—driven

[10] *Gentefication* is a play on the term gentrification ('*gente*' is Spanish for 'people') and refers to development led by the Hispanic population of a given area that ultimately contributes to the displacement of long-term residents (cf. E. E. Delgado and Swanson 2019). Typically, within the gentefication perspective, the drivers of the development are perceived to be younger, 'hip,' and economically successful Hispanic community members that have left the area for a certain amount of time to pursue educational or economic goals and then

by the (local) Hispanic population. In addition to these statements, interviewee AR-11 discusses the possibility for the residents of Barrio Logan to benefit from the current redevelopment dynamics and shares their thoughts in detail:

> "It's interesting because I'm a little bit hopeful [...] if there was, what I was hoping, which was a similar story than Little Italy where the kind of long-term rooted residents of the neighborhood were in a position to benefit. And I think that's probably pretty true in Barrio Logan, not because, so in Little Italy they could afford to move out, but the church kept them invested in the neighborhood. In Barrio Logan, I think, it's kind of an even more positive story because it's the exclusion from other neighborhoods that kept them there in devalued property [and] suddenly gives them, even though they don't have a lot of money, they are still sitting on a piece of property that's got a new sense of value [...]. I think there is some hope that the people who have remained in Barrio Logan will be able to both benefit and stay. Because I don't think that pressure that could have come, like say if Little Italy hadn't experienced what it has experienced, I think there'd be a lot more pressure to do the next East Village" (AR-11).

Besides the potential benefits for local landowners, the presented statement of interviewee AR-11 addresses the functional order of and the relationships between the different neighborhoods that have been established in the course of the overarching redevelopment and rededication trend in downtown and its eastern end. Even though the interviewee claims that the development pressure on Barrio Logan is relatively low due to the economic success of Little Italy, the statement bespeaks how the current redevelopment perspectives for downtown are increasingly including the community of Barrio Logan. This expansion of the redevelopment focus is also addressed by official planning instances and has been formalized in the current downtown community plan, which states that "downtown's relationship to its surroundings is attracting increased attention. Promoting these trends toward re-integration will be essential to making downtown a connected place, and is an objective of this plan" (CCDC 2015, p. 5.23). Thus, the planning document has set the goal to strengthen the historically and physically severed connections between downtown and its adjoining residential communities in the east. As observed and experienced by various interviewees in everyday life, this objective is already taking place at downtown's southeastern end. Accordingly, interviewee AR-11 phrases that the development "wave moves outward" from downtown, and interviewee AR-09 states that "it's definitely, it's moving south." In line with these illustrations, several other interviewees interpret the

returned to the Barrios with development ideas that are centered around certain elements of Mexican or Chicano culture or cuisine and its redefinition.

development in Barrio Logan as the result of "spillover effects from downtown" (AR-02), which are—in the perception of interviewee AR-06—"re-stitching what was once there." Interviewee AR-02 describes further that the "spillover effects from the revitalization of downtown have caused neighborhoods to become more attractive to people." Thus, the interviewee is connecting the expansion of the development area into Barrio Logan to the changing perception and image of the Chicano community in the sense of a symbolic debordering process.

In a similar vein, several interviewees link the redevelopment 'spillover' to the beginning gentrification processes in the communities of Barrio Logan and Southeastern San Diego. Not only the arriving, displaced artists from East Village, or the new restaurants and bars in Barrio Logan, but particularly the new development projects in the border area of East Village and Barrio Logan are perceived to contribute to the expansion of gentrification pressure from downtown into the traditional Chicano communities. Interviewee AR-08, for instance, describes: "So, I see it as, like, the downtown is the center of development and then the redevelopment is spreading or radiating away from downtown, that kind of beautification process. And along with that people can't afford and they have to move." Furthermore, also the planner EP-09 argues that "as development spills over [...] the rents are going up and it's starting to exclude people" (EP-09). Regarding the expansion of the development efforts into Southeastern San Diego, the urban scholar AR-02 states: "people and institutions have been moved over the freeway." In this overall process, "mostly Latino, low-income families are being pushed out and more south" (AR-04). Arguing from a more resident-specific everyday life perspective, interviewee R-03 shares that "the whole area in [Barrio] Logan around Chicano Park has changed a lot through the last five years." The interviewee is linking their observation with the direction of the development expansion and states that "you either get pushed more east or south if you live in an area that is developing and you can't afford the rent anymore" (R-03). Furthermore, the interviewee AR-08 claims that the 'displacement wave' that has forced the low-income population from East Village to Barrio Logan in the last two decades is now moving further outward and the displaced "go even further south to Shell Town and even National City" (AR-08). Similarly, the interviewee EP-04 describes: "the people who used to be able to afford the inner ring now have to move more outward."

According to the developer EP-04, the expansion and the push of the development dynamics toward the south are taking place because downtown "has hit its borders, it is coming to its max, the 5 freeway is the border, in many ways the 8 freeway and the water." The interviewee describes how the physical borders of downtown San Diego—the Bay and Interstate 5—are demarcating and thus channeling the redevelopment processes. As a consequence of the physical limitation, the target area of the redevelopment efforts has been shifted and expanded—first toward the "eastern 'frontier' of downtown, East Village" (CCDC 2015, p. 6.22) and now gradually further south and southeast into Barrio Logan and Sherman Heights. Accordingly, the urban scholar AR-09 states: "the city is growing down toward the south, you know, into Barrio Logan." Similarly, interviewee R-03 is interpreting Logan Avenue in Barrio Logan (see Figure 5.26) as the "expansion of downtown." In line with this interpretation, the interviewee EP-04 explains that "more and more spots [are] being called downtown." Ultimately, the resident R-03 from Hillcrest concludes that "the border [of downtown] is getting pushed even more south" (equivalently argued by AR-02). These interpretations correspond with the observation of interviewee EP-11 on the gradual everyday life renaming and rebranding of the western City Heights and Normal Heights area into the fictional neighborhood section of 'East North Park' (discussed in section 5.2.2), which is reflected at the southeastern downtown edge by means of "that pattern that, kind of, that East Village pushes into Barrio Logan" (EP-11). Thus, the various interviewees describe how the two communities of East Village and Barrio Logan are beginning to overlap in the course of the expansion of the development area from downtown into the inner ring. Arguing from an urban landscape and border-theoretical perspective, this overlap can be interpreted as an *ambiguous transition zone of temporal urban/rural hybridity* in which aspects that are perceived to characterize the dense and urbanized East Village area are gradually imprinted into the adjoining section of the lower density neighborhood of Barrio Logan.

In addition to the diverse interpretations on the development-led shift of the boundary of downtown, the interviewee AR-01 argues that the individually drawn boundary between East Village and Barrio Logan and its location are reflected visually by means of individual 'bordering behavior.' The interviewee stresses that "there is a visible line in perceived safety. [...] The border between Barrio Logan and East Village definitely exists, where you see the border is where the people consider to park, I have friends who would never park in Barrio Logan and walk to Petco Park" (AR-01) to watch a baseball game. Equivalently, also EP-09 from North Park discusses this phenomenon: "the barrier, the freeway, people don't want to walk under that freeway, it's scary. I always park in Barrio Logan when I go to the baseball game because there's always parking. There's always parking and it's only a ten minutes' walk. It's crazy. And no one else does it for all those reasons" (EP-09). Thus, the two interviewees are describing bordering experiences that are—from their perspective—typically avoided by certain parts of San Diego's society. From a phenomenological perspective, the avoidance of this border crossing is based on the negative perception of the relational engagement of the individual with the physical object, the freeway tunnel that is seen as the unpleasant physical border to the 'scary' urban landscape of Barrio Logan.

In summary, from a border-theoretical perspective, the various statements on gentrification and the displacement of former residents are addressing the shift of specific immaterial boundaries, such as socio-economic and socio-spatial demarcations, between East Village and Barrio Logan. Additionally, also in the symbolic-aesthetic dimension, the boundary line between the redeveloped, 'new' areas in the west and the less renovated, 'old' neighborhoods in the east and southeast is currently perceived to shift. In this course, several interviewees are experiencing an *expansion of the downtown area* into the industrialized, less dense residential neighborhood of Barrio Logan by means of new construction projects, newly opened businesses, the increase of housing prices, as well as through the influx of new residents into the Barrio. These experiences and interpretations, however, appear to be highly subjective and are thus—in parts—even contradicting each other. This is due to the reason that the individual boundaries between East Village and Barrio Logan are drawn on the basis of varying experiences, perspectives, and emotions. As a result, it can be concluded that boundary perceptions and their everyday meaning vary while never being final due to the ongoing changes in and between (the analyzed) neighborhoods. For, as argued by the interviewee EP-TJ, the southeastern downtown edge "is a place that I think is not done yet, it's still, we have to wait and see what it turns out to be." Ultimately, by addressing the uncertainty of future development, this conclusion is highlighting the processuality of neighborhoods, urban landscape, and borders as well as of perception itself.

5.4.5 Urban Landscape and Border-Theoretical Conclusion—Simultaneous Tendencies of Re- and Disconnection and the Multi-Dimensional Ambiguity of the Inner-City Boundaries

In the following, the presented quantitative and qualitative illustration of the redevelopment and neighborhood change processes in and between downtown San Diego, East Village, and Barrio Logan will be summarized. This will be done on the basis of the underlying theory of urban bordering and urban landscape processes in order to draw a theory-supported conclusion on the ambiguity, processuality, and transitional hybridity that is currently characterizing the southeastern border area of downtown San Diego. The written form of the following synthesis will be complemented by a theory-based graphic conclusion, which is presenting the gained insights in an alternative cartographic way (see Figure 5.34).

In the 1880s, the small border town of San Diego experienced a first significant development and population boom initiated by the land speculations of Alonzo E. Horton as well as through the establishment of the intercontinental railroad connection. However, shortly after its onset, in 1888, the boom has already been put to an end due to the commencement of a severe real estate crisis that led to the exodus of thousands of new residents. After the inflated real estate bubble burst, it took more than 20 years for the city to regain momentum and prestige (see Figure 4.2). Thus, in the early 20th century, the organization of the international Panama-California Exposition, the emergence of new architectural styles, as well as the rapid expansion of the streetcar network have spurred development in the neighborhoods north of Balboa Park through the attraction of new and affluent residents. Meanwhile, the areas east of the central business district and 5th Avenue—today's East Village, Barrio Logan, and Sherman Heights—and their waterfront have faced extensive industrialization, while the area south of Broadway—which is today's touristic Gaslamp Quarter—developed into a gambling and redlight district named the Stingaree. As a consequence, the local African American population and the numerous industry workers have been relocated from their homes close to the CBD to areas further east, into the large and undefined Logan Heights community. Thus, the once prestigious low-density structure of suburban Victorian homes throughout Logan Heights gave way for workers' housing and low-cost apartment buildings as well as for the growing waterfront industries. This trend has been exacerbated in the 1930s after zoning changes enabled the establishment of industrial structures in even closer vicinity to the already decaying decade-old building stock. Thus, the commencing ethnic segregation and the continuous reduction of the neighborhood qualities due to the increasing obnoxious land uses resulted in the beginning stigmatization of the undefined areas east of downtown.

In the course of the following decades, the spatially rather amorphous Logan Heights area developed into one of San Diego's most ethnically diverse communities through the influx of Japanese fishermen and Mexican as well as other immigrants—an aspects that is prevalent until today as indicated by the census data calculations above. In addition to the ongoing industrialization, also the construction of numerous freeways and bridges in the 1950s and '60s has furthered the reduction of residential use and the displacement of residents in the area east of downtown. On the one hand, this has led to the physical-spatial division of the large area into several distinct communities and neighborhoods, such as the Centre City East area (today's East Village), Barrio Logan in the south, and Sherman Heights and Logan Heights in the northeast. On the other hand, however, the construction of the physical borders has led to a social and symbolic division of the former Logan Heights community and thus the fragmentation of its common identity. Thus, in 197, with the intent to re-establish a sense of place and community and to manifest the frustration about the ongoing industrialization and contamination of the area, the marginalized residents of Barrio Logan have established Chicano Park under the pillars of the Coronado Bay Bridge.

In addition to the newly formed communities of Barrio Logan and Southeastern San Diego, also the downtown area has faced the consequences of the 'white flight' phenomenon. Particularly during the 1960s and '70s, the inner city has experienced physical decay and San Diego's suburban middle- and higher-income population has increasingly lost its excitement for downtown. As a counteraction, the City Council presented a plan to transform the decayed Horton Plaza and its surrounding area into a multi-level shopping center, which was intended to restore downtown San Diego's prestige. In a modern planning approach, the construction of the new mall has served the purpose of establishing an orderly entertainment environment by demolishing and removing the historic Stingaree redlight district, its outdated building stock, and stigmatized low-income and immigrant population. This has led to the relocation of former residents, low-cost businesses, and homeless service providers to the underutilized industrial quarter between the redevelopment project area and the newly built Interstate 5 in the east—the Centre City East community, which is today's East Village. Thus, over the years, as a consequence of (regional-)political decisions in the city and the county as well as the NIMBY-influence of higher-income suburban communities, the Centre City East area gradually became the primary destination for service providers and low-cost housing options, such as SRO rooms in San Diego.

However, in the last two decades, in the course of the ongoing public and private efforts to establish downtown San Diego as a top tourism and entertainment destination, development pressure has reached the eastern end of the inner city. The increasing limitation of developable land inside the physical borders of downtown—San Diego Bay in the south and east and Interstate 5 in the north and west—has shifted the development focus to the less developed Centre City East area, which ultimately led to the construction of the convention center in 1989, the new public library in 2013, and most importantly the mega project Petco Park in 2004. The 26 block redevelopment project includes the ancillary private development of hotels and businesses and aimed to rebrand the eastern portion of downtown San Diego into the East Village community, which was promised outstanding financial and infrastructural revitalization outcomes. However, as discussed by means of qualitative as well as qualitative data, the imagined renaissance and success of the area has not taken effect as anticipated by city officials and investors and this has limited new private development impulses in the community.

Demographically and in terms of housing-related development, the three considered communities—downtown, Barrio Logan, and Southeastern San Diego—have developed in highly different ways in the last decades (see Table 5.4 and Table 5.5). Between 2000 and 2010, the downtown area has experienced a strong population growth, but in comparison to the overall City of San Diego, it features a lower share of family households, lower median household incomes, and higher shares of white, exclusively English-speaking, as well as college-educated residents. Concerning housing-related development, the Centre City community planning area is characterized by a disproportionately high vacancy rate and median housing value as well as by a stark increase of the total number of housing units. The Barrio Logan area, in turn, features relatively low shares of white and of college-educated residents as well as of exclusively English-speaking households. Additionally, the median household income in the community appears to be particularly low. However, the calculations above are indicating a beginning influx of white, English-speaking, as well as of college-educated residents, a stark increase of the median housing value and the total number of housing units, which is accompanied by a decrease of family households in the community area. In conclusion, these developments can be interpreted as an assimilation to or expansion of the socio-spatial trend that is taking place in downtown and the East Village area. This impression is underlined even more when comparing census data on the tract level for the cohorts between 2000 and 2020, which shows that the stereotypical characteristics of redeveloping urban communities—a relatively high rate of college-educated, white, as well as English-speaking residents, a low share of family households, and a stark growth of new housing units—are increasingly prevalent in the official boundary area of East Village and Barrio Logan (tract 51). Thus, the areas that are in closer vicinity to the official downtown boundary appear to adapt more to the local demographic and housing-related urbanization trend.

Accordingly, in comparison to the development of Barrio Logan, the community of Southeastern San Diego, which is physically severed from downtown by Interstate 5, is following the respective trends of downtown and the East Village area to a much lesser degree. In contrast to the other two communities, Southeastern San Diego features the highest shares of family households and units, has not experienced significant growth in terms of the total number of housing units or the median household income, and is characterized by the lowest share of white, exclusively English-speaking, as well as of college-educated residents. Additionally, the community has faced a population decrease and features a large share of households that are not fluent in English. A comparison of the developments on the tract level, however, allows for spatial differentiation within the large community. Thus, it has been shown that the neighborhood of Sherman Heights—the community section that is closest to East Village—is undergoing a comparable, yet significantly less pronounced demographic development than the downtown and the East Village area. In comparison to Logan Heights at the eastern end of the community, the Sherman Heights area features a significantly higher share of white population and a lower share of underage residents. Thus, from a quantitative point, the Logan Heights area appears to be at an earlier stage of neighborhood redevelopment or urbanization compared to Sherman Heights, which is quantitatively also only beginning to adapt to the inner-city trends. An equivalent picture of the community of Southeastern San Diego and the adjoining Golden Hill area has been painted in section 5.3, which traces the arrival of the 'development wave' that is slowly penetrating from west to east, from the downtown and East Village area into the neighborhoods east of Interstate 5 and from north to south, from North Park over South Park into Golden Hill (cartographically illustrated in Figure 5.25).

In conclusion, the comparison of quantitative data allows for the interpretation of the official boundary area between East Village and Barrio Logan as a transition zone in which the mixed-use neighborhood of predominantly lower-income, Hispanic family households with a higher stock of single-family housing in the east blends with the growing and exclusive East Village community in the west, which comprises of a majority of white residents and a particularly low stock of single-family homes as well as family households. Thus, in quantitative terms, the boundary area between the two communities appears as a *zone of assimilation* in which the more or less suburban community is slowly but gradually developing toward a whiter, demographically stereotypical urban version in the form an urban/rural hybrid pastiche.

Equivalently, the evaluation of the qualitative interviews has demonstrated that the redevelopment dynamics are largely perceived to progress as a 'wave' from East Village eastward along the axis of Logan Avenue into Barrio Logan and partly also across the physical border of Interstate 5 into Sherman Heights (along the commercial corridor of Imperial Avenue). However, the gathered interpretations and experiences of the redevelopment in the various communities are highly subjective and thus partly diverging and contradicting. The success of the

East Village rebranding and the area's potential, the cultural value and the gastronomic development of Barrio Logan, and the socio-spatial outcomes of the change processes in East Village, Barrio Logan, and Southeastern San Diego are each perceived in diverse, individual ways and through subjective experiences. In addition to the divergence of the interpretations, the interviewees cannot be divided into two strict camps of 'believers' and 'skeptics' of the redevelopment success because even strong supporters of the development efforts have described obstacles and problems, while opponents have discussed desirable aspects or tendencies (the difficulty of categorizing interviewees will be addressed further in section 6.2). This is further highlighting the complexity and ambiguity of the state of change the redeveloping neighborhoods are currently in. Nevertheless, even though some of the gathered interpretations turn out to be rather ambiguous, three broad threads of interpretations have been identified, which are outlined in the following.

The statements that are bundled into the first interpretation thread are highlighting the success and the positive outcomes of the East Village redevelopment. Several interviewees have interpreted the redevelopment processes as improvements and corrections of a former state of predominant crime, poverty and decay, and general unpleasantness. From this interpretation, the redevelopment is facilitating a revitalization and better utilization of the area and its unused potential as the 'last frontier' or the 'last undeveloped area' of downtown. Additionally, the redesign of old physical structures (such as warehouses) and the presentation of the area as a top location for tech businesses as well as creative schools is seen as a logic continuation of the community's industrial and crafting history and its creative legacy. Besides the emphasis on the positive outcomes and the prosperous future of East Village, this interpretation line is closely linked to the perception of Barrio Logan as a community that is not attracting new development due to zoning regimentations, financial advantages of development in downtown, and the community efforts to oppose or delay constructions. Thus, Barrio Logan is perceived to successfully sustain its Hispanic community and cultural richness in its fight against development, while other interviewees are increasingly worried about the ongoing displacement of community members. Ultimately, in border-theoretical terms, the interpretations in this thread are centered around the perceived dissolution or expansion of symbolic and socio-economic everyday life boundaries of prosperous and desirable urban space from the inner-city and its successful neighborhoods toward the 'unused eastern frontier' that is finally about to utilize its potential.

In contrast to the thread of interpretations that are lauding the redevelopment efforts as successful revitalization, several interviewees have voiced concerns about displacement, gentrification, and participation. In terms of the change processes in downtown and East Village, various interviewees have described the destruction and displacement of a 'vibrant artist community' that has lived in affordable apartments of the former Stingaree district and—after the area's demolition—was forced to utilize vacant warehouses and other industrial structures in the adjoining, large East Village area. However, after the construction of Petco Park, as described by several interviewees, the settled artists as well as the transient and homeless residents of East Village have been pushed further east into the Barrio Logan area, which has been discussed as a historically underrepresented and emission-plagued community within the statements of this interpretation thread. In light of the described loss of the creative residents and their unofficial arts district in East Village, the efforts to redevelop and rebrand the eastern downtown end as a creative community are perceived as ironic and counterintuitive.

In terms of the recent development of Barrio Logan, this interpretation thread is strongly based on a distinction of white 'outsiders' that begin to access and utilize the neighborhood of the Hispanic 'insider' community. In line with this distinction, several interviewees are deploring a trend of selective appreciation in which the outsiders are perceived to romanticize the presumed authenticity of Mexican-influenced cuisine and art while maintaining longstanding prejudices and aversion against the Hispanic population itself. Furthermore, the attraction of white outsiders is perceived to both originate from as well as further the trendy and higher-priced gastronomic offers in the community, such as vegan tacos and craft beer, which are rather serving occasional visitors than community members. Within this interpretation thread, the discussed new consumption offers but also the growing number of art exhibitions in the area are perceived as 'cultural sellout' of the community and as a phenomenon of a superordinate, citywide gentrification process that radiates outward from the redeveloped downtown and ultimately displaces long-term community members to neighborhoods further east and southeast. From a border-theoretical perspective, these interpretations are painting the picture of a dissolving or shifting symbolic and socio-economic boundary that is progressing from the redeveloped East Village community into the Barrio Logan neighborhood through rebranding efforts, new development, and displacement. However, in line with the critical orientation of this thread, the statements of several interviewees can be interpreted as a contestation of the debordering tendency and thus as calls for a rebordering between East Village and Barrio Logan that are tied to the political aim of keeping out the 'white outsiders' and preventing the continuation of the Hispanic 'insider's' historic lack of spatial participation in San Diego.

The third and last compiled interpretation thread is bundling perceptions that are neither arguing for the success of the attempted East Village renaissance, nor for the onset of redevelopment or gentrification pressures in Barrio Logan. Instead, the interpretations in this thread are loosely based on an aversion against the entire downtown area and particularly against Barrio Logan and Southeastern San Diego. This aversion is expressed by avoidance—the areas are of no significance in the everyday life of a certain number of interviewees due to an array of aspects that are perceived as predominant such as homelessness, vandalism, crime, or littering. As a consequence of this interpretation, places of residence are chosen in 'safe distance' from the downtown area—the new residents of the redeveloped areas north of Balboa Park are perceived to seek a certain 'buffer zone.' Additionally, other statements of this interpretation thread are addressing the incompatibleness of the dense and transit-oriented urban landscape of the downtown area with the predominant 'San Diego lifestyle' that is based on individual mobility and suburban zoning. Concerning the communities of Barrio Logan and Southeastern San Diego, some interpretations in this last thread are negating the redevelopment processes in the two communities, emphasizing their lack of infrastructure and perceiving the areas as 'boring.'

Ultimately, the comparison of the three different interpretation threads is allowing for a two-fold conclusion. Firstly, the variety and contrariness of the various collected interpretations is reflecting the complexity and ambiguity of the areas in study and their current change. The East Village community is, on the one hand, experiencing high-rise construction projects, the demolition of large, old industrial structures, and is being promoted as a top location for new tech businesses as well as creative schools. On the other hand, however, East Village is home to large parts of the region's homeless population and service providers, empty lots and underutilized warehouses, and seems to be only partly 'activated' or frequented in the ways the development stakeholders have intended. In summary, the East Village area is oscillating between the states of its envisioned prosperous future and its industrial and neglected past. Furthermore, it is encompassing the various contradictory realities and needs of its different residents, visitors, and other user groups. This state of plurality and ambiguity ultimately presents the community as a hybrid and transitional urban landscape that is perpetuating the area's historic spatial indefiniteness (as discussed in section 5.4.1).

Secondly, the various collected interpretations allow for conclusions on the complex boundary situation between the downtown area and Barrio Logan. The variety and the contingency of the analyzed interpretations bespeaks the subjectivity of everyday life boundaries—between East Village and Barrio Logan but also beyond this example. Where the boundary between the downtown area and the inner-ring community lies and how it is perceived is highly subjective and seems to be more and more contentious. However, it is necessary to state that the current developments at the southeastern downtown edge are not promoting a full-scale transformation of Barrio Logan into a stereotypical downtown neighborhood. Instead, the ambiguously interpretated neighborhood changes are to be understood as an *expansion of the redeveloped area and its characteristics into Barrio Logan.* In this process, the transitional urban landscape in the border area of East Village and Barrio Logan increasingly eludes clear attributions or categories and is thus gradually appearing as a "postmodern pastiche" (Joassart-Marcelli and Bosco 2020, p. 47) due to its variety of influences, structures, and perceptions. For this area is currently characterized by aspects of simultaneity and contradiction; it functions as a contested in-between or "intermediate space of possible coexistence, of simultaneous inclusion and exclusion in which participation, affiliation, and accessibility are temporarily renegotiated and regulated and whose further development is uncertain" (Roßmeier and Weber 2021, p. 21). Thus, the border area between East Village and Barrio Logan constitutes a multi-dimensional (physical, functional, symbolic, and social) transition zone or, in other words, a *hybrid urban borderland* in which binary and clear distinctions dissolve and multi-layered hybridizations emerge (cartographically illustrated in Figure 5.34). Framing this with Anzaldúa (2012 [1987], p. 25), between East Village and Barrio Logan "two worlds [are] merging to form a third country," a hybrid urban borderland.

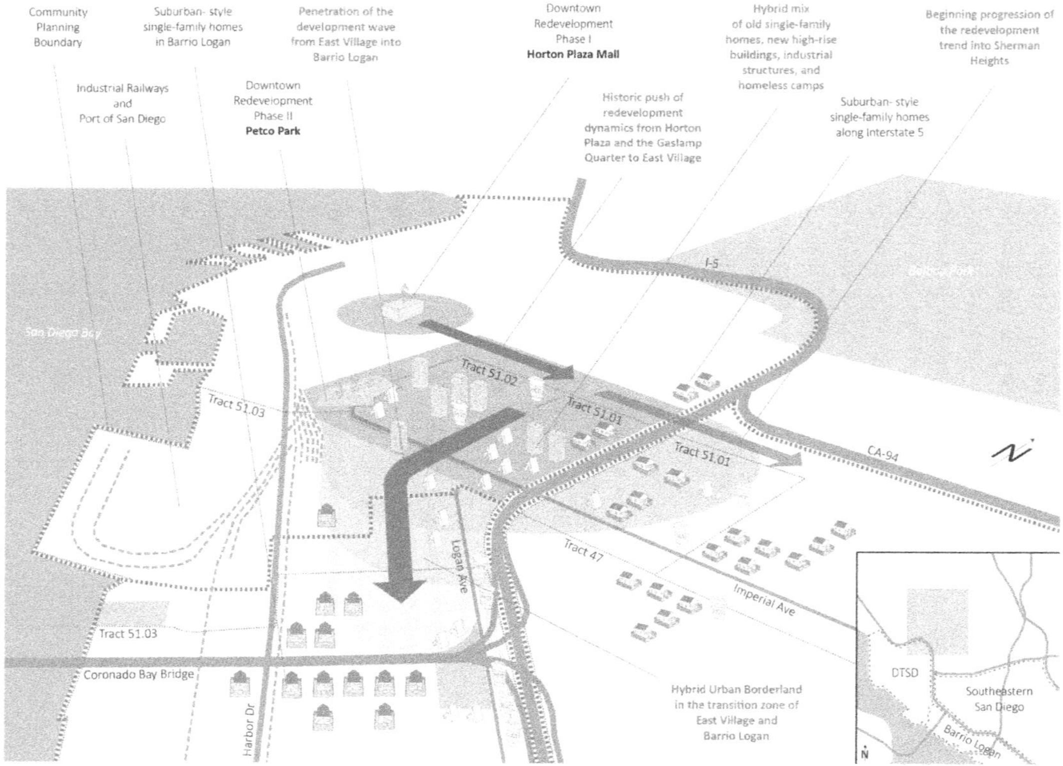

Figure 5.34 Digital 3D Sketch—the Hybrid Urban Borderland in the Transition Zone of East Village and Barrio Logan. (General information is provided in black, border-related conclusions are given in red. The red arrows are used to display the direction of the elaborated development wave. The multi-dimensional boundary stretching and dissolution, in other words, the hybrid urban borderland between East Village and Barrio Logan is displayed through the red gradient, which stretches southeast along Logan Avenue. Source: Own cartography and design)

5.5 Theory-Based Discussion and Interpretation of the Results: The Conceptualization of Hybrid Urban Borderlands and its Application in the Case Study of San Diego

In the following, before turning to the conclusion and the outlook of the present work, the main results of the study will be reflected against the background of the underlying theoretical perspective. In doing so, on the one hand, the urban landscape and border-theoretical background serves as a lens and a frame for the analyzed change processes, thus enabling a wider understanding of the recent developments in San Diego in the sense of a *theoretical interpretation*, as it will be shown below. On the other hand, the following synthesis of the theoretical foundation with selected empirical material allows for a *context-based derivation* and illustration of the conceptualization of hybrid urban borderlands (see Figure 5.35). For against the 'unattainable dream' (Paasi 2011) of developing a universally applicable border theory, the concept of hybrid urban borderlands is an approach that is developed—and tested—'in context' (cf. Wastl-Walter 2011; see section 2.4.3).

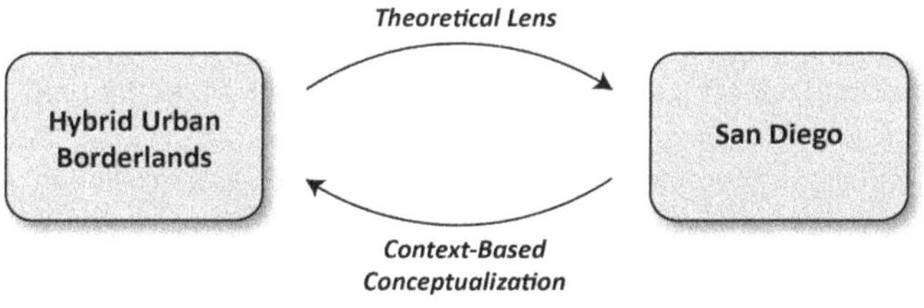

Figure 5.35 The Hybrid Urban Borderland Concept—Simultaneously an Urban Landscape and Border-Theoretical Lens for the Analysis of Neighborhood Change in San Diego and a Conceptualization Developed 'in Context.' (Source: Own cartography and design)

Following social-constructivist and post-structuralist urban landscape and border theory, the analytical focus of the present study has been directed on the reinterpretation processes and the changing experiences that are taking place in and around downtown San Diego. By means of the understanding of landscape as a 'way of seeing' (Cosgrove 1985, 1998 [1984]; Cosgrove and Daniels 1988), neighborhood change has been viewed and traced as a process that is taking place through the ongoing evolution of spatial perception and interpretation. Furthermore, from a phenomenological stance, also the material means of the various spatial perceptions have been taken into consideration—the renovated single-family homes in Golden Hill or the converted warehouse structures in East Village are each the 'external' means of the subjects' various experiences and urban landscape interpretations. In the course of the recent years and decades, the urban fabric but also the social perception of the neighborhoods of Hillcrest, North Park and City Heights, Golden Hill and South Park, as well as of East Village and Barrio Logan have changed strongly. For several interviewees, negative assumptions about and experiences of violence, vandalism, and decay in each of these neighborhoods diminished, and over time positively associated characteristics, such as vibrancy, authenticity, or creativity, have been attributed to selected areas of the inner ring. For others, positively assessed aspects of the analyzed neighborhoods, such as a strong community network or local traditional economic structures, are being destroyed in the course of the redevelopment, while a third thread of interpretations emphases the absence of significant updates in certain areas. In conclusion, these observed (re)interpretation processes demonstrate the subjectivity of urban landscape perception and interpretation or, in other words, how the 'meaningless world' (Greider and Garkovich 1994, p. 2) is constructed into a multitude of urban landscapes through the contingent attribution of symbolic meaning. Thus, it can be inferred that the 'vibrancy' of Barrio Logan, the 'hipness' of North Park, or the 'neighborhood vibe' of South Park are subjective ways of seeing and feeling—in a phenomenological-experiential way—rather than essential and innate characteristics of the various neighborhoods (cf. Burr 2005, p. 6).

The significant image change the neighborhood of Barrio Logan underwent in recent years can be understood and illustrated by means of the dialectic processes of externalization, objectivation, and internalization (Berger and Luckmann 1966, pp. 78–79; see Figure 2.2). As discussed by several interview partners, from an outsider perspective, Barrio Logan has long been perceived as a dangerous and run-down area that is suffering from gang activity, emission-heavy industrial uses, and poverty. However, in the course of the arrival of displaced artists and the redevelopment of East Village, the common perception of Barrio Logan has begun to change (as stated in various interviews). Instead of a 'no-go area,' Barrio Logan is increasingly perceived—as well as marketed—as an interesting, culturally rich neighborhood interspersed with art galleries and coffee shops and became an 'insider tip' for craft beer and 'authentic' dining experiences. In terms of the process of externalization, some interviewees have shared their impressions about the change the neighborhood underwent as well as its current 'interesting' state, while others have discussed how Barrio Logan is increasingly mentioned by their friends and in local media as an area 'worth visiting.' Thus, the subjective positive interpretation of Barrio Logan is being communicated—i.e. externalized—by several interviewees in their everyday life and is, as a result, increasingly becoming common sense. In other words, the externalized interpretation of Barrio Logan as the 'new arts district' is objectivating, which means it is integrating into the pool of common narratives and ways of seeing. As a consequence of the commencing objectivation, various interviewees have stated to recognize the changing ways in which the people around them are constructing the neighborhood of Barrio Logan and have, as a reaction to that, expressed curiosity themselves. Thus, the externalized and objectivated interpretations begin to influence subjective perceptions; the new way of seeing Barrio Logan is processually internalized into the subjective way of thinking of several interviewees—a process that can ultimately be seen as a specific form of urban landscape socialization (on landscape socialization see Kühne 2019, pp. 61–64). Ultimately, it can be concluded that the construction of Barrio Logan and its (spatial) identity is an ongoing individual and social practice that is always subject to change.

However, in contrast to the emerging interpretation of Barrio Logan as an 'opportunity area,' other constructions persist that are denying the new *way of seeing* the neighborhood. On the one hand, this is due to the infinity of possible realities, which exist next to each other and are constantly being produced (anew) in everyday life. On the other hand, however, this contrast is connected to the processuality of objectivation and socialization in general, which means that objectivating reality is not internalized at the same time by the various subjects of society. These circumstances are leading to a state of transitional simultaneity and hybridity in terms of spatial interpretations, which is particularly characterizing redeveloping neighborhoods such as Barrio Logan. Accordingly, the (perceptional) transition process and the ambiguity that is going along with it is reflected spatially in the neighborhood of San Diego's redeveloping inner ring. This is becoming particularly obvious in certain border and boundary situations between the various communities. The current ambiguous situations between the adjoining neighborhoods of North Park, Normal Heights, and City Heights or the communities of Barrio Logan and East Village are examples of the ambivalence that can characterize spaces and demarcations.

In the present study, the complexity of border and boundary situations has been met theoretically with the multi-dimensional breakdown of bordering processes. As illustrated above (see Figure 2.5), bordering in everyday life is taking place simultaneously in the physical, political, social or individual, and symbolic-aesthetic dimension. The breakdown acknowledges the existence and everyday relevance of both physical as well as immaterial demarcations and conceives borders and boundaries as practices that are—just like landscape—based on individual construction processes, constantly updated and learned, i.e. internalized, through socialization and everyday experiences. In the case context of the communities of North Park and Mid-City, the adaptability and the current ambiguity of the neighborhood boundaries is becoming particularly apparent. On the various levels of bordering, the established demarcations between the two communities are gradually dissolving, perforating, or shifting in the course of the redevelopment trend that is progressing eastward. These processes will be discussed in detailed below.

In the physical dimension, for instance, the former 'differentiation line' between the architecture and the urban fabric of the redeveloped west and the partly decayed east is gradually shifting toward the east and away from its previous location, which was Interstate 15. This can be seen as a transitional effect of the redevelopment trend that has started in Hillcrest in the 1980s and has progressed eastward into North Park from where it is now spilling over the two freeways and the official community boundaries into Normal Heights and City Heights. The commencing development in Mid-City is leading to updates of the building stock, which gradually reduces the former, strong architectural or physical-material difference between the west and the east. Additionally, new and trendy businesses are opening along the commercial corridors of Normal Heights and City Heights, and an influx of white, economically better-situated residents has set in. Thus, the Mid-City community is undergoing certain assimilations to the North Park area—physically-materially, functionally or infrastructurally, and demographically. Border-theoretically, the demographic change means a gradual dissolution of the social boundary between North Park and Mid-City. In the economic dimension of bordering, the demarcation between the communities is becoming more and more ambiguous as well. Not only in terms of the aligning housing market but also through wrongly advertised real estate listings; the areas east of Interstate 15 are frequently intentionally marketed as North Park or East North Park, which is potentially influencing the place perceptions of the local residents and leading to confusion or a kind of "dual identity" (Anzaldúa 2012 [1987], p. 63). Additionally, the newly opening businesses along University Avenue and El Cajon Boulevard, the influx of white and young urban professionals, as well as the beginning redevelopment of private homes are having a perforating effect on the symbolic and aesthetic community boundaries. For these adaptions are decreasing the means of differentiation between the redeveloped, aesthetic-oriented west and the yet still largely disinvested and functionality-oriented east.

Ultimately, North Park is perceived by several interviewees to expand eastward into neighborhoods of Normal Heights and City Heights, which are, as a result, increasingly experienced (internalized) and discussed (externalized) as 'gentrifying neighborhoods.' In conclusion, a physical, social, and symbolic debordering (cf. Liao et al. 2018) is taking place between the communities of North Park and Mid-City. This perforation contests the established spatial everyday life order and provokes a state of temporal hybridity and ambiguity in the area east of Interstate 15 that resembles the transitional, pastiche-like state of urban/rural hybrid phenomena. In the sense of Haselsberger (2014), the 'thick' and rather clear set of demarcations that have lain on top of each other along the physical border of Interstate 15—the bundle of physical, social, functional, and symbolic-aesthetic boundaries—is gradually separating. For the boundaries are dissolving or shifting away from their traditional location and begin to lie next to each other in an unordered and ambiguous way that spans an in-between zone rather than a clear demarcation line.

Lastly, building both on the empirical results and the underlying theoretical foundation, it can be concluded that the analyzed demarcations between North Park and City Heights as well as between East Village and Barrio Logan are simultaneously connecting and dividing the respective neighborhoods, they are contesting the duality of inside/outside or here/there. The demarcations have broadened from formerly clear lines into a fluid zone or a 'third room' that incorporates the two worlds it used to separate (on hybridity and the third space see Bhabha 2000, 2012; Easthope 1998; Mitchell 1997). In this sense, arguing from a relational perspective, the analyzed border areas gain an *ambiguous spatiality* (graphically illustrated in the various 3D visualizations), which underlines the theoretical and contextual connection of urban landscape and borders even more. Ultimately, it becomes possible to conceptualize these fluid in-between or transition zones in San Diego as hybrid urban borderlands (cf. Roßmeier 2020; Roßmeier and Weber 2021). In line with the remarks of Iossifova (2019, p. 4), the hybrid urban borderlands in San Diego are each a "hybrid space that contains and modifies the variety of characteristics that can be found in the core spaces which it separates or joins: old and new, traditional and modern, rural and urban."

References

Aitken, S. C. & Zonn, L. (1994). Re-presenting the place pastiche. In S. C. Aitken & L. Zonn (Eds.), *Place, power, situation, and spectacle. A geography of film* (pp. 3–25). Lanham: Rowman & Littlefield.

Amero, R. W. (1990). The Making of the Panama-California Exposition, 1909–1915. *The Journal of San Diego History, 36* (1), p. n.pag.

Anzaldúa, G. (2012 [1987]). *Borderlands/La frontera. The New Mestiza*. San Francisco: Aunt Lute Books.

Banerjee, P. & Chen, X. (2012). Living in in-between spaces: A structure-agency analysis of the India-China and India-Bangladesh borderlands. *Cities, 34*, pp. 18–29.

Berger, P. L. & Luckmann, T. (1966). *The social construction of reality*. New York: Anchor.

Bhabha, H. K. (2000). *Die Verortung der Kultur*. Tübingen: Stauffenburg.

Bhabha, H. K. (2012). Über kulturelle Hybridität: Tradition und Übersetzung. In A. Babka & G. Posselt (Eds.), *Über kulturelle Hybridität. Tradition und Übersetzung* (pp. 17–57). Wien: Turia + Kant.

Bitterman, A. & Hess, D. B. (Eds.). (2021). *The Life and Afterlife of Gay Neighborhoods. Renaissance and Resurgence*. Cham: Springer.

Burr, V. (2005). *Social constructionism*. London: Routledge.

CCDC. (2004). Downtown San Diego African-American Heritage Study. Centre City Development Corporation. https://www.sandiego.gov/sites/default/files/ccdc_african-american_heritage_study.pdf. Accessed: 22. April 2020.

CCDC. (2015). Downtown Community Plan. Centre City Development Corporation. http://civicsd.com/wp-content/uploads/2015/02/Downtown-Comunity-Plan-All-1.pdf. Accessed: 27. November 2019.

Chapple, K. & Thomas, T. (2021). Urban Displacement Project. Institute of Governmental Studies. https://www.urbandisplacement.org/. Accessed: 16 July 2021.

CofSD, P. D. (2004). Five Projects Receive Prestigious Pilot Village Designation. City of San Diego Planning Department. https://www.sandiego.gov/sites/default/files/legacy//planning/genplan/pdf/pilotvillage/040212pvsapproved.pdf. Accessed: March 5, 2021.

CofSD, P. D. (2013a). Barrio Logan Community Plan and Local Coastal Program Draft. City of San Diego Planning Department. https://www.sandiego.gov/sites/default/files/legacy/planning/community/cpu/barriologan/pdf/bl_cpu_full_w_historic_res_091913.pdf. Accessed: 5. November 2019.

CofSD, P. D. (2013b). Golden Hill Community Plan. City of San Diego Planning Department. https://www.sandiego.gov/sites/default/files/golden_hill_community_plan_2016_-_reduced_file_size.pdf. Accessed: 20. October 2021.

CofSD, P. D. (2015). Southeastern San Diego Community Plan October 2015. City of San Diego Planning Department. https://www.sandiego.gov/sites/default/files/legacy/planning/community/cpu/southeastern/pdf/sesd_community_plan_reduced.pdf. Accessed: 5. November 2019.

CofSD, P. D. (2016). North Park Community Plan. City of San Diego Planning Department. https://www.sandiego.gov/sites/default/files/amended_2016_north_park_community_plan_web.pdf. Accessed: January 11, 2022.

CofSD, P. D. (2018). General Plan. City of San Diego Planning Department. https://www.sandiego.gov/planning/programs/genplan#genplan. Accessed: 15. January 2022.

CofSD, P. D. (2019). *Uptown Community Plan.* San Diego: City of San Diego Planning Department.

Comer-Schultz, J. (2011). *History and Historic Preservation in San Diego Since 1945: Civic Identity in America's Finest City.* (PhD), Arizona State University, Tempe.

Cosgrove, D. (1985). Prospect, perspective and the evolution of the landscape idea. *Transactions of the Institute of British Geographers, 10* (1), pp. 45–62.

Cosgrove, D. (1998 [1984]). *Social Formation and Symbolic Landscape.* Madison: The University of Wisconsin Press.

Cosgrove, D. & Daniels, S. (1988). *The iconography of landscape. Essays on the symbolic representation, design and use of past environments* (Cambridge Studies in Historical Geography, Vol. 9). Cambridge: Cambridge University Press.

Coulton, C. (2012). Defining Neighborhoods for Research and Policy. *Cityscape: A Journal of Policy Development and Research, 14* (2), pp. 231–236.

Covington, D. (1993). Burlingame: The Tract of Character 1912–1914. *The Journal of San Diego History, 39* (3), p. n.pag.

Crawford, R. W. (1993). Burlingame. *The Journal of San Diego History, 39* (3), p. n.pag.

Davoudi, S. & Strange, I. (2009). Space and place in the twentieth century planning: an analytical framework and a historical review. In S. Davoudi & I. Strange (Eds.), *Conceptions of Space and Place in Strategic Spatial Planning* (pp. 7–42). London: Routledge.

Delgado, E. E. & Swanson, K. (2019). *Gentefication* in the barrio: Displacement and urban change in Southern California. *Journal of Urban Affairs*, pp. 1–16.

Delgado, K. (1998). A turning point: The conception and realization of Chicano Park. *Journal of San Diego History, 44* (1), pp. 48–61.

Dillinger, M. E. (2000). Hillcrest: From Haven to Home. *The Journal of San Diego History, 46* (4), p. n.pag.

DSDP. (2016). Downtown San Diego: The Innovation Economy's Next Frontier. A Data Driven Exploration of San Diego's Urban Renaissance. Downtown San Diego Partnership & UC San Diego Extension Center For Research On The Regional Economy. https://downtownsandiego.org/wp-content/uploads/2016/05/DSDP-Demographic-Study-2016.pdf. Accessed: 5. November 2019.

Easthope, A. (1998). Bhabha, hybridity and identiy. *Textual Practice, 12* (2), pp. 341–348.

Eddy, L. (1993). War comes to San Diego. *The Journal of San Diego History, 39* (1 & 2), p. n.pag.

Eddy, L. (1995). Visions of Paradise. *The Journal of San Diego History, 41* (3), p. n.pag.

Edler, D., Jenal, C. & Kühne, O. (Eds.). (2020). *Modern Approaches to the Visualization of Landscapes.* Wiesbaden: Springer VS.

Erie, S. P., Kogan, V. & MacKenzie, S. A. (2010). Redevelopment, San Diego Style: The Limits of Public—Private Partnerships. *Urban Affairs Review, 45* (5), pp. 644–678.

Erie, S. P., Kogan, V. & MacKenzie, S. A. (2011). *Paradise plundered: Fiscal Crisis and Governance Failures in San Diego.* Stanford: Stanford University Press.

Ervin, J. C. (2007). Reinventing downtown San Diego: A Spatial and Cultural Analysis of the Gaslamp Quarter. *The Journal of San Diego History, 53* (4), pp. 188–217.

Ervin, J. C. (2008). San Diego's Urban Trophy: Horton Plaza Redevelopment Project. *Southern California Quaterly, 90* (4), pp. 419–453.

Ford, L. (2014). The Visions of the Builders. The Historical Evolution of the San Diego Cityscape. In P. R. Pryde (Ed.), *San Diego. An Introduction to the Region* (pp. 175–190). San Diego: Sunbelt Publications.

Ford, L. & Griffin, E. (1979). The Ghettoization of Paradise. *Geographical Review, 69* (2), pp. 140–158.

Ford, L., Klevisser, F. & Carli, F. (2008). Ethnic Neighborhoods and Urban Revitalization: Can Europe Use the American Model? *Geographical Review, 98* (1), pp. 82–102.

Frank, S. (2018). Inner-City Suburbanization–No Contradiction in Terms. Middle-Class Family Enclaves are Spreading in the Cities. *Raumforschung und Raumordnung | Spatial Research and Planning, 76* (2), pp. 123–132.

Frank, S. (2020). Suburbane Räume und Lebensweisen: Von der Peripherie ins Zentrum—und nicht zurück. In I. Breckner, A. Göschel & U. Matthiesen (Eds.), *Stadtsoziologie und Stadtentwicklung. Handbuch für Wissenschaft und Praxis* (pp. 257–268). Baden-Baden: Nomos.

Gergen, K. (1999). *An Invitation to Social Construction.* Thousank Oaks: Sage.

Greider, T. & Garkovich, L. (1994). Landscapes: The Social Construction of Nature and the En vironment. *Rural Sociology, 59* (1), pp. 1–24.

Griffin, E. & Weeks, J. R. (2014). Peopling the Region. San Diego's Population Patterns. In P. R. Pryde (Ed.), *San Diego. An Introduction to the Region* (pp. 63–76). San Diego: Sunbelt Publications.

Hanlon, B. (2008). The decline of older, inner suburbs in metropolitan America. *Housing Policy Debate, 19* (3), pp. 423-456.

Hanlon, B. (2010). *Once the American Dream: Inner-ring suburbs of the metropolitan United States.* Philadelphia: Temple University Press.

Haselsberger, B. (2014). Decoding borders. Appreciating border impacts on space and people. *Planning Theory & Pratice, 15* (4), pp. 505–526.

Hennessey, G. (2000). Photographic Essay of Hillcrest. *The Journal of San Diego History, 46* (4), p. n.pag.

Hoesterey, I. (2001). *Pastiche: Cultural memory in art, film, literature.* Bloomington: Indiana University Press.

Holstein, J. A. & Gubrium, J. F. (2000). *The self we live by: narrative identity in a postmodern world.* New York: Oxford University Press.

Iossifova, D. (2013). Searching for common ground. Urban borderlands in a world of borders and boundaries. *Cities, 34*, pp. 1–5.

Iossifova, D. (2019). Borderland. In A. M. Orum (Ed.), *The Wiley Blackwell Encyclopedia of Urban and Regional Studies* (pp. 1–4). Chichester: Wiley Blackwell.

Jauhiainen, J. S. (2003). Learning from Tartu — Towards Post-Postmodern Landscapes. In H. Palang & G. Fry (Eds.), *Landscape Interfaces* (Landscape Series, Vol. 1). Dordrecht: Springer.

Joassart-Marcelli, P. (2021). *The $16 Taco: Contested Geographies of Food, Ethnicity, and Gentrification.* Seattle: University of Washington Press.

Joassart-Marcelli, P. & Bosco, F. J. (2020). The Taste of Gentrification. In A. H. Alkon, Y. Kato & J. Sbicca (Eds.), *A recipe for gentrification: Food, power, and resistance in the city* (pp. 31–53). New York: New York University Press.

Kayzar, B. (2003). Investigating Connectivity: The Vermont Street Pedestrian Bridge in San Diego. *The California Geographer, 43*, pp. 1–21.

Kayzar, B. (2006). *Analyzing Revitalization Outcomes in Downtown San Diego.* (PhD), University of California Santa Barbara and San Diego State University.

Kayzar, B. (2008a). The New Era of Mixed Uses. 1986-Present. In T. W. Paradis (Ed.), *The Greenwood Encyclopedia of Homes through American History. Volume 4 1946–Present.* (pp. 353–502). Westport: Greenwood Press.

Kayzar, B. (2008b). The Perpetual Undoing of San Diego's East Village. *Yearbook of the Association of Pacific Coast Geographers, 70*, pp. 135–160.

Kayzar, B. (2016). Ephemeral Conciliation: Community-Based Art and Redevelopment. *Yearbook of the Association of Pacific Coast Geographers, 78*, pp. 28–46.

Killory, C. (1993). Temporary Suburbs: The Lost Opportunity of San Diego's National Defense Housing Projects. *The Journal of San Diego History, 39* (1 & 2), p. n.pag.

Kühne, O. (2012). *Stadt—Landschaft—Hybridität: Ästhetische Bezüge im postmodernen Los Angeles mit seinen modernen Persistenzen.* Wiesbaden: Springer VS.

Kühne, O. (2019). *Landscape Theories. A Brief Introduction.* Wiesbaden: Springer VS.

Kühne, O. & Schönwald, A. (2015). *San Diego: Eigenlogiken, Widersprüche und Hybriditäten in und von ‚America´s finest city'.* Wiesbaden: Springer VS.

Kühne, O., Schönwald, A. & Jenal, C. (2022). Bottom-up memorial landscapes between social protest and top-down tourist destination: the case of Chicano Park in San Diego (California)—an analysis based on Ralf Dahrendorf's conflict theory. *Landscape Research*, pp. 1–17.

Kühne, O., Schönwald, A. & Weber, F. (2016). Urban/rural hybrids: The urbanisation of former suburbs (URFSURBS). *Quaestiones Geographicae, 35* (4), pp. 23-34.

Kühne, O. & Weber, F. (2019). Postmoderne Zugriffe und Differenzierungen von Stadt und Land(schaft): Stadtlandhybride, räumliche Pastiches und URFSURBS. In O. Kühne, F. Weber, K. Berr & C. Jenal (Eds.), *Handbuch Landschaft* (pp. 755–770). Wiesbaden: Springer VS.

Kühne, O., Weber, F. & Koegst, L. (2020). Der Prozess der URFSURBanisierung in San Diego (Kalifornien) aus der Perspektive von Bewohner*innen und Besucher*innen des East Vilage. In R. Duttmann, O. Kühne & F. Weber (Eds.), *Landschaft als Prozess* (pp. 475–490). Wiesbaden: Springer VS.

Le Texier, E. (2007). The Struggle against Gentrification in Barrio Logan. In R. Griswold del Castillo (Ed.), *Chicano San Diego: Cultural Space and the Struggle for Justice* (pp. 202–221). Tucson: University of Arizona Pr.

Lee, B. A., Oropesa, R. S. & Kanan, J. W. (1994). Neighborhood Context and Residential Mobility. *Demography, 31* (2), pp. 249–270.

Lee, S. & Leigh, N. G. (2005). The Role of Inner Ring Suburbs in Metropolitan Smart Growth Strategies. *Journal of Planning Literature, 19* (3), pp. 330–346.

Liao, K., Breitung, W. & Wehrhahn, R. (2018). Debordering and rebordering in the residential borderlands of suburban Guangzhou. *Urban Geography, 39* (7), pp. 1092–1112.

Lynch, K. (1992). *The image of the city* (21th ed.). Cambridge: MIT Press.

Mengers, D. W. (2017). *San Diego Trolleys. Images of Rail.* Charleston: Arcadia Publishing.

Miller, J. (2013). Excavating Golden Hill: Between "South Park" and "the Other Side of the Freeway". Retrieved from https://sandiegofreepress.org/2013/05/excavating-golden-hill-between-south-park-and-the-other-side-of-the-freeway/#.YbKYhPHML0q

Mitchell, K. (1997). Different Diasporas and the Hype of Hybridity. *Environment and Planning D: Society and Space, 15* (5), pp. 533–553.

Montes, G. E. (1977). San Diego's City Park, 1868–1902. *The Journal of San Diego History, 23* (2), p. n.pag.

Newman, D. (2006). Borders and Bordering: Towards an Interdisciplinary Dialogue. *European Journal of Social Theory, 9* (2), pp. 171–186.

Norris, F. (1983). Logan Heights: Growth and Change in the Old East End. *The Journal of San Diego History, 29* (1), p. n.pag.

Paasi, A. (2011). A *Border Theory*: An Unattainable Dream or a Realistic Aim for Border Scholars? In D. Wastl-Walter (Ed.), *The Ashgate Research Companion to Border Studies* (pp. 11–31). Farnham: Ashgate.

Pryde, P. R. & Stutz, F. P. (2014). "America's Finest City". Communities of the City of San Diego. In P. R. Pryde (Ed.), *San Diego. An Introduction to the Region* (pp. 191–208). San Diego: Sunbelt Publications.

Radeloff, V. C., Hammer, R. B., Stewart, S. I., Fried, J. S., Holcomb, S. S. & McKeefry, J. F. (2005). The wildland–urban interface in the United States. *Ecological applications, 15* (3), pp. 799–805.

Roßmeier, A. (2019). *Quo vadis ,Temporary Paradise'?* Urfsurbanisierung und räumliche Konflikte im StadtLandHybriden San Diego. In K. Berr & C. Jenal (Eds.), *Landschaftskonflikte* (pp. 591–616). Wiesbaden: Springer VS.

Roßmeier, A. (2020). Hybrid Urban Borderlands. Von der Hybridität intra-urbaner Grenzen in der internationalen Metropolregion San Diego. In B. Caesar, J. Hollstegge, F. Weber & C. Wille (Eds.), *Geographien der Grenzen. Räume—Ordnungen—Verflechtungen* (pp. 357–389). Wiesbaden: Springer VS.

Roßmeier, A. & Weber, F. (2021). Hybrid Urban Borderlands—Redevelopment Efforts and Shifting Boundaries In and Around Downtown San Diego. *Journal of Borderlands Studies*, pp. 1–27.

Roßmeier, A. & Weber, F. (2021 [2018]). Stormy times. Civic engagement in wind power development: between support and rejection. In A. Stefansky & A. Göb (Eds.), *'All change please!'—challenges and opportunities of the energy transition* (pp. 52–79). Hannover: Academy for Territorial Development in the Leibniz Association.

Rumford, C. (2014). *Cosmopolitan Borders*. Hampshire: Palgrave Macmillan.

Rumpf, L. E. (2016). *The synthesis of social reality and the evolution of Barrio Logan: A critical view of the framing and representation of urban redevelopment.* Fielding Graduate University, Santa Barbara.

San Diego State University. (2021). The San Diego Union Tribune at SDSU. SDSU. https://library.sdsu.edu/research-services/news/san-diego-union-tribune-sdsu. Accessed: 1. December 2021.

San Diego Tourism Authority. (2021). Have Fun! Hillcrest. San Diego's Hip & Cultured Uptown. San Diego Tourism Marketing District Corporation. https://www.sandiego.org/meeting-planners/regional-guides/hillcrest.aspx. Accessed: 20. August 2021.

San Diego Tourism Authority. (2022). Welcome to North Park & South Park. San Diego Tourism Authority. https://www.sandiego.org/explore/downtown-urban/north-park-south-park.aspx. Accessed: 8. March 2022.

SANDAG. (2003a). CENSUS 2000 PROFILE: Barrio Logan Community Planning Area, City of San Diego. San Diego's Regional Planning Agency. http://datasurfer.sandag.org/download/sandag_census_2000_cpa_barrio-logan.pdf. Accessed: 15. December 2019.

SANDAG. (2003b). CENSUS 2000 PROFILE: Census Tract 3.00. San Diego's Regional Planning Agency. https://datasurfer.sandag.org/download/sandag_census_2000_tract_3.pdf. Accessed: 23. August 2021.

SANDAG. (2003c). CENSUS 2000 PROFILE: Census Tract 12. San Diego's Regional Planning Agency. https://datasurfer.sandag.org/download/sandag_census_2000_tract_12.pdf. Accessed: 19. January 2021.

SANDAG. (2003d). CENSUS 2000 PROFILE: Census Tract 13. San Diego's Regional Planning Agency. https://datasurfer.sandag.org/download/sandag_census_2000_tract_13.pdf. Accessed: 19. January 2021.

SANDAG. (2003e). CENSUS 2000 PROFILE: Census Tract 14. San Diego's Regional Planning Agency. https://datasurfer.sandag.org/download/sandag_census_2000_tract_14.pdf. Accessed: 19. January 2021.

SANDAG. (2003f). CENSUS 2000 PROFILE: Census Tract 15. San Diego's Regional Planning Agency. https://datasurfer.sandag.org/download/sandag_census_2000_tract_15.pdf. Accessed: 19. January 2021.

SANDAG. (2003g). CENSUS 2000 PROFILE: Census Tract 47.00. San Diego's Regional Planning Agency. http://datasurfer.sandag.org/download/sandag_census_2000_tract_47.pdf. Accessed: 15. December 2019.

SANDAG. (2003h). CENSUS 2000 PROFILE: Census Tract 49.00. San Diego's Regional Planning Agency. http://datasurfer.sandag.org/download/sandag_census_2000_tract_49.pdf. Accessed: 15. December 2019.

SANDAG. (2003i). CENSUS 2000 PROFILE: Census Tract 50.00. San Diego's Regional Planning Agency. http://datasurfer.sandag.org/download/sandag_census_2000_tract_50.pdf. Accessed: 15. December 2019.

SANDAG. (2003j). CENSUS 2000 PROFILE: Census Tract 51.00. San Diego's Regional Planning Agency. http://datasurfer.sandag.org/download/sandag_census_2000_tract_51.pdf. Accessed: 15. December 2019.

SANDAG. (2003k). CENSUS 2000 PROFILE: Census Tract 57.00. San Diego's Regional Planning Agency. https://datasurfer.sandag.org/download/sandag_census_2000_tract_57.pdf. Accessed: 27. July 2021.

SANDAG. (2003l). CENSUS 2000 PROFILE: Census Tract 59.00. San Diego's Regional Planning Agency. https://datasurfer.sandag.org/download/sandag_census_2000_tract_59.pdf. Accessed: 27. July 2021.

SANDAG. (2003m). CENSUS 2000 PROFILE: Census Tract 60.00. San Diego's Regional Planning Agency. https://datasurfer.sandag.org/download/sandag_census_2000_tract_60.pdf. Accessed: 27. July 2021.

SANDAG. (2003n). CENSUS 2000 PROFILE: Centre City Community Planning Area, City of San Diego. San Diego's Regional Planning Agency. http://datasurfer.sandag.org/download/sandag_census_2000_cpa_downtown.pdf. Accessed: 15. December 2019.

SANDAG. (2003o). CENSUS 2000 PROFILE: City of San Diego. San Diego's Regional Planning Agency. http://datasurfer.sandag.org/download/sandag_census_2000_jurisdiction_san-diego.pdf. Accessed: 28. July 2021.

SANDAG. (2003p). CENSUS 2000 PROFILE: Golden HIll Community Planning Area, City of San Diego. San Diego's Regional Planning Agency. https://datasurfer.sandag.org/download/sandag_census_2000_cpa_greater-golden-hill.pdf. Accessed: 11. December 2021.

SANDAG. (2003q). CENSUS 2000 PROFILE: Southeastern SD Community Planning Area, City of San Diego. San Diego's Regional Planning Agency. http://datasurfer.sandag.org/download/sandag_census_2000_cpa_southeastern-san-diego.pdf. Accessed: 15. December 2019.

SANDAG. (2003r). CENSUS 2000 PROFILE: Uptown Community Planning Area, City of San Diego. San Diego's Regional Planning Agency. https://datasurfer.sandag.org/download/sandag_census_2000_cpa_uptown.pdf. Accessed: 27. July 2021.

SANDAG. (2016a). Demographic and Socioeconomic Profile 2010. CPA: Barrio Logan. San Diego's Regional Planning Agency. http://datasurfer.sandag.org/download/sandag_census_2010_cpa_barrio-logan.pdf. Accessed: 15. December 2019.

SANDAG. (2016b). Demographic and Socioeconomic Profile 2010. CPA: Downtown. San Diego's Regional Planning Agency. http://datasurfer.sandag.org/download/sandag_census_2010_cpa_downtown.pdf. Accessed: 15. December 2019.

SANDAG. (2016c). Demographic and Socioeconomic Profile 2010. CPA: Greater Golden Hill. San Diego's Regional Planning Agency. https://datasurfer.sandag.org/download/sandag_census_2010_cpa_greater-golden-hill.pdf. Accessed: 11. December 2021.

SANDAG. (2016d). Demographic and Socioeconomic Profile 2010. CPA: Southeastern San Diego. San Diego's Regional Planning Agency. https://datasurfer.sandag.org/download/sandag_census_2010_cpa_southeastern-san-diego.pdf. Accessed: 15. December 2019.

SANDAG. (2016e). Demographic and Socioeconomic Profile 2010. CPA: Uptown. San Diego's Regional Planning Agency. https://datasurfer.sandag.org/download/sandag_census_2010_cpa_uptown.pdf. Accessed: 27. July 2021.

SANDAG. (2016f). Demographic and Socioeconomic Profile 2010. JURISDICTION: San Diego. San Diego's Regional Planning Agency. https://datasurfer.sandag.org/download/sandag_census_2010_jurisdiction_san-diego.pdf. Accessed: 28. July 2021.

SANDAG. (2016g). Demographic and Socioeconomic Profile 2010. TRACT: 3. San Diego's Regional Planning Agency. https://datasurfer.sandag.org/download/sandag_census_2010_tract_3.pdf. Accessed: 23. August 2021.

SANDAG. (2016h). Demographic and Socioeconomic Profile 2010. TRACT: 12. San Diego's Regional Planning Agency. http://datasurfer.sandag.org/download/sandag_census_2010_tract_12.pdf. Accessed: 15. December 2019.

SANDAG. (2016i). Demographic and Socioeconomic Profile 2010. TRACT: 13. San Diego's Regional Planning Agency. http://datasurfer.sandag.org/download/sandag_census_2010_tract_13.pdf. Accessed: 15. December 2019.

SANDAG. (2016j). Demographic and Socioeconomic Profile 2010. TRACT: 14. San Diego's Regional Planning Agency. http://datasurfer.sandag.org/download/sandag_census_2010_tract_14.pdf. Accessed: 15. December 2019.

SANDAG. (2016k). Demographic and Socioeconomic Profile 2010. TRACT: 15. San Diego's Regional Planning Agency. http://datasurfer.sandag.org/download/sandag_census_2010_tract_15.pdf. Accessed: 15. December 2019.

SANDAG. (2016l). Demographic and Socioeconomic Profile 2010. TRACT: 47. San Diego's Regional Planning Agency. http://datasurfer.sandag.org/download/sandag_census_2010_tract_47.pdf. Accessed: 15. December 2019.

SANDAG. (2016m). Demographic and Socioeconomic Profile 2010. TRACT: 49. San Diego's Regional Planning Agency. http://datasurfer.sandag.org/download/sandag_census_2010_tract_47.pdf. Accessed: 15. December 2019.

SANDAG. (2016n). Demographic and Socioeconomic Profile 2010. TRACT: 50. San Diego's Regional Planning Agency. http://datasurfer.sandag.org/download/sandag_census_2010_tract_50.pdf. Accessed: 15. December 2019.

SANDAG. (2016o). Demographic and Socioeconomic Profile 2010. TRACT: 51. San Diego's Regional Planning Agency. http://datasurfer.sandag.org/download/sandag_census_2010_tract_51.pdf. Accessed: 15. December 2019.

SANDAG. (2016p). Demographic and Socioeconomic Profile 2010. TRACT: 57. San Diego's Regional Planning Agency. https://datasurfer.sandag.org/download/sandag_census_2010_tract_57.pdf. Accessed: 27. July 2021.

SANDAG. (2016q). Demographic and Socioeconomic Profile 2010. TRACT: 59. San Diego's Regional Planning Agency. https://datasurfer.sandag.org/download/sandag_census_2010_tract_59.pdf. Accessed: 27. July 2021.

SANDAG. (2016r). Demographic and Socioeconomic Profile 2010. TRACT: 60. San Diego's Regional Planning Agency. https://datasurfer.sandag.org/download/sandag_census_2010_tract_60.pdf. Accessed: 27. July 2021.

SANDAG & SanGIS. (2018). Parcel Lookup Tool and Geographic Boundary Viewer. SANDAG

SanGIS. (2012). About SanGIS. San Diego Geographic Information Source. https://www.sangis.org/about/index.html. Accessed: 16 July 2021.

Schimitschek, M. (2017). Golden Hill: Once-fashionable neighborhood is having a renaissance. *The San Diego Union-Tribune.*

Schütz, A. & Luckmann, T. (1973). *The Structures of the Life-World* (Vol. 1). Evanston, Illinois: Northwestern University Press.

Sedelmeier, T., Kühne, O. & Jenal, C. (2021). *Foodscapes/Nahrungslandschaften. Eine Bestandsaufnahme.* Wiesbaden: Springer VS.

SOHO. (1992). The Historic Subdivision of South Park. Save Our Heritage Organization (SOHO). http://www.sohosandiego.org/tourbooklets/historicsubdivisionofsouthpark.pdf. Accessed: 29. October 2021.

Suttles, G. D. (1972). *The social construction of communities.* Chicago: University of Chicago Press.

Szytniewski, B., Spierings, B. & Van der Velde, M. (2020). Stretching the Border: Shopping, Petty Trade and Everyday Life Experiences in the Polish-Ukrainian Borderland. *International Journal of Urban and Regional Research, 44* (3), pp. 469–483.

The San Diego Union-Tribune. (2021). About Us. The San Diego Union-Tribune. https://www.sandiegouniontribune.com/about-us. Accessed: 2. December 2021.

U.S. Census Bureau. (2021a). 2020 Census Demographic Data Map Viewer. U.S. Department of Commerce. https://mtgis-portal.geo.census.gov/arcgis/apps/MapSeries/index.html?appid=2566121a73de463995ed2b2fd7ff6eb7. Accessed: 23. August 2021.

U.S. Census Bureau. (2021b). Glossary. U.S. Department of Commerce. https://www.census.gov/glossary/#term_Censustract. Accessed: 10. April 2021.

van Houtum, H. & van Naerssen, T. (2002). Bordering, Ordering and Othering. *Tijdschrift voor economische en sociale geografie, 93* (2), pp. 125–136.

Vester, H.-G. (1993). *Soziologie der Postmoderne.* München: Quintessenz.

Vicino, T. (2008). The Quest to Confront Suburban Decline: Political Realities and Lessons. *Urban Affairs Review, 43* (4), pp. 553–581.

Wastl-Walter, D. (2011). Introduction. In D. Wastl-Walter (Ed.), *The Ashgate Research Companion to Border Studies* (pp. 1–8). Farnham: Ashgate.
Weber, F., Jenal, C., Roßmeier, A. & Kühne, O. (2017). Conflicts around Germany's Energiewende: Discourse patterns of citizens' initiatives. *Quaestiones Geographicae, 36* (4), pp. 117–130.
Wille, C. & Nienaber, B. (2020). Border and border experiences. In C. Wille & B. Nienaber (Eds.), *Border Experiences in Europe. Everyday Life—Working Life—Communication—Languages* (pp. 7–15). Baden-Baden: Nomos.
Wylie, J. (2007). *Landscape.* London, New York: Routledge.

6 Conclusion, Reflection, and Outlook on Hybrid Urban Borderlands—A Unique San Diegan Phenomenon or an Adjustable Lens?

The following paragraphs are completing the present work with reflective and conclusive remarks. The first subsection serves the purpose of summarizing and critically reflecting the study's theoretical background, the perspectival-conceptual, methodological, and methodical approaches, their scope and limitations. In the subsequent section, a 'zoomed out' conclusion and interpretation of the main results as well as an outlook on the analytical potential of the concept of hybrid urban borderlands beyond the present case context of San Diego will be given.

6.1 Methodological, Methodical, and Perspectival Reflection—Addressing Limitations and Highlighting the Analytical Scope of the Present Triangulation

The primary aim of the present study was to further the understanding of the recent redevelopment trend that has emerged in downtown San Diego and throughout selected neighborhoods of its inner ring. Within a phenomenology-oriented social constructivist methodology (section 2.1), it was of interest how individuals perceive and experience the different neighborhoods and how the everyday life meanings, the physical foundation, but also the images of the various areas change over time. However, in order to pursue this empirical goal, the aim was to develop a suitable perspectival-conceptual lens that allows to interpret the recent neighborhood change processes in a comprehensive way. Accordingly, an interdisciplinary framework of urban landscape and border theory has been established, which explores the commonalities and overlaps of the two research

A. Rossmeier, *San Diego's Hybrid Urban Borderlands*,
Räume – Grenzen – Hybriditäten, https://doi.org/10.1007/978-3-658-42667-5_6

fields (sections 2.2–2.4). Thus, on the one hand, it became possible to draw on the rich theoretical and conceptual ways of thinking, on the social constructivist, phenomenological, relational, as well as post-structuralist approaches, that have emerged in the broad field of landscape research. Within the focus on processuality and changeability in the neighborhoods of San Diego, urban landscape has been approached as an ongoing individual *practice of spatial identity creation.* Besides this understanding, the idea of *(socio-)spatial hybridization* and pluralization, which is embedded in postmodern landscape theory (cf. Kühne 2019), has been adopted as the key landscape-theoretical lens for the present study. By means of the focus on hybridization processes, attention has been brought to aspects of (temporal) ambiguity and pastiche-like spatial phenomena within and particularly between the central neighborhoods of San Diego.

On the other hand, the chosen landscape approach has been combined with border-theoretical thinking. For this purpose, it was necessary to determine a clear terminology: While the term *border* has been used for physical demarcations only, immaterial demarcations have been strictly referred to as *boundaries.* Thus, the existence not only of material but also of immaterial demarcations on the city level has been acknowledged, both of which have been shown to be of significant everyday life relevance in San Diego, as it will be illustrated further in the following conclusion. Retrospectively, the used terminological and analytical distinction of borders and boundaries has proven to be a suitable approach for the present and potentially also for future research projects. Furthermore, building on the conceptual-perspectival focus on urban landscape processes and processuality in general, an understanding of demarcations as *b/ordering* and *othering processes* has been adopted. By means of a multi-dimensional breakdown of the bordering notion—into physical, political, individual or social, and symbolic-aesthetic urban bordering processes—it became possible to take both the complexity and processuality of border and boundary situations between San Diego's inner-ring neighborhoods into account. Lastly, these unique urban landscape and border-theoretical approaches to plurality, hybridity, and multi-dimensionality have been synthesized into the conceptualization of *hybrid urban borderlands.* As applied in the present study, the unique concept and lens of hybrid urban borderlands brings into focus the temporal spatiality and ambiguity of borders and boundaries between adjacent, assimilating neighborhoods in transition.

Given the complex aim of analyzing social perceptions and experiences of neighborhood change, on the one hand, and the perspectival focus on hybridization and processes of spatial identity and thus boundary creation, on the other hand, a mixed methods approach has been chosen. In line with the post-structuralist and neopragmatic idea of combining different theories and methodologies (section 3.2; Chilla et al. 2015; Kühne and Jenal 2020a; 2020b, pp. 10–20; 2021; Kühne and Weber 2019, pp. 47–50), it became possible to triangulate qualitative interviews and participant short-term observations with rather positivist and representational approaches, such as cartographic and photographic illustration, as well as quantitative census data (section 3.3). In order to take the contingency and infinity of social reality into account, it was necessary to draw on qualitative methods, which are rich in context and allow for the expression of individual impulses and emotions. However, qualitative approaches pose certain methodological limitations. Generally, their deficit lies in the epistemological inability of fully understanding subjective meaning (Hitzler and Eberle 2004) as well as the risk of interpretative generalization. Thus, it has been crucial to highlight at different points of the analysis that social reality is—in theory and in the present case context of San Diego—contingent, processually changing, and infinite, not uniform or singular. As a result, space in general and neighborhoods in particular will never be perceived in only one specific way by all members of society. In contrast to the qualitative methods, the presented representational cartographies and photographs, but particularly the census calculations, are contextless. Furthermore, quantitative approaches are characterized by a certain rigidity, which limits their explorative potential. Nevertheless, the present quantitative approaches have created a useful background against which the contextual qualitative insights have been interpreted. Thus, the neopragmatic idea has been applied in the sense of a *meta-theory* for the triangulation of non- or more-than-representational with representational methodologies, which allows for the supplementation of diverse qualitative data with census calculations and innovative cartographies. In summary, this mixed methods approach has enabled the present study to gain a large and differentiated picture of San Diego's redevelopment trend and the concomitant b/ordering and othering processes. Furthermore, the present integration of *digital 3D sketches*, photographs, and various maps can be credited with the intent to explore alternative ways of qualitative data illustration and thus of landscape visualization beyond 2D cartography.

In terms of the interviews, a combination of semi-structured and unconstrained, rather open interviews has been conducted. While the more structured versions have mostly been prescheduled interviews with individuals of specific professional backgrounds, such as planners, developers, or local scholars, the unconstrained interviews have taken place through spontaneous encounters with individuals during the participant observations and walks. More importantly, however, this approach has followed a classification of the interviewees into the three ideal-typical perspectival groups of 'residents,' 'economy, development, and planning,' and 'academia and research.' In theory, this distinction or diversification allows for the collection of both individual everyday life experiences and expert views and thus for poly-vocality within the body of interviews. However, interviews are not always clearly classifiable into distinct groups, because, among other aspects, the interviewees have versatile (and largely unknown) biographies and are possibly narrating from various perspectives. The interviewee classified as EP-06 is a fitting example for this methodical dilemma; the interviewee is a long-term resident from Bankers Hill and has experienced the local redevelopment and change processes first hand in everyday life, however, at the same time, they are a business owner in the neighborhood as well as financially invested in local development projects. This combination of various interests results in a versatile perspective on the overall neighborhood change that is not only reflecting the experiences of a resident (in terms of changing manifold everyday life phenomena) but of an investor that is particularly arguing for the positive outcomes of the local redevelopment efforts. In addition to the issue of diverse or possibly unknown personal backgrounds, interests, and acquired special knowledge, it can be stated that the attempt to classify individuals as either lays or experts is only possible to a certain extent. As illustrated using the example of the interviewee EP-06, lay and expert but also urban scholar and planner/developer perspectives, as other examples have shown, are subject to overlaps in today's highly specialized, postmodern world. This aspect, however, is a consequential problem of postmodernity and thus rather a limitation than a shortcoming of the present methodic design.

6.2 Empirical Conclusion and Outlook on Hybrid Urban Borderlands in San Diego and beyond

Despite the illustrated difficulty of perspective classification and the quantitatively not representative number of 39 qualitative interviews, the present study allows for a brief conclusive remark on group-specific tendencies. Within the outlined lay/expert comparison, the analysis has indicated certain differences in terms of spatial conception and perspective. While the interviewed planners and in parts also scholars seem to draw on modern or Euclidian conceptions of space, the interviewees classified as residents have narrated rather from post-structuralist, relational perspectives. This has become particularly obvious in the context of spatial demarcation and the discussed urban/suburban or downtown/inner ring opposition. Especially planners have argued for the continuation of the distinction between the downtown area and the inner ring neighborhoods (based on various planning-specific characteristics, such as zoning regulations) and have emphasized the ongoing relevance of official-territorial neighborhood boundaries and borders in the social organization of everyday life in San Diego (cf. Davoudi and Strange 2009). Thus, the specific interviewees appeared to conceptualize the various neighborhoods as rigidly demarcated 'containers,' which underly planning interventions rather than social everyday processes and meanings. In contrast, residents without planning background or specialized knowledge in related fields have discussed the changing relations and the contrasts between the various communities rather on the basis of everyday life experiences, interpretations, and routines and thus by means of social b/ordering and identity construction process. These approaches to space and everyday life in San Diego are highlighting aspects of fluidity, processuality, and connection rather than rigidity and territoriality. However, the outlined contrast between lay and expert perspectives is to be understood as an 'initial claim' that needs to be addressed further by future research. This prospective agenda can be deemed particularly relevant, because modern and Euclidian conceptions of space negate the contingency and subjectivity of urban landscape and b/ordering processes and thus San Diego's multiple realities, which will never perfectly coincide with official-territorial demarcations and are always subject to change (cf. Davoudi 2012; Graham and Healey 1999).

In addition to the claimed perspective contrast, the present study has confirmed the initial expectation that the social perceptions and interpretations of the various analyzed areas have—with exceptions—changed strongly over time. Within the present spatial focus on the communities and neighborhoods of Bankers Hill and Hillcrest (section 5.1), North Park and Mid-City (section 5.2), Golden Hill and South Park (section 5.3), as well as East Village, Southeastern San Diego, and Barrio Logan (section 5.4), several interviewees have discussed architectural/physical, infrastructural, social/demographic, as well as symbolic-aesthetic changes, which they experienced over the course of the recent years and even decades. However, each of the neighborhoods has been affected differently by the redevelopment trend, as the triangulated approach of interviews, observations, census data, and cartographic and photographic illustrations has shown. Among others, this is due to the reason that the various neighborhoods of downtown San Diego and the inner ring have each underwent redevelopment processes at different times within the last decades. While Horton Plaza and the Gaslamp Quarter, for instance, are considered the first areas that underwent redevelopment in San Diego, neighborhoods such as Barrio Logan or Sherman Heights are only recently facing new constructions, increasing housing values, and demographic changes, among other updates. Further, it has been shown that some neighborhoods, such as North Park, have faced extensive redevelopment with significant demographic, physical, and symbolic changes, while the development dynamics of Bankers Hill, for instance, have appeared to be slower and less comprehensive.

Besides the temporal sequence and the varying extent, the present study has shown that the redevelopment in San Diego is taking place within a spatial pattern. As illustrated by various interviewees, the progression resembles a 'wave' that is moving away from the city center, from west to east and from one neighborhood into the other (cf. Joassart-Marcelli 2021, pp. 152–153). However, the redevelopment wave is neither affecting each community in the same ways and to the same extent, nor are the various sections of the different neighborhoods adapting equally. In this respect, it has been shown that the development trend is particularly progressing along large commercial and traffic corridors, which connect various neighborhoods across official boundaries and physical borders, such as freeways or canyons. Thus, the corridors function as 'development axes' within the city along which creative businesses, construction projects, and young college graduates in search for more affordable housing, among other aspects, are penetrating from the redeveloped neighborhoods in the west into the less invested areas in the east. These axes are the spaces where the neighborhood change processes are perceived for the first times in the less invested communities, in the form of changing storefronts and their new clientele or through renovations and reutilization.

This is particularly the case in the Mid-City community, which is currently only facing redevelopment processes along the western sections of the major corridors of El Cajon Boulevard and University Avenue, which connect the low-income community to the redeveloped major commercial intersections of North Park. Along these corridors, the trendy concept of the North Park community is being extended westward into the neighborhoods of Normal Heights and City Heights. In contrast to this section, the large residential areas in southern Mid-City as well as the eastern portion of the community appear to remain largely unaffected of the investment and reinterpretation trend. Similarly, in Barrio Logan, the influx of new residents and businesses and the arrival of trendy culinary offers, such as craft beer, specialty coffee, or vegan tacos, has begun along the traffic and commercial corridor of Logan Avenue, from which the redevelopment dynamics radiate into other areas of the community. Logan Avenue connects the lower-income community of Barrio Logan with the East Village area across their official boundary and creates a socio-spatial, functional, economic, and symbolic extension of the redeveloping central city. Besides the examples of University Avenue, El Cajon Boulevard, and Logan Avenue, the large Avenues to the West of Balboa Park—4th, 5th, and 6th Avenue—constitute a development axis between the communities of Hillcrest and Bankers Hill. Along these

Avenues, redevelopment processes are progressing from north to south, from the busy neighborhood center of Hillcrest into the northern parts of Bankers Hill. Yet, the redeveloping area in the norths stands in contrast to the less frequented southern neighborhood section of Bankers Hill, which has—until now—not been reached by the redevelopment and reinterpretation 'wave,' despite its proximity to Little Italy and the continuation of 4th, 5th, and 6th Avenue through several redeveloped neighborhoods of downtown.

However, reflecting the contingency of urban landscape and bordering processes, the present study has shown that the perceptions and experiences of the redevelopment and neighborhood changes in San Diego are by no means static and uniform but dynamic, diverse, and in parts even contrary. The divergence of interpretations can be tied to the emerging multifaceted hybridities and the ambiguity of b/ordering and othering processes, which characterize the redeveloping communities of San Diego. On the one hand, particularly the neighborhoods that have only recently entered a transition phase, such as Barrio Logan or City Heights, are facing processes of architectural or physical, social, and aesthetic-symbolic hybridization. In these communities, the physical structures and the symbolics that reflect the previous decades of low investment activity are beginning to mix with new buildings, uses, and trendy symbols and signs. In addition to these updates, also demographic change processes are setting in through the gradual outflow of lower-income residents with relatively low educational levels and the simultaneous influx of new higher-income and college-educated residents, which ultimately promotes the divergence or contrast of neighborhood experiences and (socio-)spatial interpretations further—particularly along the identified 'development axes.'

On the other hand, also neighborhoods that have already passed initial redevelopment stages, such as Hillcrest or North Park, for instance, are characterized by strong tendencies of *urban/rural hybridity*. This is due to the contrast between the gradually established residential and commercial density and transit orientation of the central corridors of University Avenue and El Cajon Boulevard and the (strong local efforts to preserve the) low-density residential structure of the surrounding, partly historic neighborhood sections. As a result, both the formerly as well as the newly transitioning communities constitute socio-spatial hybrids in the form of urbanizing former suburbs (see particularly Kühne et al. 2016), which are in the context of San Diego's official planning objectives referred to as 'urban villages' (CofSD 2004, 2019), due to their combination of stereotypically urban and suburban lifestyles, architecture and zoning, as well as modes of transportation. Ultimately, these juxtapositions and the various hybridizations they create enhance the divergence and the plurality of individual place perceptions in San Diego's redeveloping inner ring. In this sense, despite the seemingly widespread trend for 'urban village' living in San Diego's inner-ring and the officially proclaimed renaissance of downtown San Diego and East Village, it can be concluded that static as well as uniform and unambiguous interpretations of neighborhoods and development projects remain an unattained dream of modern or Euclidean urban planning—in San Diego and beyond.

Besides the urban landscape approach and the related insights, also the border-specific results allow for conclusions on the social perception and interpretation of the neighborhood changes. Generally, in line with the realization in social constructivist and relational neighborhood research that "neighborhood boundaries will always be the subject of countless debates" (Sperling 2012, p. 222), the present analysis has shown that the boundaries between the central neighborhoods of San Diego are dynamic individual and social everyday processes of distinction and spatial identity construction rather than static and given lines in space or maps. Furthermore, it has been demonstrated that the manifold hybridization processes in San Diego are not only taking place *within* the various neighborhoods but particularly also *between* the different communities, which is enhancing the plurality and ambiguity of place perceptions further. The 'redevelopment wave,' which is carrying the development and reinvestment dynamics from one neighborhood into the other, is challenging the contrasts and thus the demarcations between the affected neighborhoods in various ways. This has become particularly obvious along the 'development axes' of University Avenue, El Cajon Boulevard, or Logan Avenue, where the redevelopment is gradually undoing the physical/architectural, social and individual, as well as symbolic or aesthetic distinctions between the redeveloped and trendy areas in the west and the 'up and coming' lower income communities in the east. As a result, particularly in recent years, the socialized everyday life boundaries between the adjacent, assimilating communities are undergoing significant reinterpretation and thus redrawing processes—on the one hand, in the form of a *perforation* along the redeveloping corridors and, on the other hand, through *shifts* that push the line of differentiation further east and southeast in San Diego. Thus, ambiguous transition zones emerge, which are simultaneously part of the one and the other community, somehow here and at the same time over there. This creates a particular spatial ambiguity, a "cross-border hybridity" (Newman 2011, p. 38), which is reflected in the plurality of spatial perceptions and interpretations in San Diego—particularly between the communities of North Park and Mid-City as well as at the downtown edge, between East Village and Barrio Logan. The present study has come to refer to these undefined spaces as *hybrid urban borderlands*.

In summary, the conceptualization of hybrid urban borderlands has been established as a perspectival-conceptual approach that explores and manifests the intersections of urban landscape and border theory. The resulting unique lens has highlighted various aspects of hybridity and has helped to overcome spatial conceptions of boundedness and singularity in central San Diego's ambiguous net of simultaneously bordered and connected neighborhoods. Furthermore, this unique approach critiques "the assumption of consensus and visibility" (Rumford 2012, p. 899) of demarcations—in San Diego and potentially also beyond this case context. In line with Balibar (2004) and the idea of '*borders everywhere*,' the present study has shown that demarcations do not necessarily align with physical borders or official boundaries, they do not require physical or visual representation in order to exist and unfold their power, and they will always be drawn subjectively at various points in space and time. For San Diego and its future planning and development objectives this means that concrete redevelopment projects, neighborhoods, and their boundaries will never be of singular but of a dynamic multitude of different and equally real meanings. In this sense, also place perceptions and spatial demands in San Diego will at all times diverge, evolve further, and potentially contest each other, as several planning and development conflicts about the location and distribution of urban infill projects, affordable housing, and social service providers in the city have already shown (cf. Kayzar 2016; Roßmeier 2019).

Lastly, its multi-perspectival orientation makes the conceptualization of hybrid urban borderlands a promising approach for other case study areas that are undergoing changes related to urban redevelopment, spatial reinterpretation, and the re-negotiation of social inclusion and exclusion in neighborhoods. In this regard, it seems particularly interesting to apply the unique lens of hybrid urban borderlands also to the City of Tijuana, for instance, which faces comprehensive redevelopment processes in and around its downtown area, as the conducted observation and the interviews in Mexico indicated (Herzog and Sohn 2017, pp. 17–19). Furthermore, also other cities that are facing spatial and social transformations in the context of redevelopment and densification at the inner city edges, such as Baltimore (Hanlon 2015; Hanlon and Airgood-Obrycki 2018), Baton Rouge (Kühne and Jenal 2020a, 2021), Chicago (Charles 2013, 2018a, 2018b), Los Angeles (Roßmeier 2020c), or Paris (F. Weber 2019; F. Weber and Kühne 2017, 2020), provide contexts that should be considered potential future case studies to determine if hybrid urban borderlands are a phenomenon that is unique to San Diego or if this form of processual inner-ring neighborhood change is also found and experienced in other places. For in line with Paasi (2011, p. 19), the conceptualization of hybrid urban borderlands constitutes a unique approach that is meant to be examined, "used and re-conceptualized further in various empirical settings." Looking ahead, in addition to highlighting potential study areas, it is equally important to point toward the changes and developments that will take place in San Diego over the course of the next several years. Particularly in the context of the recent COVID-19 pandemic, it is a question for future research agendas in how far this global health emergency has and will have altered the course of the decadelong redevelopment trend. Currently, it is still subject to speculation if and to what scale the COVID-19 pandemic has changed housing and location preferences as well as private investment and public development strategies in and around downtown San Diego (cf. Lei and Liu 2022; Vicino et al. 2022). Furthermore, also the everyday consequences and local meanings of the harsh, temporary COVID-19 policy-related rebordering that has occurred on the national level between the two parts of the interconnected San Diego-Tijuana borderland has not yet been subject of in-depth research (on initial observations see Roßmeier 2020a; on COVID-19 and border closures see generally F. Weber et al. 2021; F. Weber 2022; Wille and Weber 2020).

In conclusion, it can be stated that the present study has provided a broad insight into the various realities and perspectives of central San Diego, however, it remains to be seen how and if the contested and ambiguous transition zones between the various neighborhoods—the hybrid urban borderlands—will be negotiated, constructed, and thus drawn into more or less clear boundaries in the future. In this sense, it remains to be seen which readings and interpretations of the different communities will objectivate into widespread social reality and which will lapse in the competition of meanings; and lastly, it also remains to be seen how, where, and if the decadelong wave pattern of the spatial, social, and functional expansion of San Diego's downtown into the inner-ring neighborhoods continues.

References

Balibar, É. (2004). *We, the people of Europe. Reflections on transnational citizenship.* Princeton: Princeton University Press.

Charles, S. L. (2013). Understanding the determinants of single-family residential redevelopment in the inner-ring suburbs of Chicago. *Urban Studies, 50* (8), pp. 1505–1522.

Charles, S. L. (2018a). Keeping up with the Joneses. Residential reinvestment and mansionization in American inner-ring suburbs. In B. Hanlon & T. Vicino (Eds.), *The Routledge Companion to Suburbs* (pp. 1–12). London: Routledge.

Charles, S. L. (2018b). A Typology of Mansionization in the Inner-Ring Suburbs of Chicago, Illinois, 2000–2015. *Housing Policy Debate, 28* (6), pp. 832–853.

Chilla, T., Kühne, O., Weber, F. & Weber, F. (2015). ‚Neopragmatische' Argumente zur Vereinbarkeit von konzeptioneller Diskussion und Praxis der Regionalentwicklung. In O. Kühne & F. Weber (Eds.), *Bausteine der Regionalentwicklung* (pp. 13–24). Wiesbaden: Springer VS.

CofSD, P. D. (2004). Five Projects Receive Prestigious Pilot Village Designation. City of San Diego Planning Department. https://www.sandiego.gov/sites/default/files/legacy//planning/genplan/pdf/pilotvillage/040212pvsapproved.pdf. Accessed: March 5, 2021.

CofSD, P. D. (2019). Why Pilot Villages Are Important. City of San Diego Planning Department. https://www.sandiego.gov/planning/genplan/pilotvillage/important. Accessed: 5. November 2019.

Davoudi, S. (2012). The Legacy of Positivism and the Emergence of Interpretive Tradition in Spatial Planning. *Regional Studies, 46* (4), pp. 429–441.

Davoudi, S. & Strange, I. (2009). Introduction. In S. Davoudi & I. Strange (Eds.), *Conceptions of Space and Place in Strategic Spatial Planning* (pp. 1–6). London: Routledge.

Graham, S. & Healey, P. (1999). Relational concepts of space and place: Issues for planning theory and practice. *European Planning Studies, 7* (5), pp. 623–646.

Hanlon, B. (2015). Beyond Sprawl: Social Sustainability and Reinvestment in the Baltimore Suburbs. In K. B. Anacker (Ed.), *The New American Suburb* (pp. 1–20). London: Routledge.

Hanlon, B. & Airgood-Obrycki, W. (2018). Suburban revalorization: Residential infill and rehabilitation in Baltimore County's older suburbs. *Environment and Planning A: Economy and Space, 50* (4), pp. 895–921.

Herzog, L. A. & Sohn, C. (2017). The co-mingling of bordering dynamics in the San Diego–Tijuana cross-border metropolis. *Territory, Politics, Governance, 7* (2), pp. 177–199.

Hitzler, R. & Eberle, T. S. (2004). Phenomenological Life-World Analysis. In U. Flick, E. von Kardoff & I. Steinke (Eds.), *A Companion to Qualitative Research* (pp. 67–71). London: Sage.

Joassart-Marcelli, P. (2021). *The $16 Taco: Contested Geographies of Food, Ethnicity, and Gentrification.* Seattle: University of Washington Press.

Kayzar, B. (2016). Ephemeral Conciliation: Community-Based Art and Redevelopment. *Yearbook of the Association of Pacific Coast Geographers, 78*, pp. 28–46.

Kühne, O. (2019). *Landscape Theories. A Brief Introduction.* Wiesbaden: Springer VS.

Kühne, O. & Jenal, C. (2020a). Baton Rouge (Louisiana): On the Importance of Thematic Cartography for 'Neopragmatic Horizontal Geography'. *KN—Journal of Cartography and Geographic Information*, pp. 1–9.

Kühne, O. & Jenal, C. (2020b). *Baton Rouge–the Multivillage Metropolis. A Neopragmatic Landscape Biographical Approach on Spatial Pastiches, Hybridization, and Differentiation.* Wiesbaden: Springer VS.

Kühne, O. & Jenal, C. (2021). Baton Rouge–A Neopragmatic Regional Geographical Approach. *urban science, 5* (17), pp. 1–23.

Kühne, O., Schönwald, A. & Weber, F. (2016). Urban/rural hybrids: The urbanisation of former suburbs (URFSURBS). *Quaestiones Geographicae, 35* (4), pp. 23–34.

Kühne, O. & Weber, F. (2019). *Hybrid California. Annäherungen an den Golden State, seine Entwicklungen, Ästhetisierungen und Inszenierungen.* Wiesbaden: Springer VS.

Lei, L. & Liu, X. (2022). TheCOVID-19pandemicandresidentialmobilityintentionsinthe United States: Evidence from GoogleTrends data. *Population, Space and Place, 28* (6), pp. 1–18.

Newman, D. (2011). Contemporary Research Agendas in Border Studies: An Overview. In D. Wastl-Walter (Ed.), *The Ashgate Research Companion to Border Studies* (pp. 33–47). Farnham: Ashgate.

Paasi, A. (2011). A *Border Theory*: An Unattainable Dream or a Realistic Aim for Border Scholars? In D. Wastl-Walter (Ed.), *The Ashgate Research Companion to Border Studies* (pp. 11–31). Farnham: Ashgate.

Roßmeier, A. (2019). *Quo vadis ‚Temporary Paradise'?* Urfsurbanisierung und räumliche Konflikte im StadtLandHybriden San Diego. In K. Berr & C. Jenal (Eds.), *Landschaftskonflikte* (pp. 591–616). Wiesbaden: Springer VS.

Roßmeier, A. (2020a). COVID-19 in the borderland of San Diego and Tijuana: between rebordering processes and contradictory realities. In C. Wille & R. Kanesu (Eds.), *Borders in Perspective UniGR-CBS Thematic Issue. Bordering in Pandemic Times: Insights into the COVID-19 Lockdown* (Vol. 4, pp. 57–61).

Rumford, C. (2012). Towards a Multiperspectival Study of Borders. *Geopolitics, 17* (4), pp. 887–902.

Sperling, J. (2012). The Tyranny of Census Geography: Small-Area Data and Neighborhood Statistics. *Cityscape: A Journal of Policy Development and Research, 14* (2), pp. 219–223.

Vicino, T., Voigt, R. H., Kabir, M. & Michanie, J. (2022). Urban Crises and the Covid-19 Pandemic: An Analytical Framework forMetropolitan Resiliency. *Urban Planning, 7* (3), pp. 4–14.

Weber, F. (2019). Von divergierenden Grenzziehungen und Konflikten im StadtLandHybriden des Grand Paris. In K. Berr & C. Jenal (Eds.), *Landschaftskonflikte* (pp. 637–663). Wiesbaden: Springer VS.

Weber, F. (2022). Cross-border cooperation in the border region of Germany, France, and Luxembourg in times of Covid-19. *European Societies, 24* (3), pp. 1–28.

Weber, F. & Kühne, O. (2017). Hybrid suburbia: New research perspectives in France and Southern California. *Quaestiones Geographicae, 36* (4), pp. 17–28.

Weber, F. & Kühne, O. (2020). Hybride Grenzen. Stadt-Land-Entwicklungsprozesse im Großraum Paris aus der Perspektive der Border Studies. *Geographische Rundschau, 72* (1–2), pp. 48–53.

Weber, F., Theis, R. & Terrolion, K. (Eds.). (2021). *Grenzerfahrungen. COVID-19 und die deutsch-französischen Bezieungen. Expériences transfrontalières—Les relations franco-allemandes à l'heure de la COVID-19.* Wiesbaden: Springer.

Wille, C. & Weber, F. (2020). Analyzing border geographies in times of COVID-19. In G. Mein & J. Pause (Eds.), *Self and Society in the Corona Crisis. Perspectives from the Humanities and Social Sciences* (Vol. 2, The Ends of Humanities, pp. 1–17). Esch-sur-Alzette: Melusina Press.

www.ingramcontent.com/pod-product-compliance
Lightning Source LLC
Chambersburg PA
CBHW030522080425
24757CB00001B/25
* 9 7 8 3 6 5 8 4 2 6 6 6 8 *